Sculpting in ZBrush Made Simple

Explore powerful modeling and character creation techniques used for VFX, games, and 3D printing

Lukas Kutschera

‹packt›

Sculpting in ZBrush Made Simple

Group Product Manager: Rohit Rajkumar

Publishing Product Manager: Chayan Majumdar

Senior Editor: Hayden Edwards

Technical Editor: Reenish Kulshrestha

Copy Editor: Safis Editing

Project Coordinator: Shagun Saini

Proofreader: Safis Editing

Indexer: Subalakshmi Govindhan

Production Designer: Nilesh Mohite

Marketing Coordinators: Anamika Singh and Nivedita Pandey

First published: March 2024
Production reference: 2140324

Published by Packt Publishing Ltd
Grosvenor House
11 St Paul's Square
Birmingham
B3 1RB

ISBN 978-1-80323-576-9

www.packtpub.com

Contributors

About the author

Lukas Kutschera is a 3D artist known for his digitally sculpted characters and creatures for high-end clients in the VFX, games, and collectibles industries. These clients include Platige Image, Pixomondo, and Hot Toys. He has also earned recognition for his contributions to the Golden Globe-winning TV show *House of the Dragon*, where he sculpted the show's largest dragon, Vhagar. Lukas's interest and expertise in anatomy is evident in his personal artwork, which features digitally sculpted anatomy studies and portraits. Besides his professional work, he shares his techniques and knowledge with up-and-coming artists in his mentorship program.

I want to thank the editing team and reviewers for their diligence and attention to detail in helping me bring this book to life.

About the reviewers

Rasmus Angeria is a 3D character artist with four years of experience. Specializing in ZBrush and proficient in various 3D programs, Rasmus is a dedicated professional currently contributing to the studio Griever Games. Having been an integral part of the studio for three years, Rasmus has played a crucial role in shaping the characters of *Forlorn Outcast*, a souls-like RPG game, which is scheduled for an early 2024 release. With a keen eye for detail and a commitment to excellence, Rasmus's sculpted characters stand as a testament to their artistic skill and contribution to the studio. As an adaptable and collaborative team member, Rasmus continues to make 3D characters in the ever-evolving world of digital art.

Mahmoud Hady is an experienced 3D character artist from Egypt, specializing in realistic/hyperreal characters/creatures. He also has experience in polygon modeling (high-poly and low-poly), texturing, shading, and lookdev for assets. Mahmoud has worked on games such as *The Lord of the Rings: Return to Moria* and *Sammy and the Little Girl*, as well as having worked on multiple commercials/ads for big names in the Middle East, including Uber, Pepsi, Valorant, Big Chips, Lay's, KFC, Egyptian Telecom (WE), and so on. He has also instructed character modeling for games at the Information Technology Institute in Egypt.

Damiaan Thelen started his ZBrush journey as a hobbyist in 2014. After three years of self-study, he started working as a character artist in the video games industry, where he uses ZBrush almost daily as one of the most integral parts of the character creation pipeline. He is currently the lead character artist for the German game development studio Egosoft.

Maxime Forveille is a French 3D artist currently working as a freelancer. He's been in this industry for more than 10 years. During those years, he has worked in multiple fields such as archviz (Australia), movies (*Justice League* and *Mowgli*, in Canada), and the luxury industry (France). Sharing his knowledge has always been something pleasant for him; that's why he also teaches 3D software in a few French schools.

ZBrush is one of his favorite tools if not his favorite. Besides the freedom ZBrush brings to sculpting very dense meshes, Maxime also likes the technical aspect and does not hesitate to write his own plugins for ZBrush when he needs to. Some of his biggest missions using ZBrush are to make models that are then printed at a human scale.

Table of Contents

3

Exploring the Gizmo, PolyGroups, and Masking   73

4

Exploring Brushes and Alphas   93

5

Creating an Optimized Mesh Using ZRemesher and ZProject   115

Part 2: Creating Characters from Scratch: A Comprehensive Guide

8

9

Creating Costumes, Armor, and Accessories with Classic Modeling Techniques   263

10

Preparing and Exporting Our Model for 3D Printing   323

Part 3: Sculpting a Female Head: Tips and Techniques

11

12

13

Building a Portfolio and Leveraging Social Media   457

Preface

Perhaps the most effective way to articulate what ZBrush is about and what it offers is by envisioning scenes such as Godzilla engaging in a fierce battle with King Kong, colossal dragons soaring over fictional cities, or digital doubles of actors executing extraordinary stunts. These scenarios, with their full glory and intricate details, would be impossible to conceive without the capabilities of this software, and its ability to create incredibly detailed 3D models.

Although several digital sculpting programs exist, ZBrush stands out as the unquestionable leader in VFX, games, and numerous other industries. This is not only because of its vast amount of unique and powerful tools, constant updates, and excellent customization options, but especially due to its large and passionate community. This collective enthusiasm makes ZBrush the optimal choice for anyone wanting to start their journey in digital sculpting.

When I first tried ZBrush, it was remarkable to see how fast I could achieve interesting results, without having had much drawing practice or even a formal art background. Since then, I've been hooked on the digital sculpting experience. In writing this book, I did so with certainty and joy, knowing that others may become just as captivated, unlocking diverse and rewarding career paths in the process.

As you navigate through this book, I hope the hands-on examples make all those tools and workflows seem less scary and inspire you to start your digital sculpting journey. The idea is to keep things fun so that learning new tools doesn't become a chore, but instead, you see it as opening up possibilities for creating new and better things.

Who this book is for

This book is for 3D artists, digital sculptors, modelers, and anyone looking to learn about the ZBrush sculpting software. It's also helpful for professionals switching to ZBrush or expanding their skill set.

While prior ZBrush experience and artistic abilities can aid you, they're not prerequisites to understanding the book. The book covers the most common and most useful ZBrush workflows and can benefit both beginner and intermediate artists looking to tap into the vast possibilities of ZBrush.

What this book covers

In *Chapter 1*, *Getting Started with ZBrush*, you will learn how to navigate ZBrush and get started by loading a model and doing a simple sculpting practice.

In *Chapter 2*, *Sculpting a Demon Bust with DynaMesh*, you will explore one of ZBrush's most powerful sculpting tools, DynaMesh. With this tool, you can follow along in the creation process of a demon bust sculpture, while learning about character art and design principles.

In *Chapter 3*, *Exploring the Gizmo, PolyGroups, and Masking*, you will learn about the most essential functionality and tools in ZBrush: the Gizmo tool, Polygrouping, and Masking. These tools will be used for many workflows, allowing you to create your models more efficiently.

In *Chapter 4*, *Exploring Brushes and Alphas*, you will become familiar with ZBrush's vast selection of brushes, Alphas, and customization options. With this information, you will be able to create custom brushes to add detail to the demon bust from *Chapter 2*.

In *Chapter 5*, *Creating an Optimized Mesh Using ZRemesher and ZProject*, you will learn about ZBrush's most popular retopology tool, ZRemesher, as well as ZProject, which lets you transfer detail between meshes. This is an essential part of many character creation workflows, and you will use it to push the quality and level of detail of the demon sculpture.

In *Chapter 6*, *Texturing Your Sculpture with Materials, Polypaint, and UVs*, you will get familiar with ZBrush's Materials, and its painting tool, called Polypaint. These lessons will be applied in the practical example of your demon bust, so you can put them into action immediately.

In *Chapter 7*, *Lighting and Rendering Your Model*, you will learn about lights and how you can create a custom light setup, so you can then render and present your model to your clients or audience.

In *Chapter 8*, *Sculpting Human Anatomy*, you will begin the creation process of a gladiator sculpture that will be optimized for 3D printing. This chapter shows how you can establish realistic anatomy for your character, which is a key skill set in many ZBrush and sculpting-related job opportunities.

In *Chapter 9*, *Creating Costumes, Armor, and Accessories with Classic Modeling Techniques*, you will proceed to create the gladiator character, while learning various modeling techniques, which will make you a more effective and versatile ZBrush artist.

In *Chapter 10*, *Preparing and Exporting Our Model for 3D Printing*, you will finalize the gladiator model, making it water-tight and splitting it into pieces, so that it can be printed without issues.

In *Chapter 11*, *Sculpting a Female Head*, you will begin the third and last project of the book, which is one of the most popular sculpting subjects: sculpting a female head. You will learn about various blockout options, anatomy fundamentals, and last but not least, some tips for sculpting a likeness.

In *Chapter 12, Adding Skin Detail, Sculpting Hair, and Using FiberMesh*, you will proceed with the head sculpture from the previous chapter, adding realistic skin detail and sculpting hair. The last part of the chapter focuses on ZBrush's hair system, FiberMesh, as an alternative way of creating hair.

In *Chapter 13, Building a Portfolio and Leveraging Social Media*, you will get tips for building your portfolio and using social media, so you can advertise your ZBrush and sculpting skills to get job opportunities and commissions.

To get the most out of this book

Having some background in digital sculpting or 3D modeling will help make some of the concepts clearer a little bit faster. If you have already used ZBrush before, you may experience a bit less frustration, since ZBrush can be a bit unintuitive at first. However, experience is not necessary to effectively follow the examples provided in the chapters, and with some persistence, you should do just fine.

Software/hardware covered in the book	Operating system requirements
ZBrush 2023 (though older versions work too – I wrote this book using 2022.0.7)	Windows 10 or 11 (64-bit edition), or macOS 11.5 or above

Having a tablet is strongly recommended, as sculpting with the mouse is far inferior and cannot give the same results as using a pen with pressure sensitivity.

Conventions used

There are a number of text conventions used throughout this book.

Bold: Indicates a new term, an important word, or words that you see onscreen. For instance, words in menus or dialog boxes appear in **bold**. Here is an example: "Then, navigate to the **Project** or **Tool** menus inside the LightBox editor and find a suitable model for your needs."

> Tips or important notes
> If it is possible to collect multiple angles of your reference subject, make sure you do so – it will help you visualize the three-dimensional shape of your subject better. This holds especially true when working on a portrait or character likeness, where it is essential to have a high attention to detail.

Get in touch

Feedback from our readers is always welcome.

General feedback: If you have questions about any aspect of this book, email us at customercare@ packtpub.com and mention the book title in the subject of your message.

Errata: Although we have taken every care to ensure the accuracy of our content, mistakes do happen. If you have found a mistake in this book, we would be grateful if you would report this to us. Please visit www.packtpub.com/support/errata and fill in the form.

Piracy: If you come across any illegal copies of our works in any form on the internet, we would be grateful if you would provide us with the location address or website name. Please contact us at copyright@packt.com with a link to the material.

If you are interested in becoming an author: If there is a topic that you have expertise in and you are interested in either writing or contributing to a book, please visit authors.packtpub.com.

Share Your Thoughts

Once you've read *Sculpting in ZBrush Made Simple*, we'd love to hear your thoughts! Scan the QR code below to go straight to the Amazon review page for this book and share your feedback.

https://packt.link/r/1-803-23576-4

Your review is important to us and the tech community and will help us make sure we're delivering excellent quality content.

Download a free PDF copy of this book

Thanks for purchasing this book!

Do you like to read on the go but are unable to carry your print books everywhere?

Is your eBook purchase not compatible with the device of your choice?

Don't worry, now with every Packt book you get a DRM-free PDF version of that book at no cost.

Read anywhere, any place, on any device. Search, copy, and paste code from your favorite technical books directly into your application.

The perks don't stop there, you can get exclusive access to discounts, newsletters, and great free content in your inbox daily

Follow these simple steps to get the benefits:

1. Scan the QR code or visit the link below

https://packt.link/free-ebook/978-1-80323-576-9

2. Submit your proof of purchase
3. That's it! We'll send your free PDF and other benefits to your email directly

Part 1:
The Adventure Begins:
Sculpting in ZBrush

In *Part 1* of this book, you will be quickly introduced to ZBrush and its UI, after which you will jump right into sculpting. *Chapters 2* to *7* explore the creative process of sculpting a demon bust, and you are encouraged to follow along. This will let you try out the tools and techniques of each chapter in a fun and goal-oriented way.

This part includes the following chapters:

- *Chapter 1, Getting Started with ZBrush*
- *Chapter 2, Sculpting a Demon Bust with DynaMesh*
- *Chapter 3, Exploring the Gizmo, PolyGroups, and Masking*
- *Chapter 4, Exploring Brushes and Alphas*
- *Chapter 5, Creating an Optimized Mesh Using ZRemesher and ZProject*
- *Chapter 6, Texturing Your Sculpt with Materials, Polypaint, and UVs*
- *Chapter 7, Lighting and Rendering Your Model*

1
Getting Started with ZBrush

Congratulations on taking your first step into the world of ZBrush. In this first chapter, we'll introduce you to the **user interface** (**UI**), navigation, and some other fundamentals of ZBrush so that you can start sculpting as quickly as possible!

ZBrush is known in the **computer-generated imagery** (**CGI**) community as a tool that can handle a very high polycount, letting you work on highly detailed models, something that was not possible in other 3D software for a long time. This ability made it a popular choice in the games and VFX industry to create highly detailed, life-like characters, creatures, and environments, and ZBrush remains the number one sculpting tool for characters and creatures up until today.

Another reason for ZBrush's popularity comes from the developers' willingness to add tools and find solutions for specialized industries. This ensures that almost every user will find tools that fit their exact needs for any scenario. However, this can result in ZBrush having a lot of tools that most users will never need. To keep this book easily digestible, we'll only cover what is needed for the main tasks of the average digital sculptor, but it is also important to understand how to navigate through the software.

Compared to other 3D software, ZBrush has a steep learning curve when it comes to its UI and navigation, which can be frustrating at first. However, if you persevere and get past the initial difficulties, you will find that ZBrush will become very intuitive very quickly, and the vast customization options will allow you to streamline your workflow.

In the first section of this chapter, you will learn how to load a basic shape so that you can start experimenting with different sculpting brushes and get a feel for the sculpting experience, as well as learn how to navigate around the 3D space.

The next section will give you an overview of ZBrush's UI, its various menus, and the various tools and functions that ZBrush offers. Finally, you will explore different camera settings, as well as ZBrush's file types and when they should be used.

In this chapter, we are going to cover the following topics:

- Loading models, navigation, and creating your first sculpture
- Exploring and customizing the ZBrush UI
- Understanding ZBrush file types

Technical requirements

To follow this book, you'll need to install ZBrush with a valid license. You can find a download link on Maxon's website: `https://www.maxon.net/en/ZBrush`.

You do have the option to use ZBrushCore if you want a cheap alternative to get started, although it is missing many of the full features of ZBrush, so most of this book would not be relevant to you.

To have the best possible user experience with ZBrush, make sure to fulfill the following minimum requirements:

- **OS**: 64-bit editions of Windows 10 or 11 (32-bit operating systems are no longer supported)
- **CPU**: Intel i7/i9 technology and newer or AMD Ryzen and newer
- **RAM**: 4 GB (16+ GB is strongly recommended)
- **HDD**: 20 GB of free hard drive space for ZBrush and its scratch disk
- **Pen tablet**: Mouse or compatible pressure-sensitive tablet, which must support the WinTab API (a tablet is strongly recommended)
- **Monitor**: 1280×1024 monitor resolution with 32-bit color
- **Video card**: Must support OpenGL 3.3 or higher and Vulkan 1.1 or higher

> **Important note**
> If you are a professional artist, you will want to get a tablet. Although you can use a mouse and achieve some results, the lack of pressure sensitivity that comes with using a mouse will make certain results impossible or take an unreasonable amount of time.

Loading models, navigation, and creating your first sculpture

When you first get started with ZBrush, it might take a while to get used to the unusual UI and navigation, and experiencing some frustration is not uncommon. This can detract from the awesome experience of using a digital sculpting tool for the first time and getting to see all of its creative possibilities.

Therefore, before diving into too many functions and technical aspects, let's get sculpting first so that you can see how fun and rewarding it is.

Loading a starting model

After launching ZBrush, there are a couple of starting points to choose from. A versatile and powerful one is ZBrush's content browser called LightBox. Alternatively, you can also load a basic shape, such as a sphere or a cylinder. Let's look at both options.

LightBox

LightBox has a great variety of models, including everything from simple shapes to faces and full human and animal bodies. You can access it by pressing the , key. Then, navigate to the **Project** or **Tool** menus inside the LightBox editor and find a suitable model for your needs:

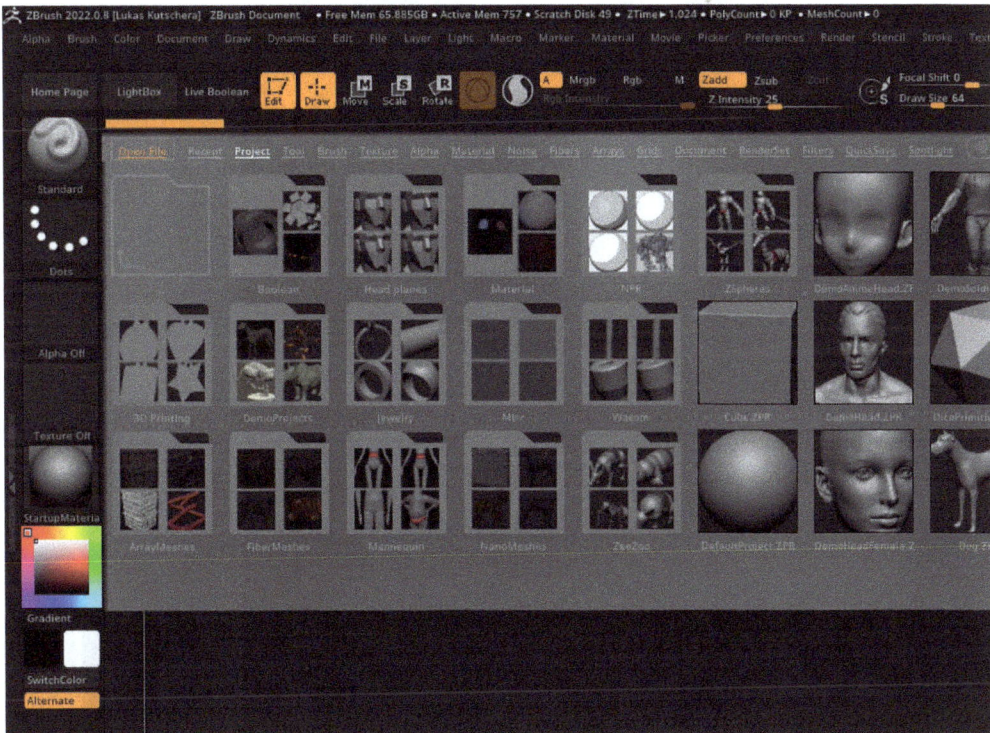

Figure 1.1 – LightBox and its many resources

Some of the models might be similar to what you want to create, saving you a lot of time and providing you with a great starting point.

Now, let's look at loading a basic shape, which we will use to start our first sculpting project.

Using a basic shape for your sculpture

Basic shapes can be a good choice if they have a similar shape to your desired result. For example, you might want to sculpt a wedding ring, which makes a **Ring3D** model a suitable shape to start the modeling process.

Since we just want to test out some sculpting brushes, we will be using the **Cube3D** shape. Follow these steps:

1. Navigate the menu on the right side of the canvas, which is the **Tool** palette. There, you can see a list of icons, including a star. If you click on the star, an overview of 3D models and 2.5D brushes appears (the latter are irrelevant to the focus of this book). In this example, we will pick a **Cube3D** shape and drag it onto the screen:

Figure 1.2 – Picking Cube3D from the basic shape menu

2. However, when you drag the shape onto your screen, you will only see it as a flat 2D shape. Here, our cube just looks like a square. This is because when you launch ZBrush, you are not automatically in **Edit** mode:

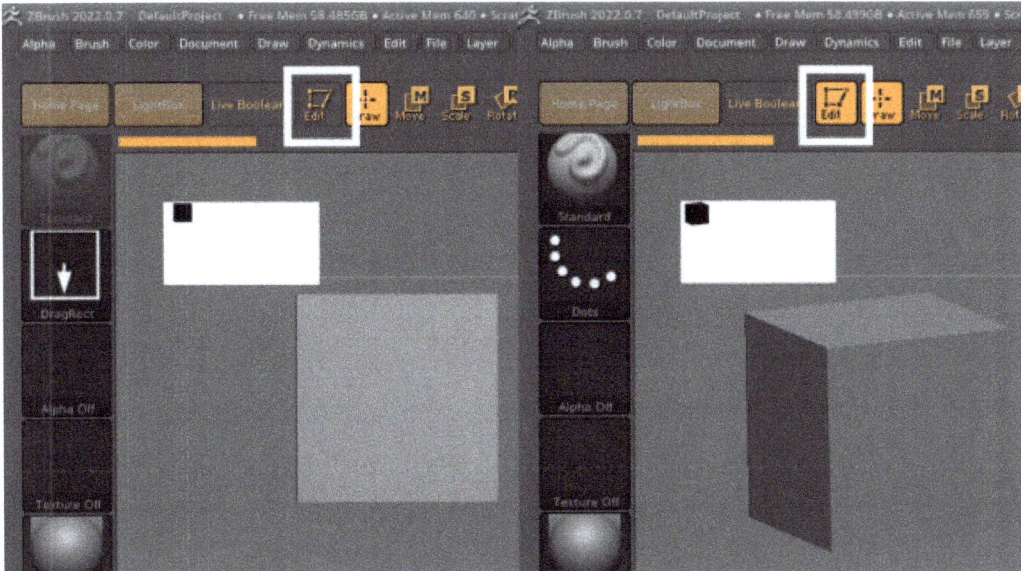

Figure 1.3 – Edit mode disabled (left) and enabled (right)

In order to switch to **Edit** mode (also known as **3D** mode) and start modeling/sculpting, you need to hit *T* on the keyboard. You can now left-click and rotate around the model in 3D space.

> **Important note**
>
> ZBrush 2.5D launches in **2.5D** mode, which means you can't modify any shape, model, or ZTool until you switch to **Edit** mode by hitting *T* on your keyboard. This often causes frustration because you won't be able to start sculpting until you enter **Edit** mode. The **Edit** mode button, above the **Undo** slider, will appear orange if it is enabled.
>
> Also, if you accidentally draw something on the canvas in **2.5D** mode, you can clear the canvas again by pressing *Ctrl + N*.

3. Now, before you can actually modify the shape, there is one more thing you need to do, which is click **Make PolyMesh3D** in the **Tool** palette. After you use this command, the name of your subtool changes from **Cube3D** to **PM3D_Cube3D**:

Figure 1.4 – Using the Make PolyMesh3D command to enable the shape for modeling/sculpting

At this point, you should have a cube on your canvas, but you now need to know how to navigate the scene and how to move around the cube and view it from any angle.

Navigating around your sculpture

Navigating in ZBrush is not always intuitive and can be frustrating, but with practice, you can move around your model smoothly and work more effectively.

Here are some of the basic movements:

- In order to zoom in and out, do the following: press *Alt* and hold, then right-click and hold; now, release the *Alt* key and drag the mouse while holding the right-click to zoom in and out.

- In order to rotate around your model, right-click and move the mouse.

- In order to move horizontally and vertically, do the following: press *Alt* and hold, then right-click and hold, then move your mouse at the same time.

You can perform all these movements with left-clicking, instead of right-clicking, as well. However, when left-clicking, you have to click on a space in the canvas to be able to navigate, while right-clicking works also when your mouse is hovering over a mesh.

> **Important note**
>
> If you use the right mouse button to navigate, you do not have to click on a space in the canvas, but you can also click on subtools in the scene and navigate regardless.

Another essential function is the **Frame** command – if you ever want to bring your camera back to your model, you can hit *F* on the keyboard or click on the **Frame** button on the **Icon** toolbar on the right side of your canvas:

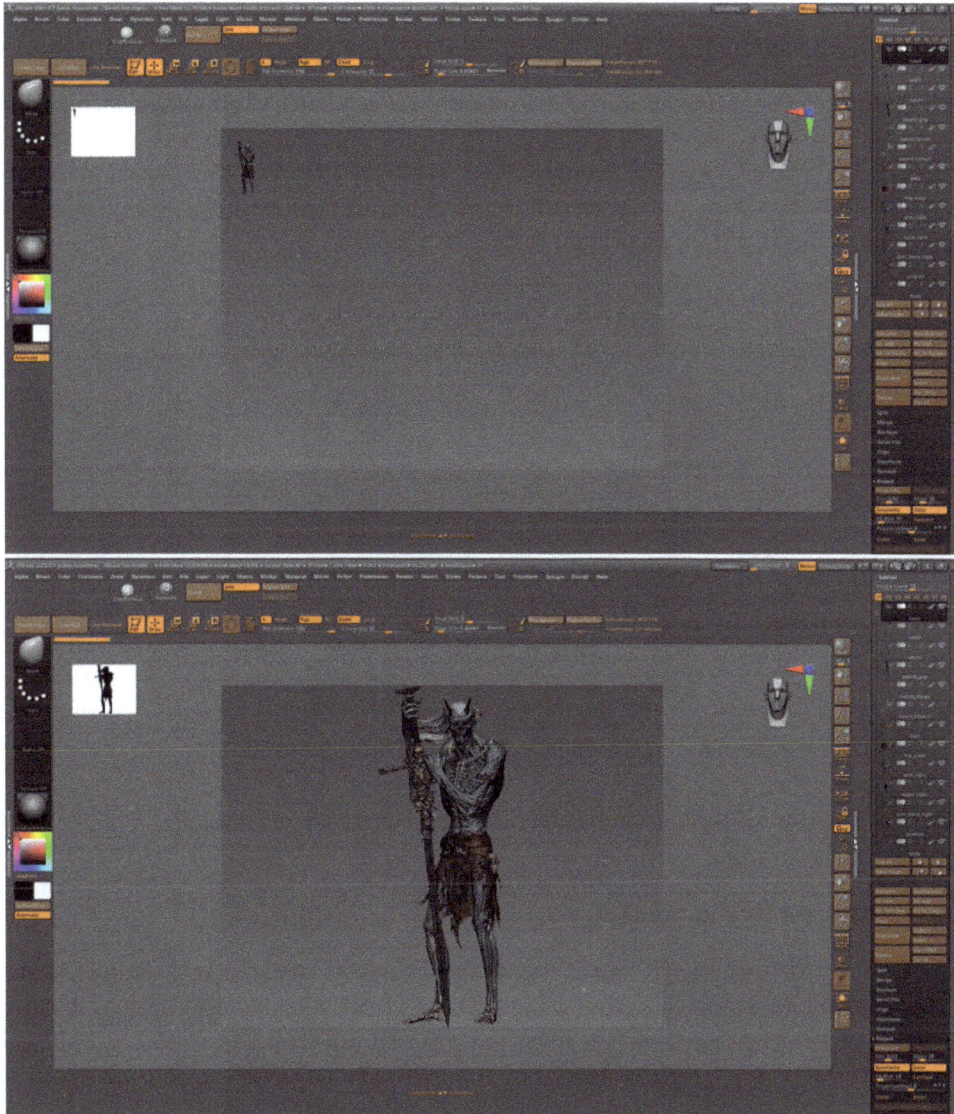

Figure 1.5 – Using Frame to bring the model back to full size in the center of the canvas

As frustrating as ZBrush's way of navigating can be at the very beginning, it will become intuitive quickly, so keep at it! Next, you can finally start sculpting!

Creating your first sculpture

At this point, you can start sculpting. Some things will probably still be confusing, and a lot of the tools and functions will be explained a bit later, throughout the book, but for now, try to focus on getting creative with the tools you have. Let's start by experimenting with some of ZBrush's many brushes.

Using sculpting brushes

Open the **Brush** menu by pressing *B* and select any brush. You can pick a **ClayBuildup**, **Standard**, or **DamStandard** brush if you want a very basic sculpting brush, but of course, you can experiment with all of the available brushes. To use the brushes, simply apply pressure with your pen on the model, and you will see the effect. Now, you can get a feel for the sculpting experience in ZBrush and see how you can get interesting results very quickly:

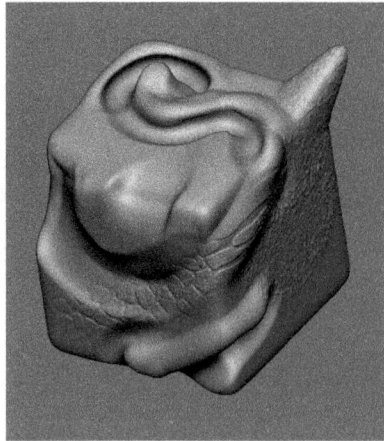

Figure 1.6 – Experimenting with different brushes in ZBrush

Brushes will be introduced throughout the book, but they are covered in detail in *Chapter 4* if you are interested.

Adding resolution to your sculpture

You may press *Ctrl + D* a couple of times to subdivide your model, adding resolution to it. Make sure that your **ActivePoints** (mesh resolution) count does not get too high, though, since it will affect performance. Anything beyond 10 million points will make ZBrush increasingly slow, depending on your PC's hardware:

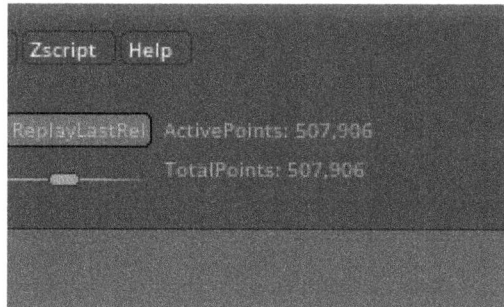

Figure 1.7 – ActivePoints count, showing the density of your mesh

You can find more information about subdividing in *Chapter 5*.

Getting creative

At this point, you can spend any time you like sculpting, experimenting, and getting used to working in ZBrush.

When you hold the *Alt* key while sculpting, you will put your brush into **Subtract** mode, carving away from it, instead of building up forms.

Additionally, you can hold *Shift* to enable the **Smooth** brush, which lets you smooth out the surface. Again, this will be covered in more depth in *Chapter 4*, but for now, these commands will give you a lot of possibilities for sculpting something unique and interesting.

Also, you can undo your last brush stroke(s) by pressing *Ctrl* + *Z* or dragging on the **Undo** slider (above the canvas) that stores the last steps:

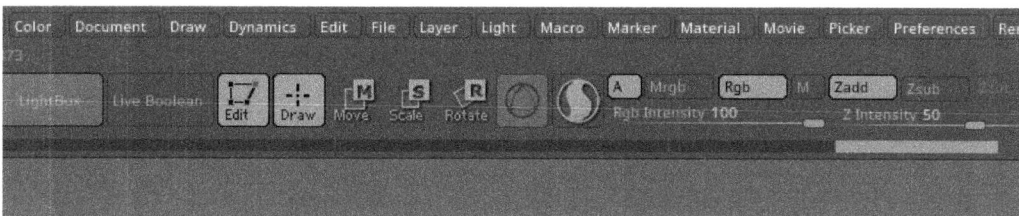

Figure 1.8 – Undo slider for restoring an older state of the model

This concludes the first section of this chapter. You learned about LightBox and primitive shapes as starting points for sculpting in ZBrush. After that, you practiced navigating around your model and finally tested some of ZBrush's sculpting brushes on a cube model. Next, we will take a closer look at the UI with its numerous menus and tools.

Exploring and customizing the ZBrush UI

The ZBrush UI contains a 3D canvas, where you can view and work on your models. By default, it should look like this (though I have changed the UI color to allow for better readability in print):

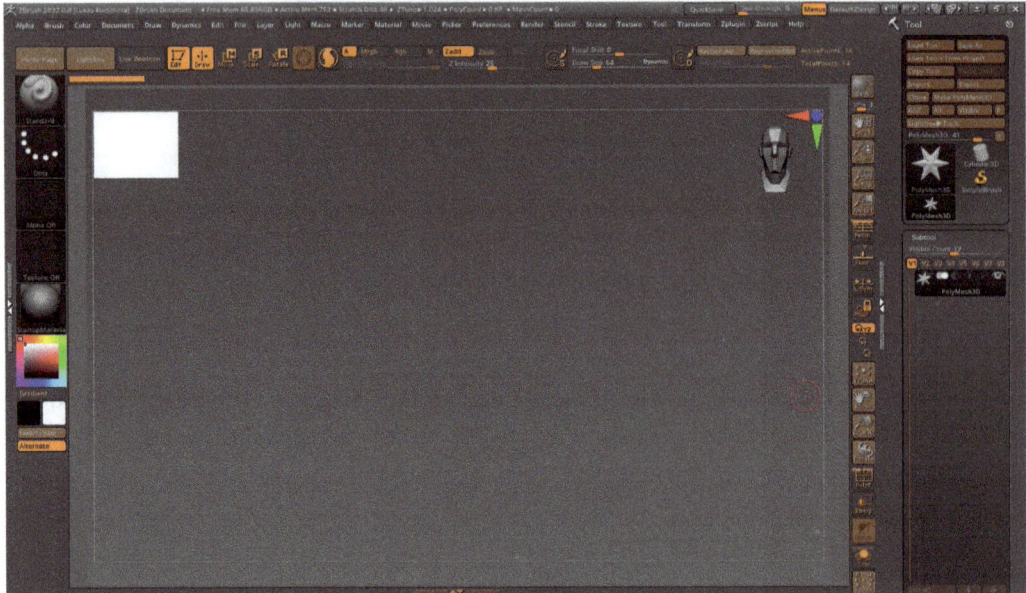

Figure 1.9 – The ZBrush UI

The buttons around the canvas are assembled there to be easily accessible; however, these same buttons can also be found in the ZBrush menus. If you click on a menu, it will open and show multiple sub-menus for you to explore.

Clicking on the **Divider Space** icon on the left and right sides of the screen (marked with a circle in the following screenshot) will create additional space, allowing you to drag and dock menus onto this space and keep them open and easily accessible:

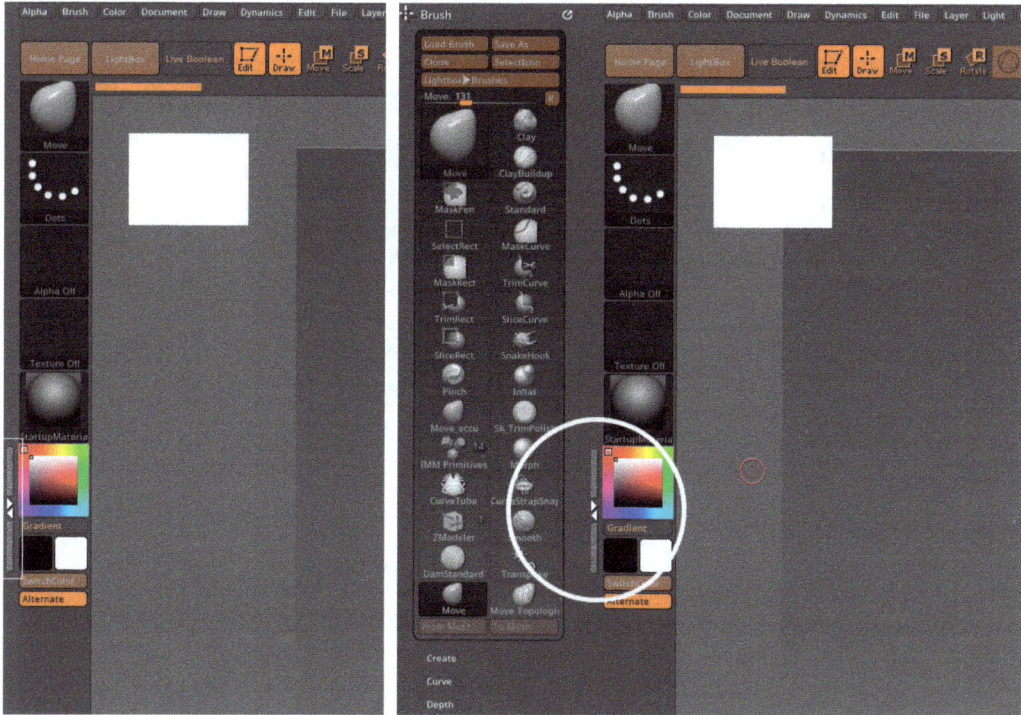

Figure 1.10 – Before (left) and after (right) adding space on the left side of the canvas

To do so, left-click and drag the circle symbol in the upper-left corner of the menu, move them to the side of the canvas, and then release to dock them there.

Let's look at the menus and see the wide range of features ZBrush has to offer.

Exploring the menus and palettes

All of ZBrush's functionality is organized in menus at the top of the screen. You can move each menu to the space around the canvas to keep the menu open for faster access to its tools; this allows for greater efficiency when frequent access to a particular tool is needed.

In the following list, all of ZBrush's menus are briefly explained. For the sake of a complete overview, no menu is left out, regardless of its importance, but you may skip parts of this section for now and come back to it when you want to look up specific information.

That being said, the menus are located in the top bar, as marked here:

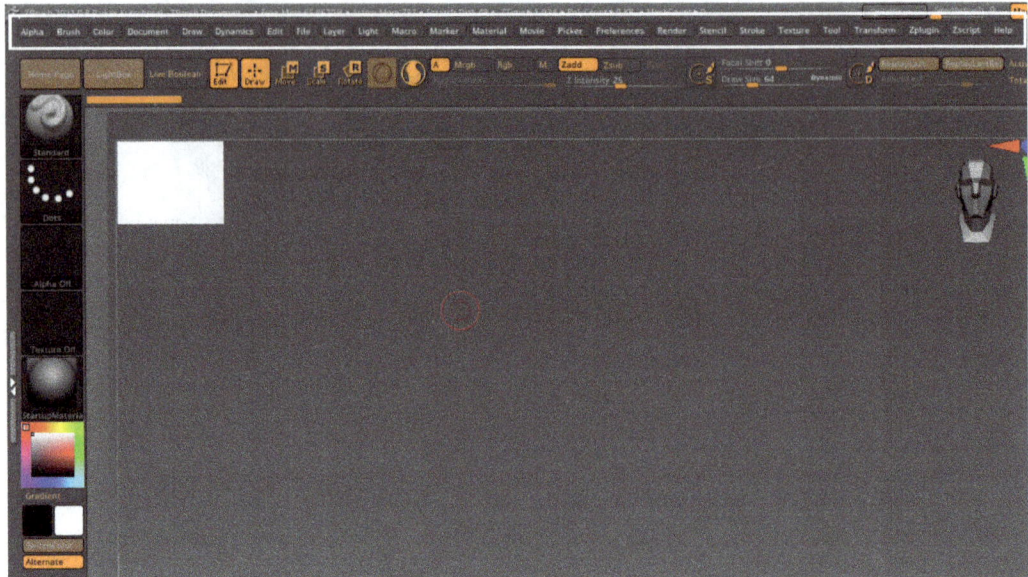

Figure 1.11 – Top-bar menus

Here are the menus and their tools and benefits:

- **Alpha**: These are grayscale images that you can use with your brushes to achieve specific effects on the surface of a sculpt (for example, **Skin Detail**, **Rock Surface**, **Fabric Weave**). The **Alpha** menu lets you create your own Alphas, modify existing ones, and import and export them. Here, we have the same brush, but equipped with different Alphas:

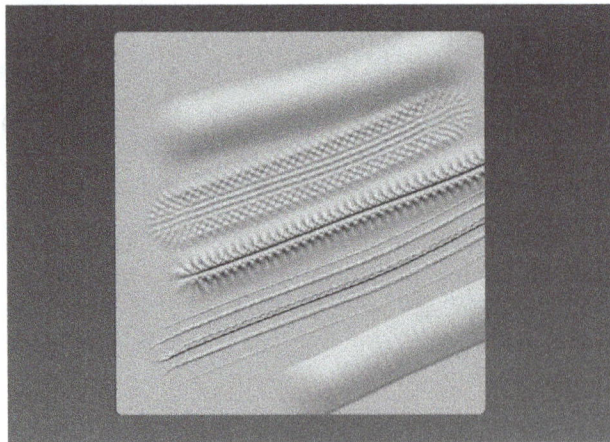

Figure 1.12 – The effect of different Alphas applied to a plane

- **Brush**: These are the main tools used for sculpting and modeling. ZBrush has a wide variety of brushes; some of them are more common, while others have unique attributes. As you can see in this screenshot, brushes can produce very diverse results:

Figure 1.13 – The effect (and versatility) of different brush strokes applied to a plane

The **Brush** menu lets you create, modify, import, and save brushes. While there are many modification options in this menu, the **Stroke** menu has even more options; we will go into detail about brushes later on in the book.

- **Color**: This menu allows you to choose and apply colors to your models. This can elevate the visual, as you can see in the following screenshot:

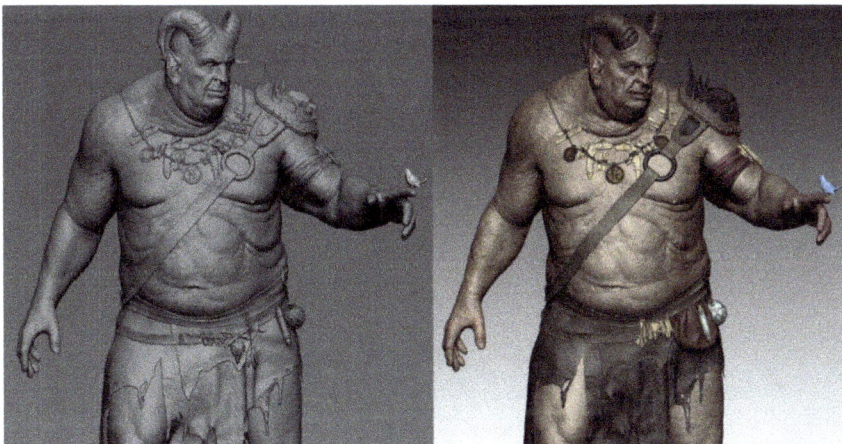

Figure 1.14 – The effect before and after color is applied to a model

- **Document**: This menu lets you modify the document – that is, the canvas: the main working space in ZBrush. You can set dimensions, import images, and export them as a `.zbr` file (which will be explained later in this chapter). The canvas is the area inside the rectangle:

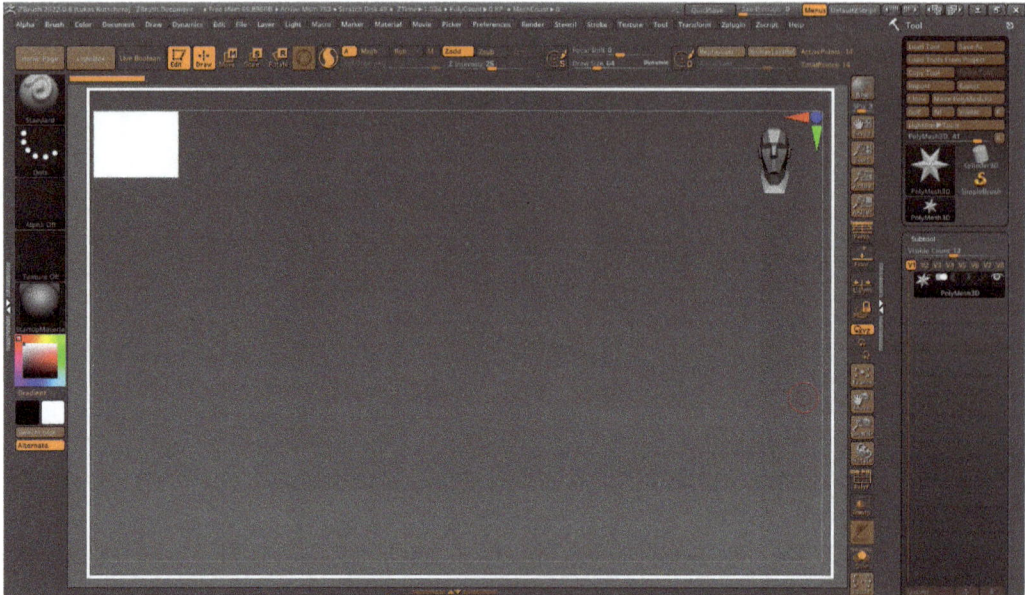

Figure 1.15 – ZBrush's canvas

- **Draw**: This menu contains many of the most important settings for how tools in ZBrush behave. Because of their importance and frequent use, some of the tools are accessible around the canvas. The most important features are **Radius**, **Intensity**, and **Falloff** brushes, as well as the **Color** and **Material** mode buttons (**RGB, M,** and **MRGB**). There is also the **Perspective (Focal Length)** option and other essential camera options.

Here are strokes with the same brush with the same Alpha but with different brush options from the **Draw** menu:

Figure 1.16 – The result of different brush functions available in the Draw menu

- **Dynamics**: With ZBrush 2021, a new feature called **Dynamics** was introduced. This tool allows you to run simulations on meshes, which can be used to create different physical effects. One common use for this tool is creating realistic-looking clothing through the simulation of wrinkles. The wrinkles of the following dress were not sculpted but simulated by the press of a button:

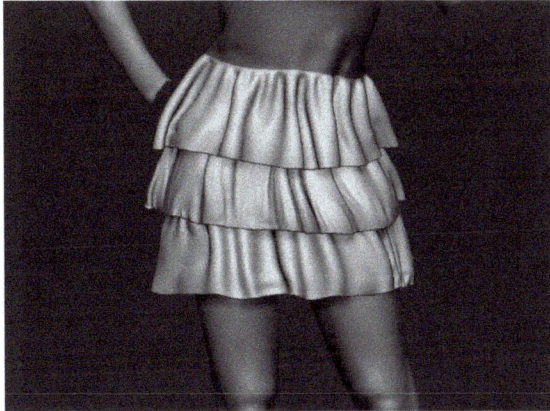

Figure 1.17 – Creating wrinkles with the Dynamics functionality

- **Edit**: This menu contains the important **Undo** and **Redo** functions. An **Undo/Redo** slider is also visible in the default UI, containing the same functionality:

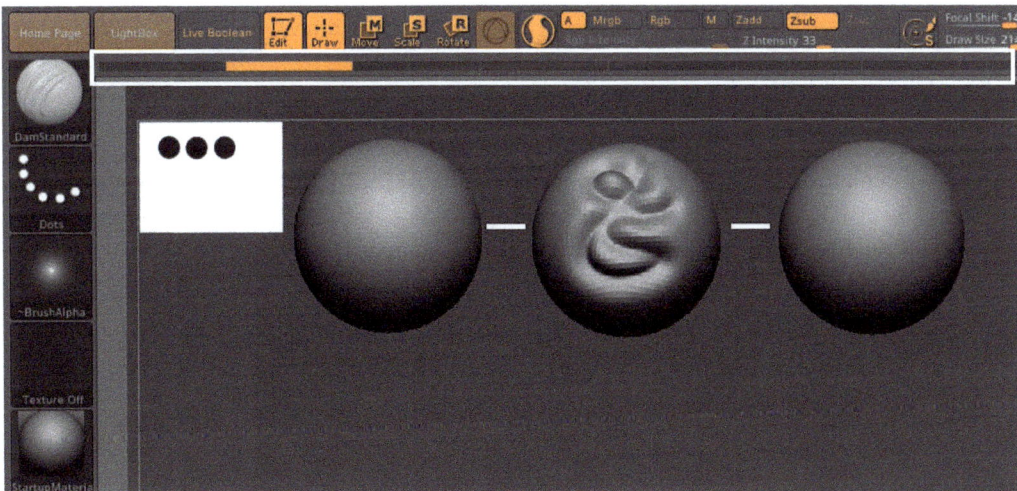

Figure 1.18 – Modifications of a sphere undone with the Undo slider

- **File**: Navigate to this menu to load, save, and export different file types from ZProjects to ZTools, canvases, and more.

- **Layer**: This palette contains information about the canvas and lets you create multiple canvases, which you can blend together to create a final image. This feature is not relevant for classic modeling or sculpting projects but might be useful for 2D artists.

- **Light**: This palette lets you adjust lighting in your scene by allowing you to create and orient multiple lights. These can be useful when creating renders inside ZBrush to present your model in the best possible way. You can see the effect of a custom light setup in a portrait sculpt here:

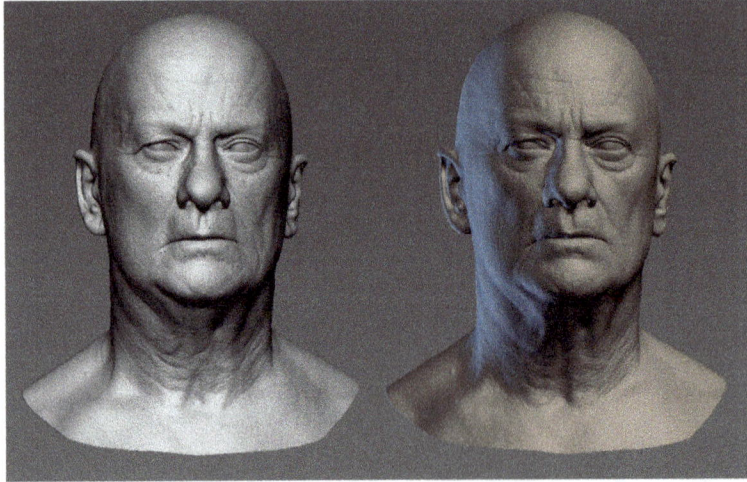

Figure 1.19 – A sculpture with default lighting (left) and the same sculpture
with lighting of different colors and directions (right)

- **Macros**: Allow you to record a sequence of actions, helping you speed up repetitive tasks.

- **Marker**: Markers let you record the position, size, and orientation of an object. Once you place an object on the canvas, it becomes converted to pixols (`https://docs.pixologic.com/getting-started/basic-concepts/the-pixol/`) and can't be edited after another tool has been applied. Using markers lets you recall an object later so that you can redraw and edit it.

- **Material**: This palette allows you to create, modify, and save materials that determine how your 3D models are displayed on the canvas. Materials affect how your model reacts to light and can have a strong visual effect. Here is the same model with different materials applied:

Figure 1.20 – The result of different materials on the same model

- **Movie**: This palette allows you to record videos and create simple animations. This is a great way to create and export previews of your model for yourself or clients.

- **Picker**: This palette gives you options for adjusting how your brushes function in ZBrush. This is a useful way to create very specific effects, but it may not be needed much in regular workflows; however, it is another tool to cover any sculpting scenario.

- **Preferences**: This menu contains a wealth of settings for all areas of ZBrush. From optimizing performance, managing quicksaves, and customizing the UI to saving hotkeys, this menu covers a lot. We will go over the most useful options later in this chapter.

- **Render**: This palette allows you to set up render settings for your 3D model so that you can maximize the quality of the images and present them according to your intention. You can create different render passes such as **Depth**, **Ambient Occlusion**, or **Subsurface Scattering** and export these to composite a final image in Photoshop or your image-editing software of choice. Rendering your model will calculate things such as shadows in a more realistic way, like so:

Figure 1.21 – Before rendering (left) and after rendering with more realistic shadows (right)

- **Stencil**: Stencils let you mask out areas of your model based on grayscale images called Alphas. This lets you sculpt certain areas without affecting others, creating the desired look.

- **Stroke**: This menu, as with the **Brush** palette, gives you options for customizing your brush. You can create a variety of brushes that function in different ways. Some of these options will be covered in more detail later in the book.

- **Texture**: This menu lets you import, modify, and export textures. This way, you can apply color information from textures created outside of ZBrush to your models and get them to the next level. You can also add textures to the **Spotlight** tool here, which is handy for manually projecting color onto your models. Instead of painting a model in ZBrush, you can also import textures and apply it to a model:

Figure 1.22 – A model with imported textures from the Substance 3D Painter texturing software

- **Tool**: This menu contains a list of all the individual 3D models in your scene, called **subtools**. Different functions help keep these subtools organized, but there are also options that allow you to duplicate, delete, or insert 3D models, which makes this one of the most essential parts of the **Tool** menu.

  It also features the **DynaMesh** modeling tool, which is a great tool for concept creation and a more dynamic way of sculpting. Plus, the **NanoMesh** and **Deformation** menus offer unique ways to create certain effects on your 3D models beyond the more basic sculpting tools by adding meshes on the surface of your model or by simply applying deformations such as bending, inflating, or polishing.

Other useful parts of this menu are ZBrush's **Polypaint** color system and its **FiberMesh** hair system – these are very powerful tools that will be introduced throughout this book, showing their effect in a practical application:

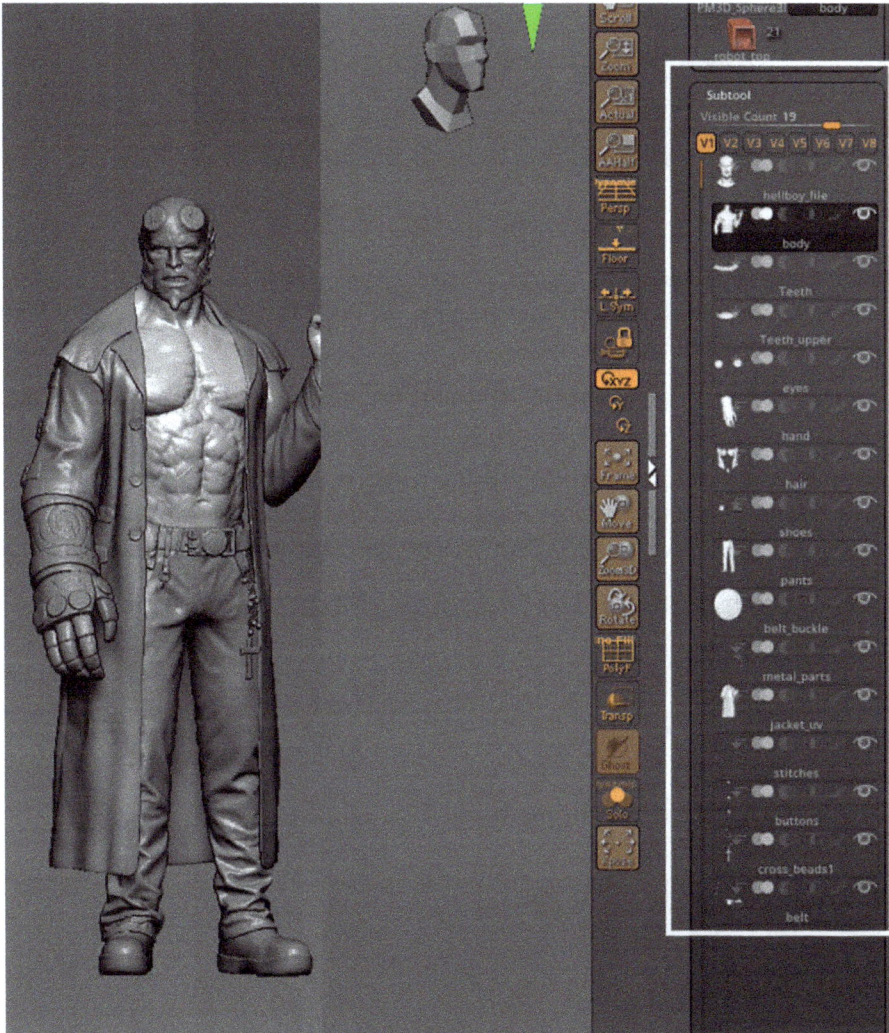

Figure 1.23   The Subtool palette, containing all models of the ZTool

- **Transform**: This palette contains options to move, scale, and orient your model, as well as navigate the canvas. Other essential options are displayed on the right side of the screen (if you are working with the standard UI). One noteworthy option is **Symmetry Settings**, which allows you to work on your model symmetrically and is one of ZBrush's most important functionalities when it comes to character and prop creation. Symmetry can be applied along multiple axes, as well as in a radial fashion:

Figure 1.24 – Symmetry along one axis (left) and radial symmetry (right)

- **ZPlugin**: This palette comes with an array of tools and plugins that cover a number of important functions for the professional artist (for example, **3D Print Hub**, **Scale Master**, **UV Master**). These will be introduced in later chapters.

- **ZScript**: ZScript is a scripting language built into ZBrush and can be used for tutorials, macros, and plugins. It's a great way to automate repetitive tasks and add new functionality via ZPlugins.

- **Help**: In this palette, you can find links to online resources, such as documentation and tutorials for new functionality. Other links give access to customer support, such as ZBrushCentral (a community for passionate sculptors) and ZClassroom (ZBrush's learning resource):

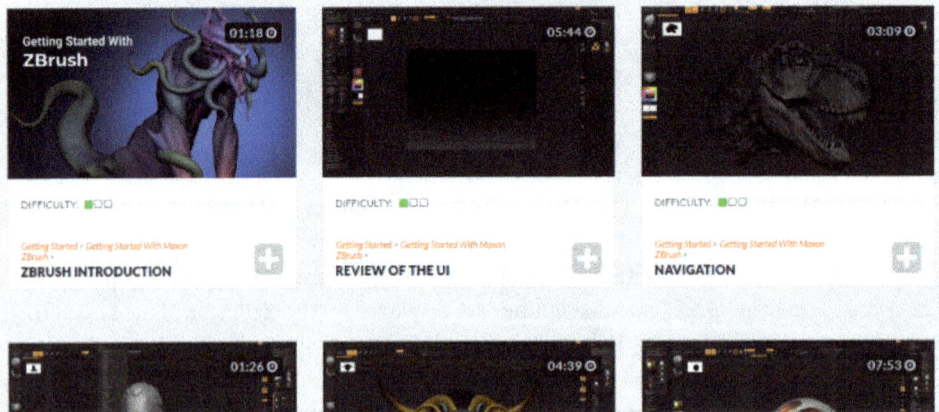

Figure 1.25 – ZClassroom

Some of the functions that are available in these palettes are also included in the vertical toolbar on the right side of the canvas, as seen in *Figure 1.26*. Let's take a look at some of the more important ones:

Command	Function
1: BPR (Best Preview Render) + **SPix** (SubPixel) slider	Renders the model, adding shadows and other properties, based on the render settings, set up in the **Render** palette in the **Render Properties** menu. The **SPix** slider improves the quality of the render.
2: **Scroll**	Lets you move the document/canvas.
3: **Zoom**	Lets you zoom in on the canvas.
4: **Actual**	Repositions the canvas in the center of the window, resetting any zoom that was applied.
5: **AAHalf**	Halves the size of the canvas.
6: **Perspective Distortion**	This button enables the **Dynamic Perspective** mode, which creates perspective distortions, making closer items appear larger. The degree of distortion depends on the focal length set up in the **Draw** palette.
7: **Floor Grid**	Lets you see the floor, allowing you to assess the position of your 3D model in 3D space.
8: **Local Symmetry**	Shifts the anchor point of **Symmetry** mode from the center of the scene to the center of the object, which lets you work with **Symmetry** mode on objects that are not located in the center of the ZBrush scene.
9: **Lock Camera**	Locks the camera, which can be useful if you want to preserve a certain viewing angle on your model.
10: **XYZ/Y/Z**	**XYZ** lets you rotate around your models in a normal way on all axes, **Y** lets you rotate only around the *y* axis, and **Z** lets you rotate only around the *z* axis.
11: **Frame**	Positions the selected subtool in the center of the canvas.
12: **Move**	Lets you move around the canvas as you normally would using your pen or mouse.
13: **Zoom3D**	Lets you zoom in on your models as you normally would using your pen or mouse.
14: **Rotate**	Lets you rotate around your models as you normally would using your pen or mouse.
15: **Draw Polyframe**	Turns on **Wireframe** mode, which lets you see the topology of the selected subtool.
16: **Transparency**	Lets you see the selected subtool through other subtools in the scene, similar to an X-ray effect.

Command	Function
17: **Ghost**	This is an extra feature for transparency, changing the shading of the active subtool from a darker to a brighter shading. This can help visualize your models in a different way and becomes a useful alternative to the default **Transparency** mode.
18: **Solo**	This will isolate your active subtool and make every other subtool on the canvas disappear. This is great for working on an isolated subtool without other subtools obstructing the view.
19: **X-Pose**	This function "explodes" your subtools, moving them away from each other to be distributed around the center of the canvas. This can be useful for visualizing your subtools individually.

Figure 1.26 – Vertical toolbar buttons and explanations

At this point, you should have a rough overview of the tools at your disposal. As previously mentioned, ZBrush has a wide variety of different tools that serve different industries, so some of them will be irrelevant for character creation and sculpting workflows. This list should serve as an overview that you can come back to at any point; the practical use of the most important tools will be introduced in the following chapters, up until the end of this book.

Setting some useful preferences

There are a few preferences that can help you get the most out of your ZBrush experience:

- Go to **Preferences** | **Interface** | **Button Size** and try lowering the size value. With a smaller **Button Size** value, more space will become available to maximize the canvas, allowing you to see more of the model and its detail.

- Go to **Preferences** | **Mem** and increase **MaxPolyPerMesh** to a higher number (up to 100, depending on your PC's capabilities). This will allow you to create meshes in higher resolution for projects where a high level of detail is needed.

- Go to **Preferences** | **QuickSave** and adjust the following:

  - **Maximum Duration**: The length of time after which ZBrush will automatically save your project. This can help avoid the need to manually save files frequently. You can adjust this value, but it is recommended that you do not set the value too high, or you can disable it altogether; this is because ZBrush will crash in certain circumstances without quicksaves, which could potentially lead to you losing a lot of work and time.

  - **Rest Duration**: In case you interrupt your work in ZBrush, this is the number of minutes after which ZBrush will make a quicksave.

These preferences are a good starting point for maximizing your canvas size, removing limits on mesh density, and preventing loss of work and time through quicksaves. You may need to adjust more preferences according to your specific needs as you become more familiar with the software and establish your workflows.

Creating your custom ZBrush UI

One of ZBrush's strengths is the amount of customization available and how easy it is to build a custom UI. A custom UI will allow you to save time by making frequently used commands more accessible without needing to open menus. While it may seem like a small difference, this will eventually add up to a lot of saved time, and the sculpting experience feels smoother through it as well.

In order to build and save your custom UI, do the following:

1.   Go to **Preferences** | **Config** | **Enable Customize**:

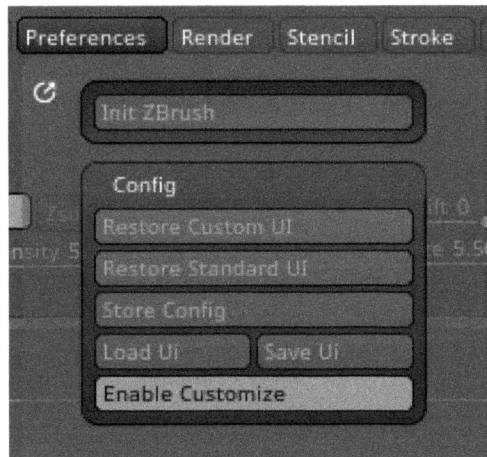

Figure 1.27 – Enable the Customize option

2.   Hover over any button (almost all buttons are customizable) and select it using *Ctrl* + *Alt* + left-click. Then, drag it to any free space around the canvas. As you place new buttons and your custom UI grows, ZBrush will automatically reduce the size of the canvas, so be mindful to only pick what you frequently need:

Figure 1.28 – Before (left) and after (right) adding elements to the UI

3. Once you are done building your UI, go to **Preferences | Config** and turn off **Enable Customize**.

4. Now, you can hit **Store UI** – this will save your UI as your new custom UI, and ZBrush will launch with this UI from that point forward. Alternatively, you can also select **Save UI** and simply save it and load it at a later point again.

Here, you can see my custom UI. If you compare it to the standard UI with the same button size, the canvas is smaller. If you work with a large computer screen, it might not be an issue, though, and the improved efficiency might make it worth it:

Figure 1.29 – My custom UI

> **Important note**
>
> If you are just getting started with ZBrush, there is little benefit in downloading custom UIs or spending a lot of time planning and building one. Rather, add commands and buttons over time as you access the same tools frequently. That way, you keep your UI slim and don't overcrowd your workspace.

You should now feel comfortable navigating the ZBrush UI and creating your own custom UI. While you will only start to learn about most of the tools in this book in future chapters, you can start building your custom UI from the very beginning of your journey. Eventually, you will find ZBrush to be very intuitive and efficient, allowing you to spend the most time sculpting and producing results.

In the next section, we will explore the file types that ZBrush uses – the purpose of each type can be unclear for beginners, so clarifying them now will help you to save everything you need while avoiding saving unwanted content.

Understanding ZBrush file types

In this section, we will examine the main file types used in ZBrush: ZTool, ZProject, and ZDoc. This topic can be confusing in the beginning, so hopefully, an overview of the properties, as well as the pros and cons, will allow you to use them effectively. You will be able to store all the information you need but not in excess, reducing file size and saving hard drive space on your PC.

ZTools (.ztl)

ZTool files are the main ZBrush tool files used to store the sculpting, painting, and texture data associated with a 3D model in ZBrush. Layers, Morph Targets, and everything applied to your 3D models (subtools) will be saved.

You can save your file as a ZTool by clicking **Tool| Save as (ZTool)**.

ZTools will save the active tool, which consists of every subtool displayed in the subtool list. The subtools saved are the ones shown in the area highlighted in the following screenshot:

Figure 1.30 – All the content saved in a ZTool file

Note that newer ZBrush versions are able to open old `.ztl` files, but you cannot open a ZTool created in a newer ZBrush version than the one you are using.

ZProjects (.zpr)

When you save a project, ZBrush creates a single file that includes all the loaded ZTools. You can work on one ZTool as your main project, but if you are working with multiple ZTool files at the same time, **ZProjects** will save every loaded ZTool in the session. Beyond that, the ZProject will also save canvas (document) information, timeline animation, and the camera:

Figure 1.31 – All the content saved in a ZProject file, marked with rectangles

ZProjects are a great format to save your whole ZBrush session so that at a later time, you can reopen it and continue your work exactly where you left off.

The ZProject file size will be considerably larger than a ZTool file, so unless you need that additional information saved, choosing ZTool over ZProject will make more sense. You can save your ZProjects by clicking **File| Save as (ZProject)**.

ZDocs (.zbr)

ZDocs store information that is visible on your canvas (the canvas being the area on the screen where all the sculpting and painting is happening). Sculptors and modelers can export ZDocs and the associated canvas information to save images of their 3D models to be presented to their clients or to be further edited in software such as Photoshop:

Figure 1.32 – All the content saved in a ZDoc file, marked with a rectangle

This format will not store any 3D model, so avoid the mistake of trying to save your sculptures using the ZDoc format. It will simply save the pixels displayed at the time on the canvas. This can include your visible subtools, as well as any modification you made to the canvas itself, such as background colors, gradients, and dimensions. You can save your ZDocs by clicking **Document | Save as (ZProject)**.

The **Document** menu also contains **Import** and **Export** options, allowing you to import and export flat images. This is useful for creating previews or work-in-progress shots of your models for yourself or clients.

Now that you understand the properties of different file types, you'll be able to use them in many ways. The biggest takeaway is to avoid saving files as ZProjects if all the needed information is already included in the ZTool; this is because ZProjects are large files, and more storage space is required for busy sculptors.

Summary

Congratulations on completing the first chapter! If you are new to ZBrush, getting used to its unique navigation and UI might take a while, but you can feel great about having taken the first step in learning new software that has set up countless artists for a rewarding and fulfilling career.

In the first section, you learned how to load a model, navigate around the canvas, and create a simple sculpture. This allowed you to experiment with brushes and get a sense of ZBrush's creative potential.

Then, you got an overview of the ZBrush UI and learned how you can create your own custom UI, which increases your workflow speed, making you a more effective and valuable artist. Finally, you learned about different file types in ZBrush and when you should use which one.

In the next chapter, you will get to create your own concept sculpture using ZBrush's powerful **DynaMesh** functionality and learn about character design principles, which help you create better-looking models.

Further reading

- ZBrush's official tools documentation: `https://docs.pixologic.com/`
- Michael Pavolich's (artist and ZBrush instructor) YouTube channel, which contains videos covering ZBrush's UI and tools in detail: `https://www.youtube.com/channel/UCWiZI2dglzpaCYNnjcejS-Q`

2

Sculpting a Demon Bust with DynaMesh

In this chapter, we will start sculpting and using our creativity and imagination.

We will begin by looking at the purpose and goals of a successful concept sculpt before jumping into a useful tool called DynaMesh. It is a dynamic tessellation tool, which means that it can generate new topology on your mesh as you work on it, and allows for a particularly dynamic and flexible way of sculpting. You will learn the very basics of DynaMesh, using this to sculpt a demon bust.

Then, we will discuss character art and design principles, which are useful for ZBrush users who want to produce visually appealing sculptures for personal or professional use.

After completing this chapter, you will be able to sculpt a concept model in ZBrush from a reference using DynaMesh and the appropriate brushes and tools.

So, in this chapter, we will cover these topics:

- Concept sculpting with DynaMesh
- Sculpting a demon bust
- Exploring character art and design principles

Technical requirements

For the best experience, it is recommended that you have a strong PC that meets the minimum requirements described in the *Technical requirements* section of *Chapter 1*. However, you can work on this chapter with just a mouse, a functional PC setup, and a ZBrush license.

You can download the reference on collecting and viewing software for free here: https://www.pureref.com/download.php. This will allow you to follow along more closely with the practical sculpting project of this chapter.

Concept sculpting with DynaMesh

In this section, I will explain the goals and benefits of concept sculpting, as well as introduce you to DynaMesh, which is the perfect tool for this process.

Understanding concept sculpting

Concept sculpting involves creating visual concepts or blueprints that can be used to develop ideas and serve as a guide when creating larger, more detailed sculptures. ZBrush is used by many creative industries to do this, allowing you to create sculptures that explore and refine ideas in a loose, versatile, and quick fashion.

Here are some of the characteristics of a concept sculpt:

- They are rough and tend to look "unfinished" as they show relatively little precision and attention to detail. The purpose is to iterate fast so that the "bigger picture" can be developed.

- While concept sculptures shouldn't waste time on detail, they should communicate the ideas behind the design clearly so that the clients can imagine how the finished design could look.

- They are often created in a series as the artist goes through iterations to explore different angles and implement feedback along the way.

- They are versatile, allowing for quick changes and experimentation with different ideas.

Now that we understand the idea behind concept sculpting, let's look at ZBrush's most effective tool for concepting: DynaMesh.

Introducing DynaMesh

When you need to sculpt a concept, **DynaMesh** is a great option. At its core, DynaMesh is a dynamic tessellation tool that updates a mesh's topology, creating evenly sized faces throughout the whole mesh.

To understand this, we must understand what **topology** is: the structure of a mesh and the organization of faces, edges, and vertices that make up an object. Let's break down those terms:

- **Vertex**: A single point in three-dimensional space
- **Edge**: A line segment that connects two vertices
- **Face**: A flat surface defined by three or more vertices that are connected by edges

These elements are depicted in the following figure:

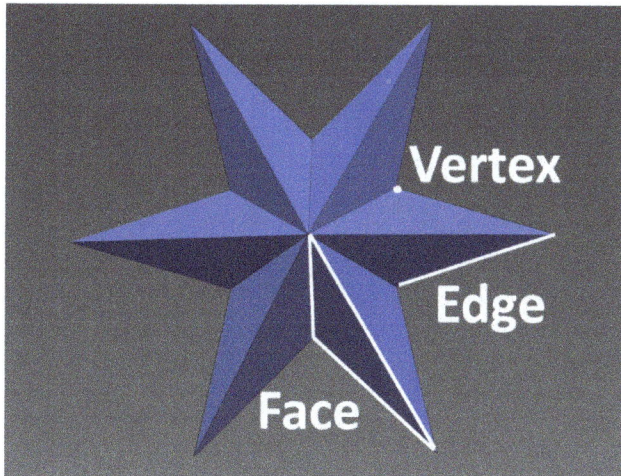

Figure 2.1 – Components of a polygonal mesh

When sculpting, you may find that the topology has become so distorted that sculpting brushes can no longer give you the desired results. At this point, DynaMesh comes in handy by quickly creating a new topology in a matter of seconds.

Here are some of the benefits of using DynaMesh for concept sculpting:

- It is very flexible since you are not restricted by the topology. Once your manipulation of the model distorts and stretches the topology to a point where you can no longer sculpt on it properly, you can create a new evenly dense topology and continue to refine your sculpture until you decide to repeat the process.

- Since you do not have to commit to a certain topology, you will not hesitate as much to make bigger changes to the model at any stage of the sculpting process. However, when working with a base mesh or classic retopology tools a large change in proportions would distort the topology too much or require more time-consuming work to restore the topology through poly modeling or a retopology workflow.

- It is easy to use and has great performance power. If your computer meets the minimum requirements, you will find that higher resolution DynaMesh models can be worked with smoothly and they create new topology quickly as well. However, keeping the polycount of your model below 1 million polygons will offer the best sculpting performance.

- DynaMesh allows users to merge multiple meshes into one continuous mesh. This usually requires time-consuming manual modeling or error-prone Boolean operations, but with DynaMesh, this happens automatically when two meshes surpass a certain proximity and a "DynaMesh operation" is performed.

The following figure shows the effect of a DynaMesh operation on a mesh with stretched polygons:

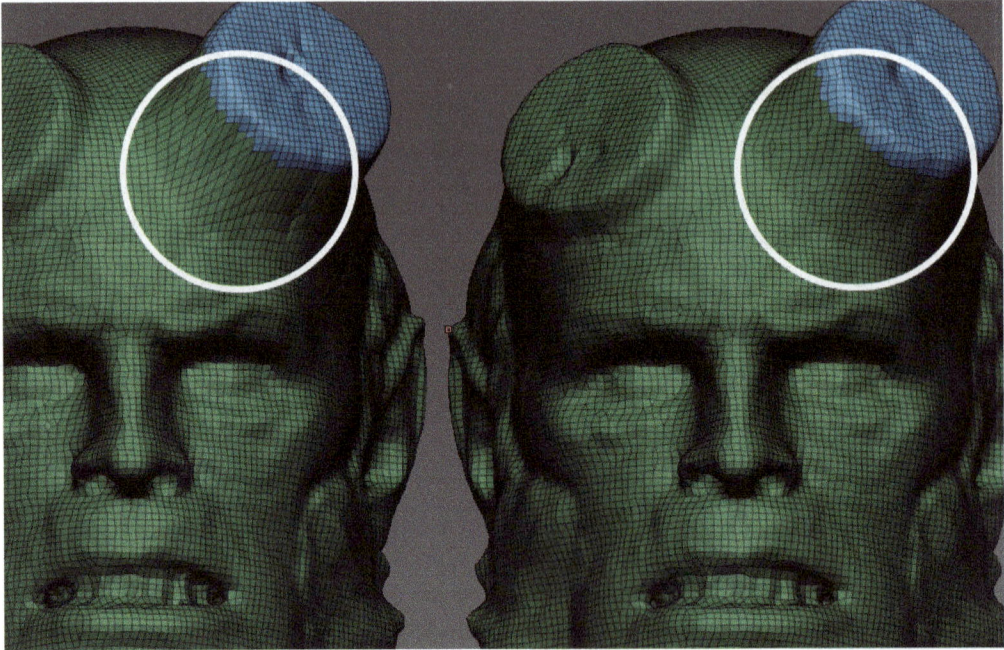

Figure 2.2 – Mesh with stretched faces before (left) and after (right) using DynaMesh

- It allows for a smooth, seamless sculpting experience. By removing many of the technical barriers to sculpting, an artist can focus exclusively on translating their vision into 3D and achieve a "flow" state.

Now that we know what DynaMesh is and how it can be beneficial in the sculpting process, especially for concept sculpting, let's put this information to practical use by starting to sculpt a demon bust using DynaMesh.

Sculpting a demon bust

In this section, we will begin creating a demon bust, which we'll ultimately finish in *Chapter 7*. This section focuses on blocking out the sculpture and establishing its overall proportions and design:

Figure 2.3 – Our demon concept sculpture

To prepare for this project, we will start by gathering references. Then, we will enter **DynaMesh** mode and use a couple of basic sculpting brushes to work on our sculpture. Along the way, we will look at how to set up hotkeys and some sculpting workflow tips.

Gathering references for your sculpture

The first step in creating a successful sculpture is often to gather reference images. This is especially important when the subject is new to an artist, but even highly experienced and skilled sculptors use a lot of references for their work. It presents the artist with unique ideas and allows them to identify common characteristics of successful artwork that can be adapted, modified, and included in their artwork.

Considering that the world's best artists use references, it would be a wasted opportunity for an aspiring artist not to take advantage of this phase of creation. Of course, how important it is depends on the specific circumstances and goals of any individual artist; as a general rule for beginners, it can be stated that any minute spent on this first phase of the project will pay off in the final result.

It is good practice to gather references for different aspects of the sculpting project. Here are some useful topics to consider:

- **Anatomy**: Sculpting a human or creature requires anatomical knowledge to produce a believable and visually appealing sculpture. Although some anatomical details seem subtle, we have evolved to pick up on inconsistencies quickly – especially when it comes to the human face. This means there is little room for error, so anatomy references are among the most important references to collect:

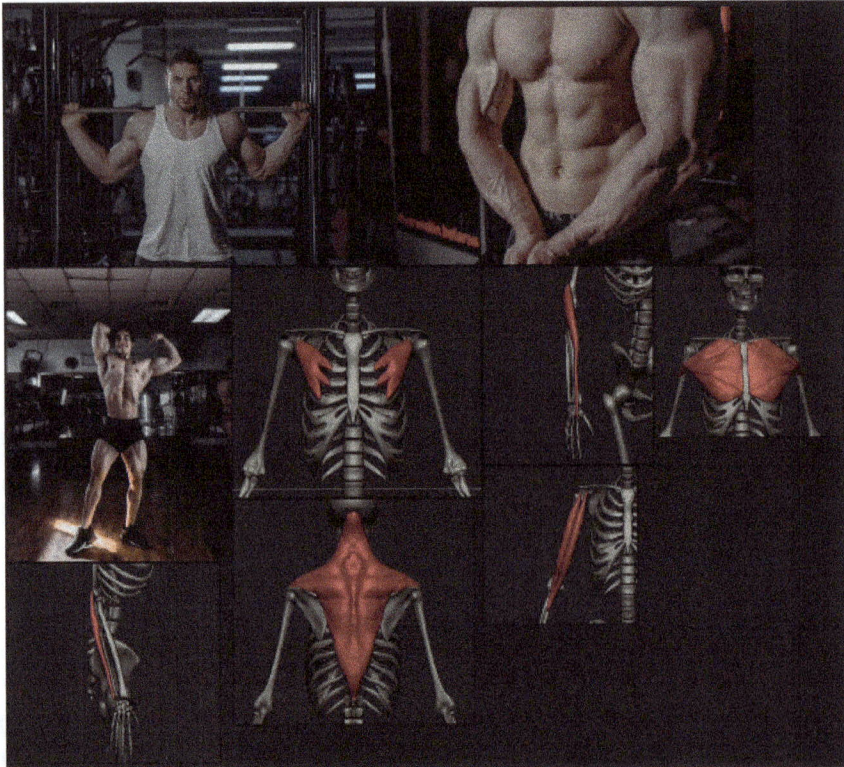

Figure 2.4 – Anatomy reference

- **Pose**: The pose or facial expression of a character or portrait sculpture can greatly affect the artwork. A character in a neutral pose or expression will look boring, even if the sculpture is of high quality. Taking inspiration from movies, illustrations, and fellow sculptors can help take your sculpture to the next level:

Figure 2.5 – Pose reference

- **Costume**: Character artists usually create characters with costumes, armor, and accessories. When designing these parts of the character, a lot of knowledge and experience are needed as it is an individual skill. If you do not know about designing costumes or armor, you should gather reference material and try to incorporate elements of existing designs or identify design principles to apply to your artwork:

Figure 2.6 – Costume reference

- **Lighting**: Lighting can have a big effect on the appearance and effect of a sculpture, so looking at different light setups can help you experiment more and create a lighting scheme that brings out the best in your sculpture:

Figure 2.7 – Lighting reference

- **Texture and color**: In many cases, you want your sculpture to have a certain texture and color. Although texture is not very important in the hierarchy of design elements, it can still have a powerful visual effect and allow for more storytelling. Photographers are aware of this and many popular pictures focus purely on interesting surface patterns. Different parts of your model, such as skin, cloth, and metal, can be enhanced with a realistic and proficiently applied surface structure. For that reason, it is a good idea to collect reference images for this aspect alone:

Figure 2.8 – Texture reference

Important note

If it is possible to collect multiple angles of your reference subject, make sure you do so – it will help you visualize the three-dimensional shape of your subject better. This holds especially true when working on a portrait or character likeness, where it is essential to have a high attention to detail.

Now, let's look at some resources you can use to gather references:

- **Google's Advance Image Search** (`https://www.google.com/advanced_image_search`): Here, you can specify large image sizes, which makes it a great resource for detailed reference. It is especially useful for finding reference material when creating likenesses:

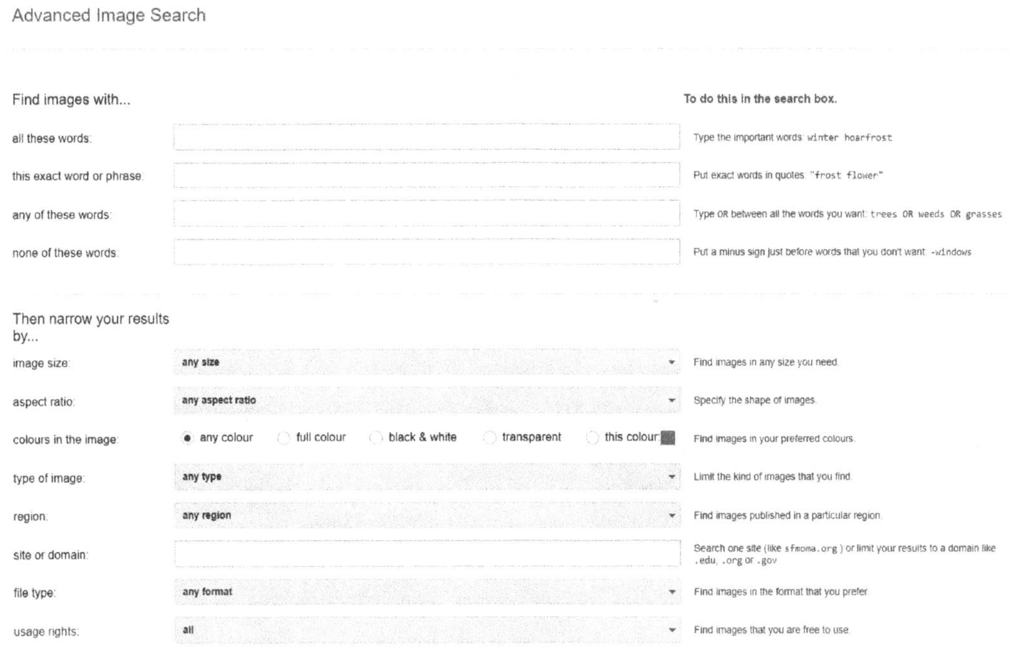

Figure 2.9 – Google's Advanced Image Search

- **Pinterest** (`https://www.pinterest.com/`): Pinterest's algorithm favors visually appealing and popular images, making it an ideal source of inspiration for artists and designers:

Figure 2.10 – Pinterest

- **ArtStation** (https://www.artstation.com/): ArtStation is an excellent resource for (aspiring) professional artists in the VFX, games, and collectibles industries. This website features much of their competition, helping you to judge your portfolio quality. ArtStation is also a great source of inspiration and reference images:

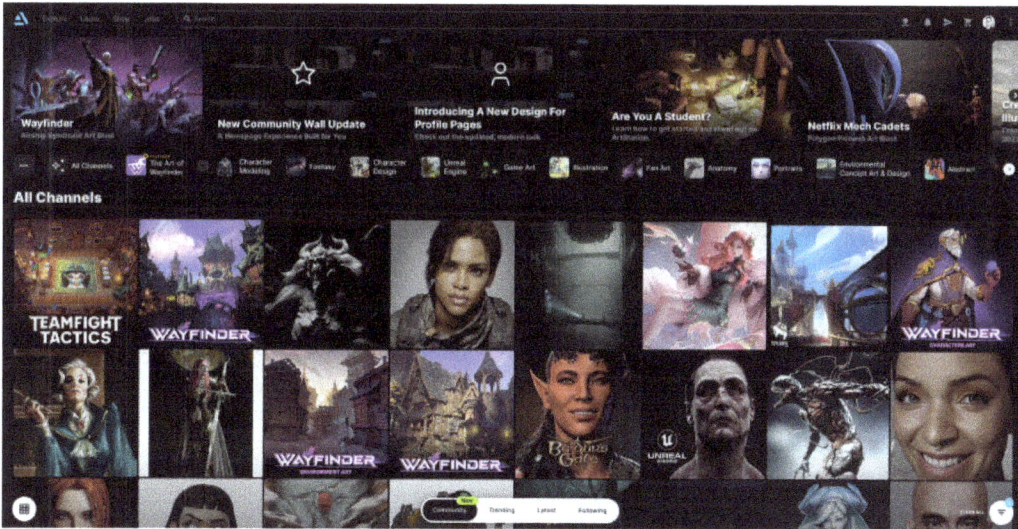

Figure 2.11 – ArtStation

Once you have gathered your references, you'll want to collect them somewhere. **Pureref** is a free reference viewing software that's used by many digital artists. It stores images conveniently and lets you make simple modifications to group and assemble them in your preferred way. You can download it here: https://www.pureref.com/download.php.

As a reference for our demon sculpture, we can use skulls and human faces for anatomy references and different sculptures and paintings for design inspiration. I have assembled a board of images that include some of the mentioned aspects, as well as some of my previous projects:

Figure 2.12 – Reference images in Pureref

At this point, you should have gathered some references that will set your project up for success. In the next subsection, we'll start sculpting.

Blocking out the demon

In this section, we'll begin blocking out the rough proportions and silhouette of the demon. Here, we want to create a suitable starting point from which we can refine the model more.

We can start this process by picking a sphere in the **Tool** palette and hitting **Make PolyMesh3D**:

Figure 2.13 – Accessing the Sphere3D model

I will introduce you to the most useful sculpting brushes throughout this chapter, but first, we will start with the **Move** brush. You can open the **Brush** menu by pressing *B*. Once you've opened the **Brush** menu, look for the **Move** tool and select it.

If you have a lot of brushes in your **Brush** menu, it can be inconvenient to have to browse all the brushes to find the one you are looking for. If you want to find brushes faster, you can also start typing *M* on the keyboard, which grays out brushes that do not begin with "M" – this makes it easier to find them.

The **Move** brush is used to move large parts of your model, which can help you to make drastic changes to the proportions and silhouette of your model. It is also frequently used for smaller changes that require precision, such as dragging small parts or even single vertices in 3D space.

To sculpt a symmetrical humanoid head, enable **Symmetry** mode by pressing *X*; once you've done this, you can use the **Move** tool to deform the sphere into something that looks closer to a skull or human head (you'll learn more about **Symmetry** mode in *Chapter 3*):

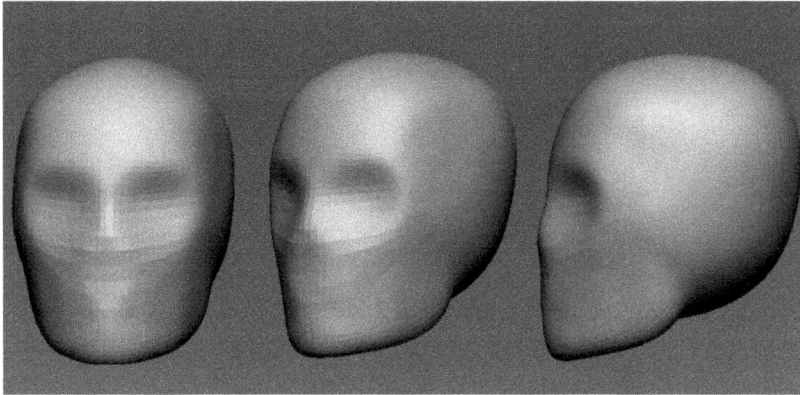

Figure 2.14 – Blocking out the head with the Move brush

Next, we can do the same for the upper body. Go to **Tool | Subtool | Append | Sphere3D**, then enable **Symmetry** mode and use the **Move** brush to shape the sphere into an upper body. Note that the result does not have to look perfect and refined at all, but you'll want to try and get the overall proportions right, like so:

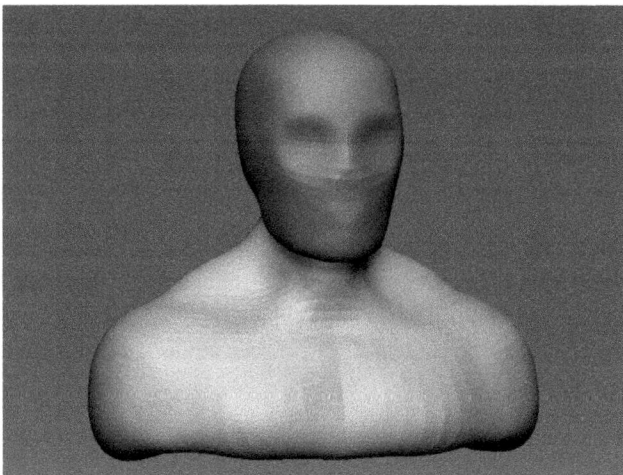

Figure 2.15 – Blocking out the head and upper body

From here, we can combine both subtools by selecting the head, which is the subtool in the list above the body, and then going to **Tool | Subtool | Merge | MergeDown**. You will learn more about subtools, the **Subtool** menu, and its functions in the next chapter.

Now, we can use DynaMesh and turn this subtool into a single object that shares one continuous surface. Then, we can sculpt the transition from neck to head, which will give us a seamless result.

Entering DynaMesh mode

Now, we can turn on **DynaMesh** mode, which can be found in **Tool** | **Geometry** | **DynaMesh**:

Figure 2.16 – DynaMesh menu

We can adjust the **Resolution** value to determine how many polygons our final mesh will contain. A higher value produces a higher polygon mesh; however, finding a good resolution requires some experimentation. As a general rule, it is good to keep the resolution as low as possible while being able to sculpt the level of detail needed. This will not only keep DynaMesh performing faster but will also help the artist focus on the bigger picture rather than commit to detail too early.

The default **Resolution** value of 128 produces a 180k polygon mesh for my sculpt, which is a good resolution to start working with. Hitting the large **DynaMesh** button will then launch **DynaMesh** mode.

DynaMesh not only updates the topology but also merges vertices that fall under a certain proximity. This results in two close meshes being merged into a continuous shape that shares one surface and has a shared volume. In the case of our demon bust, applying DynaMesh has now merged the head and body parts, creating one continuous mesh. Now, we can smooth out the model by switching to the smoothing brush (by holding *Shift*).

Now, we have a rough bust sculpture in **DynaMesh** mode. Next, we can start refining it more by using basic sculpting brushes.

In the following subsections, we'll look at two useful sculpting brushes: **ClayBuildup** and **DamStandard**. They are among ZBrush's most essential brushes and can be used to sculpt almost anything. You will also learn how to set up hotkeys, which will allow you to access brushes faster and increase your efficiency.

Sculpting the face with the ClayBuildup brush

To quickly add or remove the digital clay we have created so far, there are several useful brushes that we can use, but one of the most popular options is the **ClayBuildup** brush. This brush is perfect for adding and removing the digital clay of your model quickly, and it can be used to sculpt almost anything, making it a great brush for beginners who are sculpting for the first time; you may find that you prefer a different brush or different alphas eventually, but to start, you can't go wrong with this one.

By default, the **ClayBuildup** brush is equipped with a rectangular alpha. To obtain smoother results while sculpting, simply click on the **Alpha** icon to the left of the canvas and select **Alpha 06**. This brush and alpha combination allows for quickly building forms while keeping them smooth-looking.

Then, to sculpt with this brush, use your tablet pen to draw on the model By default, this brush is in addition mode (known as **ZAdd**), which means that this brush will add volume to the surface, building it up. You will see forms build up on your sculpture based on how much pressure you apply with your pen.

Alternatively, you can put your brush into subtract mode (**ZSub**) by holding down the *Alt* key while sculpting. This will allow you to remove digital clay and create concave shapes. You can also switch to this mode by enabling **ZSub** in the **Draw** palette or by right-clicking on the canvas and selecting it from a menu that pops up.

The result of sculpting with the **ClayBuildup** brush in **ZAdd** and **ZSub** modes can be seen here:

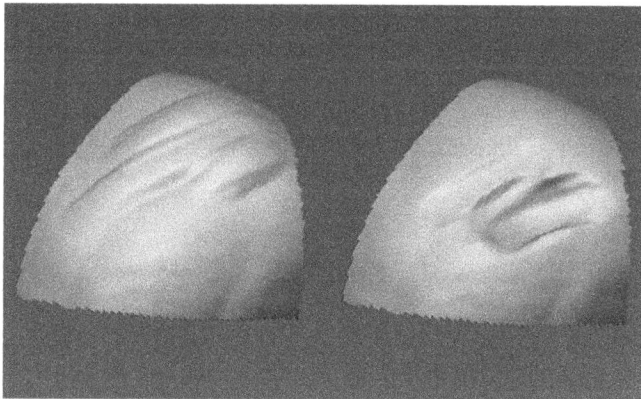

Figure 2.17 – The ClayBuildup brush in ZAdd mode (left) and ZSub mode (right)

Now that you know the basic functionality of the **ClayBuildup** brush, you can use it to build up some forms of your character's face. Here is a typical progression, in which you work from the overall proportions to slightly more refined facial features:

1. First, sculpt in the main proportions. A good way to start this is to sculpt a rough skull around the eyes. This will build the brow region and cheekbones, and ensure that you have sufficient

depth when adding the eyes and eyelids. You can also add a shape for the ears, and block out rough shapes for the nose and lips. Photos of both faces and the human skull – from various angles – can help establish these proportions more accurately. You can also sculpt the major upper body and neck muscles, as well as the Adam's apple.

Simply apply overlapping brush strokes with the **ClayBuildup** brush, building up the forms to resemble your reference. The result of this first sculpting pass should look similar to this:

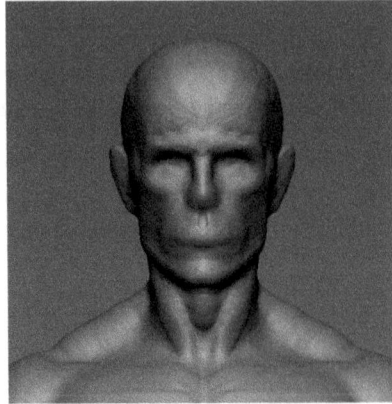

Figure 2.18 – First stage of blockout

2. Next, refine each of the facial features a bit more, applying careful brush strokes with the **ClayBuildup** brush. This way, you can sculpt the lips and add holes and more refinement to the nose and ears. As a beginner, it's not always very easy to get the result you intend to sculpt, but with time, this will become better and your brush strokes will be more precise and effective. For the eyes, you can add spherical shapes, out of which you can later make the eyelids. Now, the head should look like this:

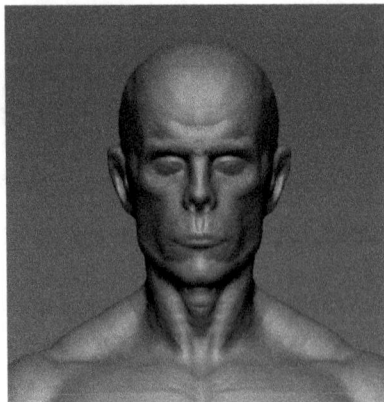

Figure 2.19 – Refining the facial features

Chapter 10 will cover how to sculpt heads and the individual features in more detail if you are interested in that.

3. Now, you can refine the whole face even more. Make sure you add space between the eyelids by using the **ClayBuildup** brush in subtract mode so that you can add the eyeballs later, and add more detail to any other area as well.

This stage can take a while, and you need both patience and good reference photos of faces to develop the facial shapes and make the face appear believable and well-balanced. Sculpting a realistic face is a challenging subject, so if you are a beginner and the first attempt is not looking too great, it is normal; you just have to keep practicing:

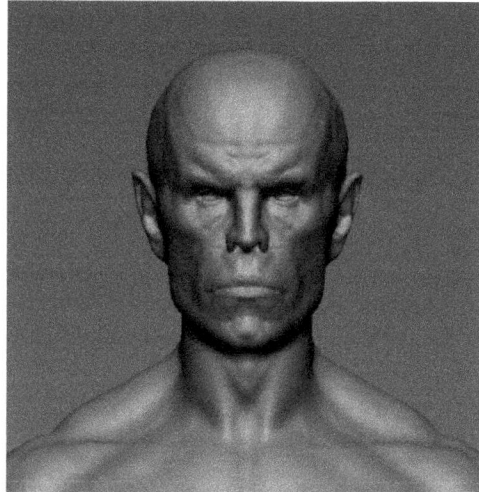

Figure 2.20 – Refining the face even more and adding space for the eyeballs

---

Important note

Whenever the polygons of your model become too stretched or compressed, and you want to update the topology, hold *Ctrl* and left-click, then drag a rectangle into an empty space. This makes DynaMesh generate new even topology and you can continue to sculpt.

---

4. Every once in a while, you can smooth out your mesh by holding *Shift*, which will equip the **Smooth** brush. With this brush, you can remove some of the artifacts that the **ClayBuildup** brush leaves on the mesh's surface. However, ensure you apply the **Smooth** brush with a small radius and with moderation so that you do not destroy the forms too much:

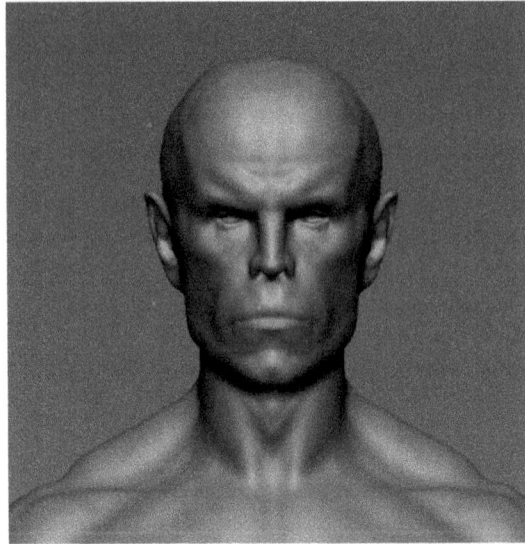

Figure 2.21 – Using the Smooth brush to give the mesh a more polished look

Using only the **ClayBuildup** and **Smooth** brushes can give you a lot of results, but when it comes to the finer detailing, ZBrush has the perfect brush: the **DamStandard** brush. We'll take a look at this brush next!

Defining shapes with the DamStandard brush

With the **DamStandard** brush, you can define certain shapes with more precision, which allows you to cut lines or make sharp protrusions by switching to **ZAdd** mode. While there are brushes that can achieve similar results, the **DamStandard** brush has proven to be the most popular brush among ZBrush users and is a great tool for any beginner sculptor.

As shown here, the **DamStandard** brush is great for adding sharp lines, creases, and wrinkles, which makes it a must-have tool for organic sculpting. Here, it was used to define many of the shapes in our demon's skull:

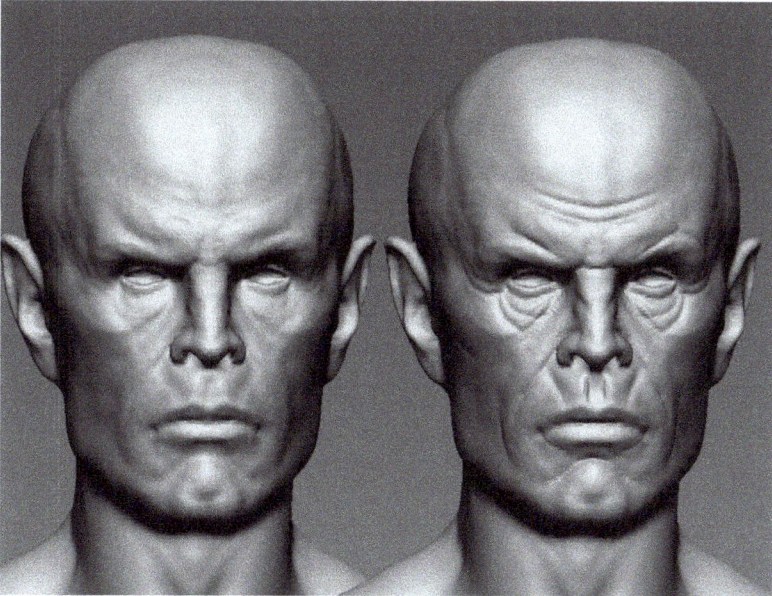

Figure 2.22 – Before (left) and after (right) using the DamStandard brush to add sharp lines

If you exaggerate the skull slightly, your demon bust will be more menacing:

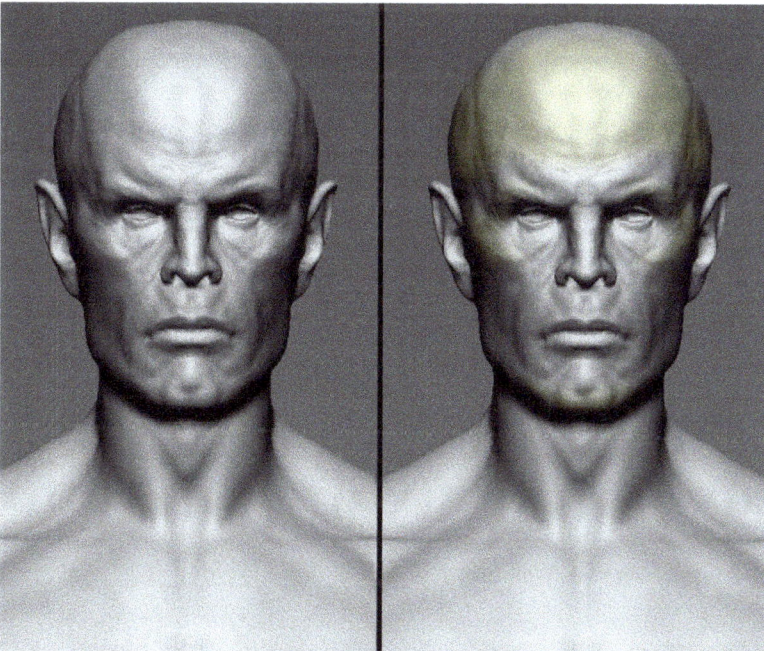

Figure 2.23 – Low-resolution sculpture (left) and skull structure highlighted (right)

With that, you know how to use the basic brushes to sculpt the demon head using a simple sculpting workflow. Next, you will learn how to speed up this workflow by creating hotkeys that allow you to access brushes faster, therefore saving a lot of time.

Using hotkeys to access brushes

Once you have been sculpting your demon bust for a while, you will notice that you frequently need to switch between brushes such as **Move**, **ClayBuildup**, and **DamStandard**. If you access these brushes by navigating to the **Brush** menu and selecting a brush, the time spent on this will accumulate and disrupt your sculpting flow.

To speed up the process, ZBrush allows users to create **hotkeys** to access frequently used features. If you want to create a hotkey for a tool, navigate to the appropriate brush or button, hold *Ctrl + Alt*, then left-click on the item. ZBrush will display a message, asking you to press any key combination. Now, you can assign your hotkey.

> **Important note**
>
> If you want to keep using hotkeys, make sure you save them. To do this, navigate to **Preferences | Hotkeys**. Now, you can either hit **Save** to save a hotkey file, which you can load at a later point, or you can select **Store**, which will make your current hotkey setup the default of ZBrush, so that it will be available every time you launch ZBrush.

Everyone has a slightly different workflow and uses different brushes and tools, but for whatever tools you use the most, creating a hotkey will save time and improve the sculpting experience, so make use of that option. For example, you may save your main sculpting brushes on numbers *1* to *5* on your keyboard, although this depends mostly on personal taste.

At this point, you are familiar with the most important brushes, and you can access them quickly, allowing you to focus on the sculpting and design process involved in it. Next, let's look at some tips for sculpting – specifically, for sculpting our demon's face.

Sculpting tips

At this stage, very limited technical knowledge or tools are required, and you can do most of the work using a handful of commands and sculpting brushes. Using these tools to their maximum potential and creating a great-looking sculpture will now depend more on the quality of reference you have, paying attention to design rules, and your approach to sculpting in general.

Before going back to refining your sculpture, let's take a quick look at some general tips for creating better sculptures in ZBrush:

- Make sure that your reference collection contains real photo references, as well as inspiring artwork, that you can find on ArtStation, Pinterest, or Google. Using other artists' work as a reference does not mean you have to copy exact parts, but rather you can create your own twist on their style, or just analyze the projects to see what makes them successful and find ways to use that for your project.

- Look at your model from every angle to make sure that you don't miss anatomy, proportion, or other mistakes. A low angle and side view can be especially useful to evaluate the proportions of a face and find flaws that are not as noticeable from the usual front view, which is what you work with the majority of the time:

Figure 2.24 – Looking at a model from different angles

- Take a step back from time to time to evaluate your model and see if the overall design is working well. Taking a break for a day or more lets you look at the model with fresh eyes so that you can judge it more objectively.

- To ensure realistic human proportions, you can check if a human skull would fit inside your model that matches the head shape without excessive intersection. You can find a skull in Anatomy ZTool in LightBox:

Figure 2.25 – Using a skull model to check facial proportions and anatomy

- If you are a beginner, it is better to sculpt a more humanoid demon as opposed to a "fantasy" design because it will be easier to create a believable and appealing design. If you decide to create a fantasy creature, make sure it is still based on human and/or animal anatomy to some degree.

> **Important note**
>
> Beginners often make the mistake of assuming that stylized and alien-looking characters do not require much anatomy foundation, or that you can sculpt those designs more easily. However, to achieve a successful stylization, you must first master human or animal anatomy and study realism – that is the foundation that allows you to take certain properties and exaggerate or alter them into a fantasy design that looks believable.

- Be patient with yourself: it can take time to see satisfying results, as with any craft. You may find it especially frustrating that you cannot translate what you have in your mind or your reference into the 3D model. Eventually, manual sculpting will not be the limiting factor anymore and your ideas, imagination, and knowledge become the bottleneck. At this point, sculpting becomes more fun and rewarding, so try persevering in the initial challenges.

- For more inspiration and to have a more visual and detailed demonstration of a DynaMesh blockout, YouTube has many timelapse videos. While these videos are usually sped up quite a bit, it can be useful to see the character creation process with no part missing. Here is a great example of such a video by Sergey Gricay: https://www.youtube.com/watch?v=3jxxfSAMD6Q.

These tips should help you sculpt the head more effectively. Once you have established the face of the demon, you can add extra details, such as horns, teeth, and eyes. We will do this next.

Adding horns

Every demon needs a pair of mean-looking horns. To add them, follow these instructions:

1. Paint a circular mask on each side of the demon's head by holding *Ctrl* and left-clicking. By default, this will enable the **MaskPen** brush, which lets you paint a mask on your mesh (if you want to learn more about masking and the various masking tools, take a look at the next chapter).

2. Invert the mask by holding *Ctrl* and left-clicking in an empty space on the canvas.

3. Press *W* to switch to **Gizmo** mode and translate the unmasked part away from the head, creating a horn shape.

4. Use the **Smooth** brush to taper off the sharp end of the horn.

5. Hold *Ctrl* while left-clicking and dragging a mask into an empty space to remove the mask.

6. Once again, hold *Ctrl* while left-clicking and dragging a mask into an empty space to let DynaMesh create a new topology.

Roughly, this process will look like this:

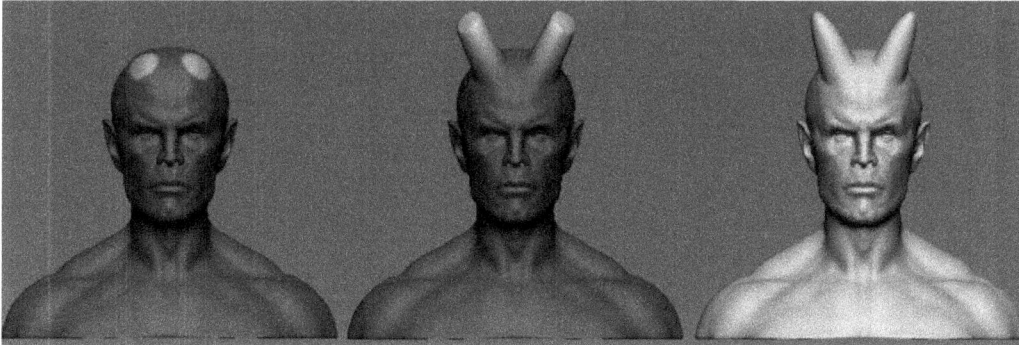

Figure 2.26 – Creating horns using masking and Gizmo mode

At this point, you can use the **Move** brush, or other sculpting brushes, to make more changes to the horns until you are satisfied with the shape.

Alternatively, you can use **Gizmo** mode to "push out" the horns.

> **Important note**
>
> **Gizmo** mode's main function is to move, rotate, and scale objects in ZBrush, but it also has some more specific functions, as well as additional deformation tools, that let you change an object's shape in various ways. You will learn more about all of its tools in *Chapter 3*.

Simply press *W* to access the gizmo and move it by holding *Ctrl* while left-clicking and dragging the white arrows. This way, you can push out the unmasked part. You can also scale it with the same **Gizmo** menu – that is, by holding *Ctrl* while left-clicking and dragging the yellow rectangle in the center of the gizmo. You can use this scale tool to get a tapering effect on the horn.

You can also rotate an unmasked part of the tip of the horns with the gizmo to bend the horn. After that, you can DynaMesh the result and repeat the process. This workflow will still need some adjustment with the **Move** and other sculpting brushes. It should look like this:

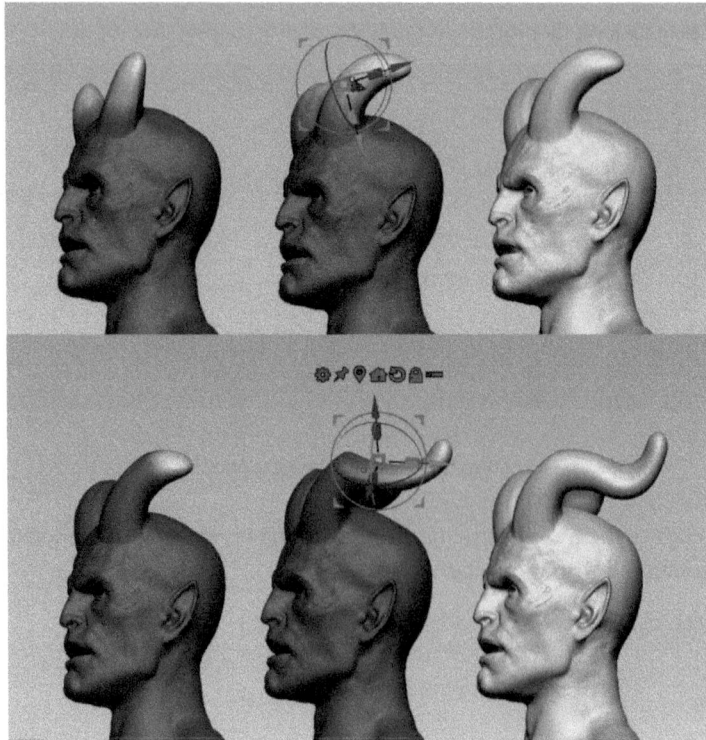

Figure 2.27 – Creating horns using gizmo rotation and sculpting brushes

I decided to make the horns on my demon model a bit longer and give them some curvature, as well as sculpt in some flat areas, which resulted in this look:

Figure 2.28 – Refining the horns

Now that our demon has horns, let's add teeth.

Adding teeth

To create a mean-looking demon, we should probably have some nasty-looking teeth showing.

First, we need to open the demon's mouth. We can do this by using the **Move** brush to pull the lips apart. Then, we want to use **Masking** and **Gizmo** to create a cavity – this is similar to how we created the horns, but pushing inward instead of outward (again, we will look at these two topics in more detail in the next chapter). This is the process:

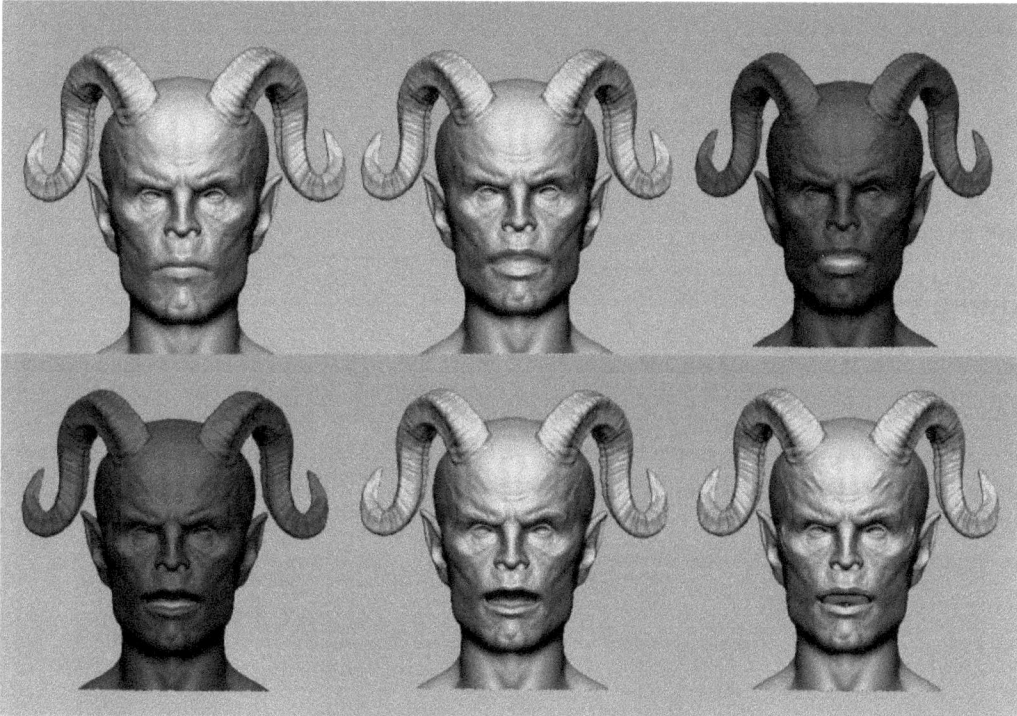

Figure 2.29 – Creating the mouth cavity

For the teeth, I recommend creating a new subtool instead of sculpting them. Our DynaMesh resolution is not quite high enough, and we do not want to risk merging the teeth. Saying that, we might want to keep space between them and modify each tooth individually.

You can turn any shape into teeth – I will append **Cylinder3D** and taper one end with the **Smooth** brush, like so:

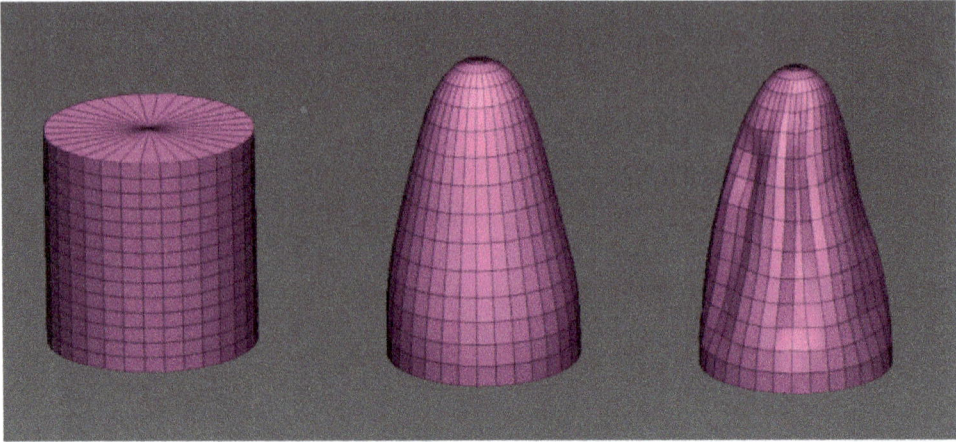

Figure 2.30 – Turning Cylinder3D into a tooth

Now, you can duplicate the teeth and place them in the mouth, using **Gizmo** mode's transforms to move, rotate, and scale the teeth.

Adding eyes

Eyes are the most important part of any character or creature design. This is because, as humans, we are naturally drawn to focus on the face, but even more specifically, on the eyes.

There are several ways to add eyes, but let's look at an easy way. Navigate to **Tool** | **Subtool** | **Append** and pick a Sphere3D model. Now, we can sculpt in some detail with the **ClayBuildup** brush in **Subtract Mode**:

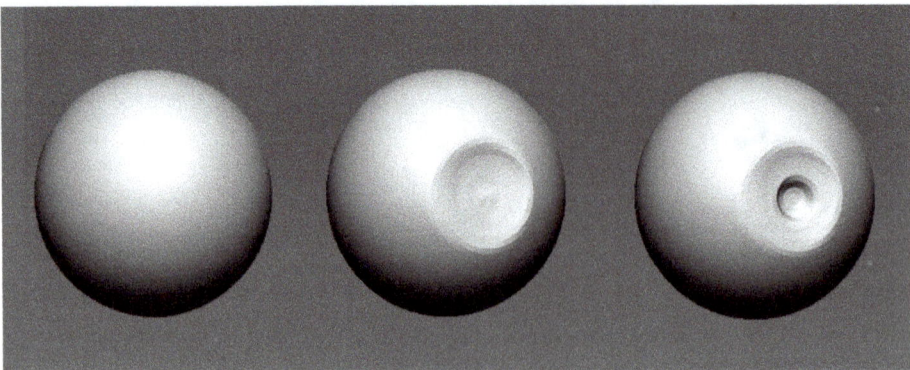

Figure 2.31 – Turning a Sphere3D into an eyeball

Position that eyeball in the left eye socket, then go to **Tool** | **Geometry** | **Modify Topology** | **Mirror and Weld**. With that, you have another eyeball mirrored across the *X*-axis. Make sure **Local Symmetry** is not enabled when you use the **Mirror and Weld** tool since this would weld the eye in its center instead of producing a second, mirrored eye.

This step requires that your model is both symmetrical and centered in the 3D space. If this is not the case, you need to duplicate the subtool and position it manually. After that, you can merge both eye subtools via **Tool | Subtool | Merge | MergeDown**.

Instead of duplicating the eye and then merging both subtools, you can duplicate the eye while staying in one subtool.

> **Important note**
>
> To quickly duplicate a mesh, hold *Ctrl* while moving the gizmo. Only unmasked parts of the active subtool will be duplicated as part of the same subtool. This only works if every continuous mesh in the scene is either 100% unmasked or 100% masked; otherwise, the gizmo will move unmasked parts instead of duplicating them. This operation can't be performed on a subtool with subdivision levels.

With that, the overall model has been established and we have a demon bust with horns, teeth, and eyes:

Figure 2.32 – Demon bust with eyes and teeth

Equipped with the **ClayBuildup** and **DamStandard** brushes, you can go a long way, and DynaMesh ensures a flexible and smooth workflow, allowing you to update the topology quickly at any point. However, DynaMesh also comes with some downsides that you need to pay attention to.

At this point, you should have blocked out your demon character and established rough proportions for it. You can spend as much time as you want to refine the sculpt and experiment with different proportions and shapes; if you're new to digital sculpting, this will also give you the chance to get more familiar with the tools and sculpting process itself.

Fixing bad DynaMesh results

DynaMesh's strength lies in its flexibility and ability to merge and deform parts quickly to create new shapes and explore ideas. However, when you are using DynaMesh on thin objects, it can give poor results. Let's take a look at how to avoid this.

If the thickness of a mesh is below the vertex-merging threshold of DynaMesh, it will create problems for the resulting model. In this case, you will end up with a one-sided surface without thickness and with holes in it, as shown here:

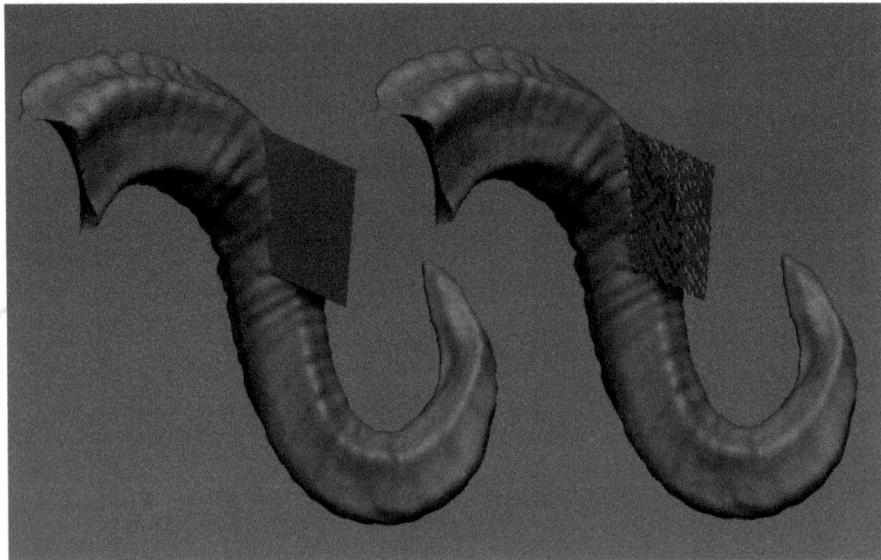

Figure 2.33 – Thin or one-sided surfaces do not work well with DynaMesh

When your mesh is not a solid, water-tight mesh, DynaMesh interprets it as a simple plane with no thickness or volume. This can happen by accident if there are holes in your mesh. In some cases, DynaMesh will simply close the hole and it will work, but other times, there will be a Swiss-cheese-type result, as shown in the preceding figure.

To fix these areas, use the **Inflate** brush (which is in your **Brush** menu by default). Inflate the area around the holes until there is sufficient thickness and no more holes are visible. After using DynaMesh again, holes should no longer appear, and you can continue sculpting:

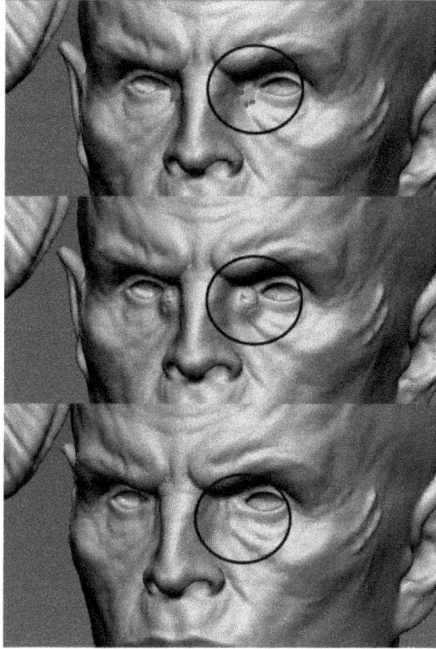

Figure 2.34 – Fixing holes in a DynaMesh mesh with the Inflate brush

That concludes the sculpting part of this chapter. You've learned about useful types of reference pictures, blocked out your demon bust, and refined it, adding horns, teeth, and eyes. At the same time, you've learned about hotkeys and other ZBrush tools, plus some general sculpting tips that allow you to sculpt more effectively.

In the next section, we will examine some of the most important design principles that will help us to improve our demon bust as we refine it.

Exploring character art and design principles

In this section, we'll examine character art and design principles. Having a good grasp of ZBrush's tools and being proficient in their use is great, but for many users, who sculpt characters and creatures for entertainment and collectibles industries, this is not enough. As a digital sculptor and modeler, you need to know what makes a great, convincing 3D character – and why. Understanding design principles will help improve the quality of any asset.

It should be stated that unlike learning how to use software, developing a good sense of design takes time, and results do not show as fast. The good news, however, is that it can be learned and improved just like any other skill. Analyzing successful designs and testing your designs, along with receiving feedback from others, are key ingredients to becoming a more proficient artist and designer.

Let's examine a few of the most important design principles for creating character art:

- **Shapes**: The shape language of a character design can greatly affect how we perceive that character. The silhouette is the most important element in this regard, but even smaller forms and shapes can have an effect. Square shapes signal strength, reliability, and trustworthiness; round shapes make a character appear more friendly, welcoming, and jolly; sharp or angular shapes create a more dangerous or evil-looking character.

  The following figure illustrates how modifications to the proportions of the face can make a character appear different in nature and temperament:

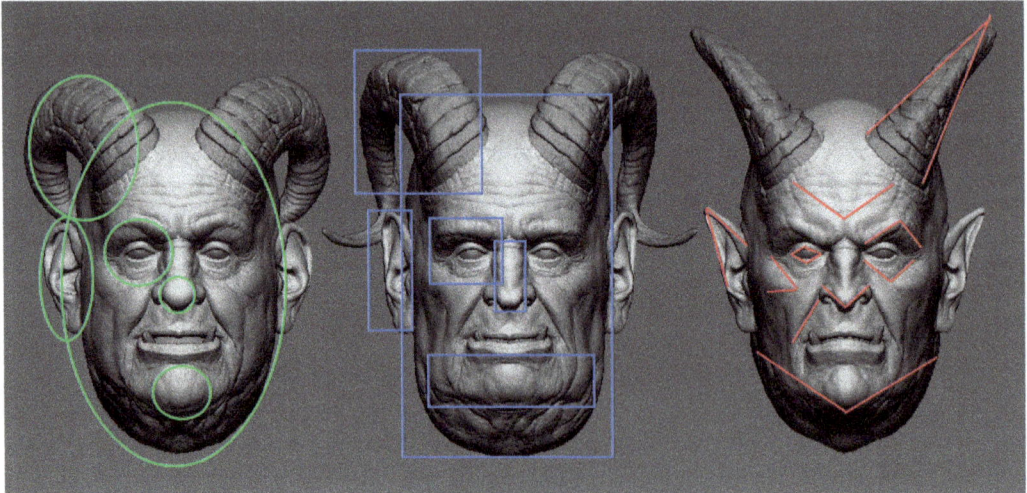

Figure 2.35 – The effect of shape language

- **Form**: Form is a three-dimensional shape that can give artwork depth. While positive forms protrude from the sculpture, negative forms are concave. Lighting also plays a major role in emphasizing form, either emphasizing or de-emphasizing it. Both positive and negative forms need to work together well to create an appealing sculpture with depth.

  If you take anatomy as a subject in art, you'll note how it lacks storytelling compared to other art subjects, and it usually does not present a very unique concept. However, the correct use of form in such anatomic sculptures, displaying depth and structure through shadows and highlights, makes for a visually appealing subject with a strong visual effect:

Figure 2.36 – An example of anatomy sculpting using form

- **Silhouette**: The same principles of shape language apply to silhouettes, where rounded shapes appear friendly, triangular shapes seem menacing, and square shapes appear stable and strong:

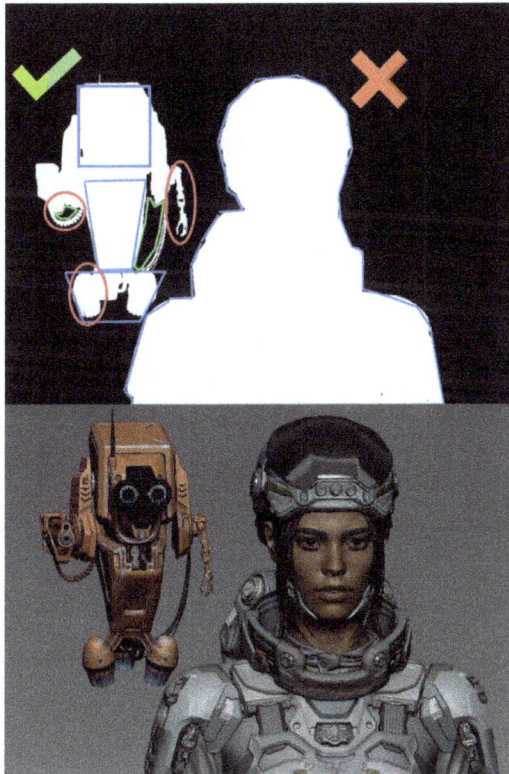

Figure 2.37 – Strong and weak silhouettes

The silhouettes in the preceding figure demonstrate a clear difference between a weak and a strong silhouette.

On the left, we have a robot, with a silhouette made up of rectangular and tapered rectangular shapes that are simple and easy to read. The silhouette also shows a bullet chain, a mechanical arm, and engine exhausts, which indicate the functionality and purpose of the robot. Plus, the use of negative space (marked in green) helps make the silhouette more interesting.

The character on the right looks minimalist, but the shapes can't be identified as ellipses, squares, or rectangles, but more like a mix. Furthermore, the silhouette does not provide any information or storytelling, and there is no use of negative space.

- **Pose**: Posing a character interestingly and dynamically can help create a strong and appealing silhouette. The pose may also tell a story about the character or emphasize what they are about. Here, the pose on the left suggests arrogance and contempt, with the King's hand pose possibly representing his desire for control. The hunched-over pose of the troll on the right shows age, while his interaction with the bird shows a good-natured and curious side to the character:

Figure 2.38 – Storytelling through poses

- **Guiding lines**: Guiding lines direct the viewer's attention to various aspects of the sculpture, retaining people's interest in the artwork longer. This photograph illustrates the use of multiple elements to guide the viewer's eye toward the center of the image:

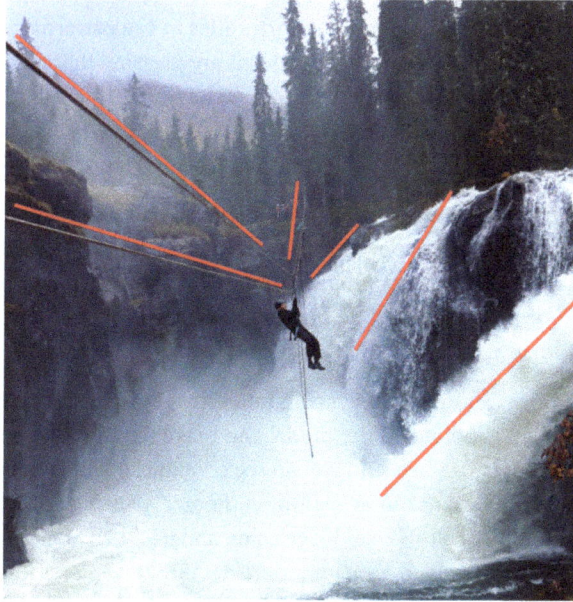

Figure 2.39 – Guiding lines in photography

- **Patterns**: Patterns are repeated shapes, objects, or colors that can be arranged in an organized or scattered way. Patterns are pleasing to the eye and there are many opportunities to incorporate them into your artwork:

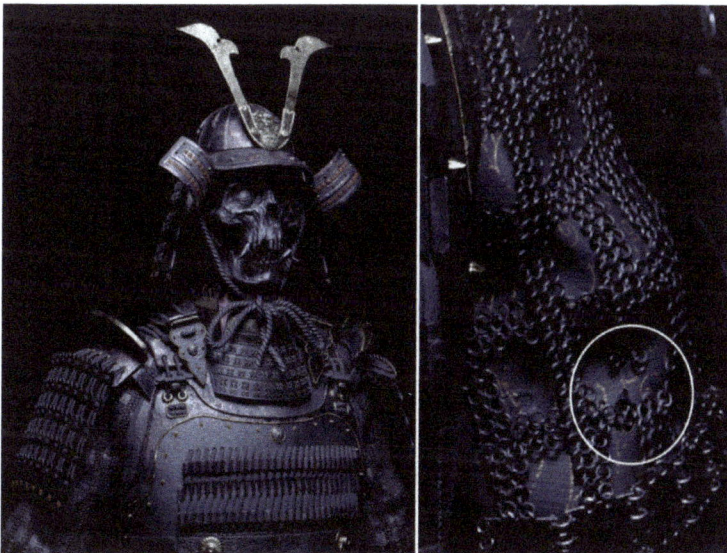

Figure 2.40 – Multiple patterns, including broken patterns (right) on a Samurai's armor

Costumes and armor are especially good opportunities to use patterns, as in the Samurai suit in the artwork here. While regular, ordered patterns are pleasing to look at, you can also break the pattern by interrupting the repetition of similar elements. This will create a focal point on your subject and can be taken advantage of by artists – damaged chainmail is an example of this technique.

- **Rule of thirds**: This rule refers to a ratio of elements in a work of art that can be applied to color, material, or level of detail. For example, it suggests that 2/3 of your sculpture should have minimal detail, while the remaining 1/3 should be highly detailed. As another example, it suggests that 2/3 of a model could be painted with something colorful and the other 1/3 could simply be painted white.

However, this rule does not mean that you should only apply this ratio. Uneven distribution can be used to create focal points on the model and make it easier to look at. In the following figure, which shows a girl in a sci-fi suit, we can see several elements with uneven distribution, which makes it possible to focus our attention on them – we can see hard armor parts versus soft (fabric) parts, turquoise color versus beige color, and low detail parts versus highly detailed parts. Including the robot, we even have low, more desaturated colors (turquoise, beige) versus saturated colors of the robot (orange), and while the girl takes up the majority of the frame, the robot only occupies a smaller space:

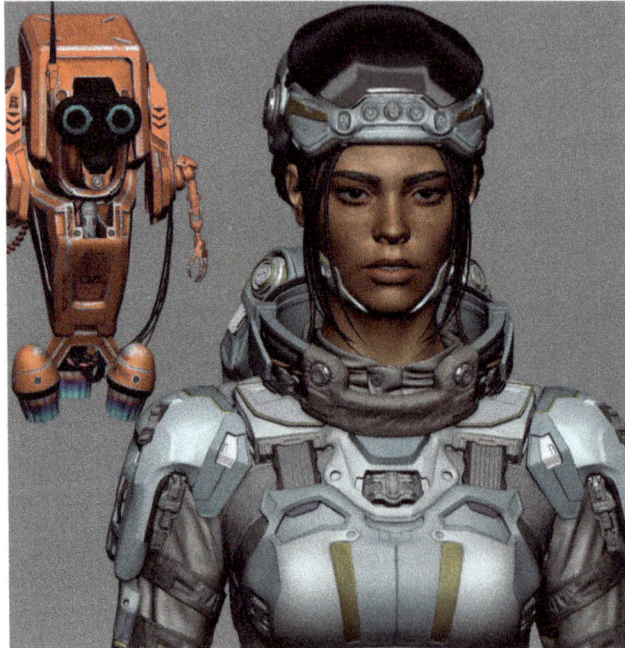

Figure 2.41 – The rule of thirds applied to multiple elements of a sci-fi design

- **Areas of rest**: This relates to the rule of thirds and states that there must be a balance between highly detailed and less-detailed portions of a sculpture. Areas with fewer details allow the eye to rest, which is why it is important to include those areas, especially in highly detailed sculptures. Adding details everywhere will not be pleasing to the eye as there are too many competing details competing for attention. However, none of these rules are absolute and you can also make a sculpture that's covered in detail if color or other elements are used to simplify the model and guide the eye:

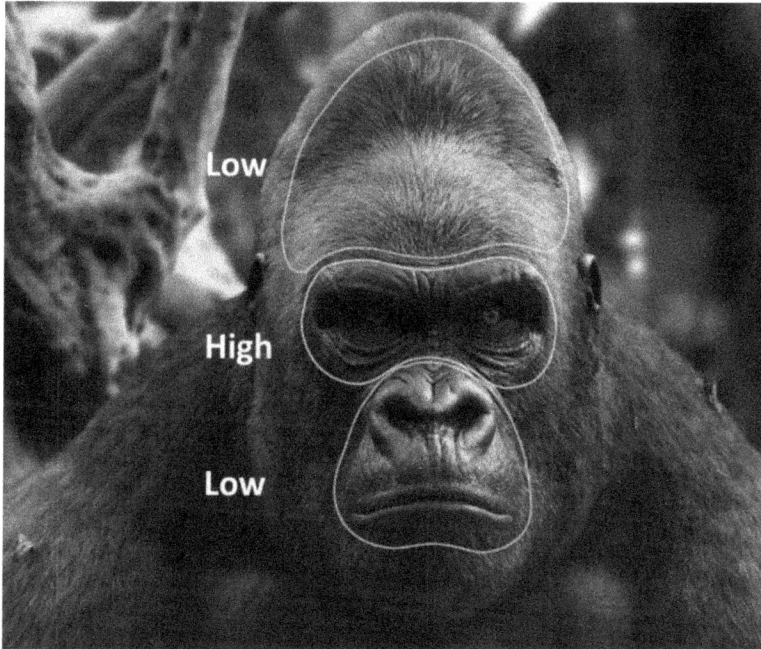

Figure 2.42 – A balance of more and less detailed areas

For this gorilla model, I made the area around the eyes highly detailed but kept other areas less busy. This created multiple focal points and made the composition more pleasing to look at than if I had covered everything in a lot of detail.

Now that we've looked at some of these guidelines' definitions and examples, let's see which ones we were about to include in our demon sculpture.

Design principles in the demon bust sculpture

Let's consider our demon sculpture:

- **Shapes**: The demon is rendered with triangular and sharp shapes to give it a more vicious look. The spherical shape of the upper head contrasts with this and helps the model to remain anatomically valid:

Figure 2.43 – Using shapes in our demon bust

- **Form**: The forms here catch highlights and create shadows, resulting in a sculpt with a sense of depth:

Figure 2.44 – Using form in our demon bust

- **Silhouette**: With the characteristic pointy ears and the easily identifiable horns, the silhouette of the demon is simple and easy to read:

Figure 2.45 – Using silhouettes in our demon bust

- **Guiding lines**: Several features, such as the horns, ears, or neck, create lines that lead the viewer's eye toward the center of the head, as shown in the following figure:

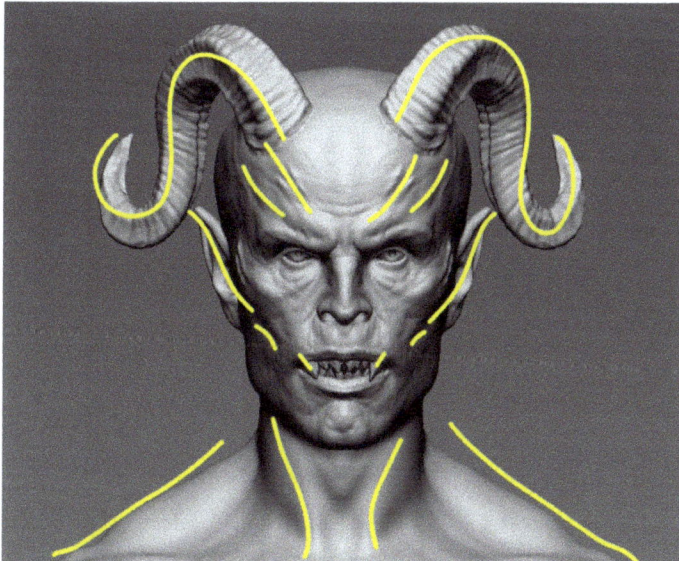

Figure 2.46 – Using guiding lines in our demon bust

- **Patterns**: The forehead and cheekbones have repeated bumpy structures, while the surface of the horns has repeated ridges:

Figure 2.47 – Using patterns in our demon bust

- **Areas of rest**: The sculpture includes two areas of rest, one on the forehead and one on the trapezius, creating a balance between high-detail and low-detail areas. This creates focal points on the sculpture and makes it more visually appealing:

Figure 2.48 – Using areas of rest in our demon bust

Later, we will add two additional design principles to our demon sculpture:

- **Pose**: Once we finish the detailing work on the head, we can disable the symmetry, then rotate and tilt the head to give it a more natural and interesting pose.

- **Rule of thirds**: When we add detail and color to the demon, we will apply these using a ratio that is based on the rule of thirds, making our design easy to read

That concludes this section, in which I introduced design principles such as silhouette, pose, and guiding lines to help improve the design quality of your ZBrush sculptures. These guidelines and rules break down the sculpture into elements and shapes, and allow you to approach the sculpting process more methodically. This is especially useful for beginners, who can create visually appealing results without a lot of experience and put them to use in their portfolio or start creating for clients and employers.

Summary

This chapter concludes our discussion of DynaMesh and digital sculpting.

We began this chapter by discussing the purpose of concept sculpting and why DynaMesh is the perfect tool for it before using the tool to create a bust of a demon. After creating a rough sculpture, I went on to introduce some of the most important design and character art principles with examples from my own projects. With this information, you can refine your sculpture further while aiming to implement these principles in your work.

You should now be confident in picking any concept – whether it be your own or one chosen by someone else—and starting to recreate it in digital clay inside of ZBrush.

In the next chapter, you will dive deeper into some of ZBrush's most powerful functionality, including **Gizmo** mode, **Symmetry** mode, and the **Masking** and **Selection** tools. This will allow you to handle a much bigger variety of modeling and sculpting tasks, and improve your overall efficiency.

3

Exploring the Gizmo, PolyGroups, and Masking

In this chapter, you will explore some of ZBrush's most powerful functionality, demonstrated on the **DemoSoldier** 3D model from ZBrush's content library, LightBox. We will examine ZBrush's individual 3D model units called subtools, and how to modify and move these subtools in 3D space.

Then we'll introduce ZBrush's powerful **Masking** and **Selection** tools, using them to make additional modifications to our model, as well as looking at PolyGrouping, an efficient way to organize your models, make local changes, and maximize your efficiency as a modeler or sculptor.

After completing this chapter, you will be able to navigate ZBrush's project structure and create your own 3D model consisting of multiple subtools, which you can deform and transform using various techniques.

So, we're going to cover the following topics:

- Managing your 3D model
- Using the Gizmo to move objects
- Understanding PolyGroups
- Using Selection and Masking tools

Technical requirements

For the best experience, it is recommended that you have a strong PC that meets the minimum requirements described in the first chapter's *Technical requirements* section. However, you can work on this chapter with just a mouse, a functional PC setup, and a ZBrush license.

Managing your 3D model

To begin this chapter, we are going to load the demoSoldier.zpr ZProject file from LightBox, in order to have a practical example while I introduce the next topics.

To do this, press , on your keyboard to open LightBox and navigate to the **Project** sub-menu. Then double-click on demoSoldier.zpr and the project will load:

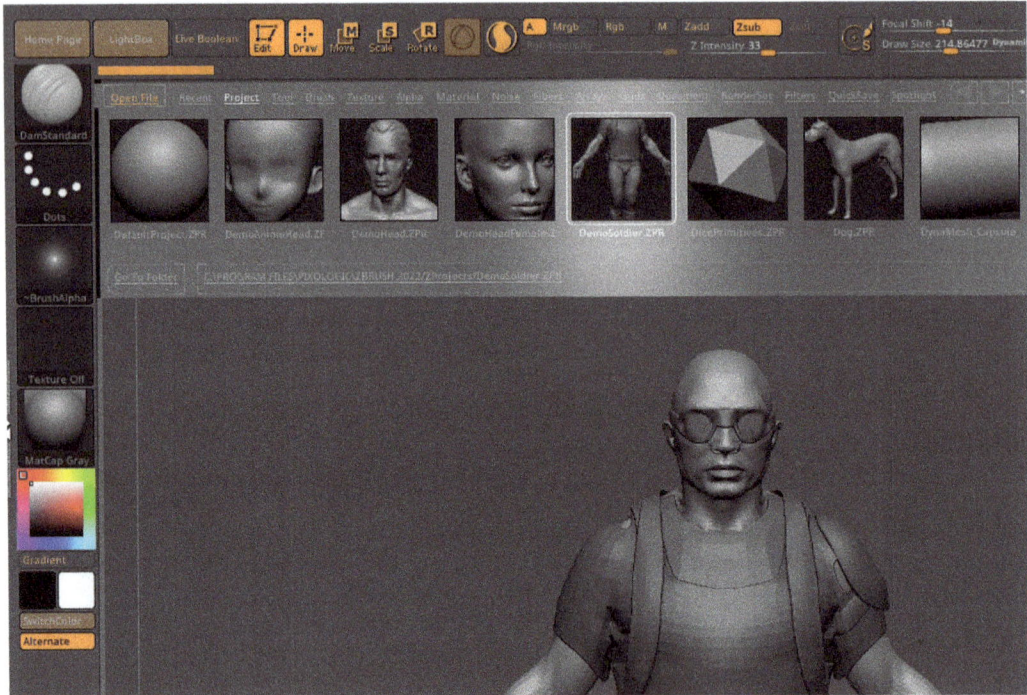

Figure 3.1 – DemoSoldier, available in LightBox

With the project open, we will now look at subtools, the individual 3D models within a tool, introducing you to several useful operations for managing these subtools. We will also learn about **Symmetry** mode, a powerful functionality that most ZBrush users could not live without – as its name suggests, it allows you to achieve symmetric results by mirroring your sculpting on one side to the other side.

Subtools

Subtools are individual meshes (objects) in your 3D scene. In our solider scene, you can see how different clothing pieces such as **shirt** or **backpack** are listed in the **Subtool** menu:

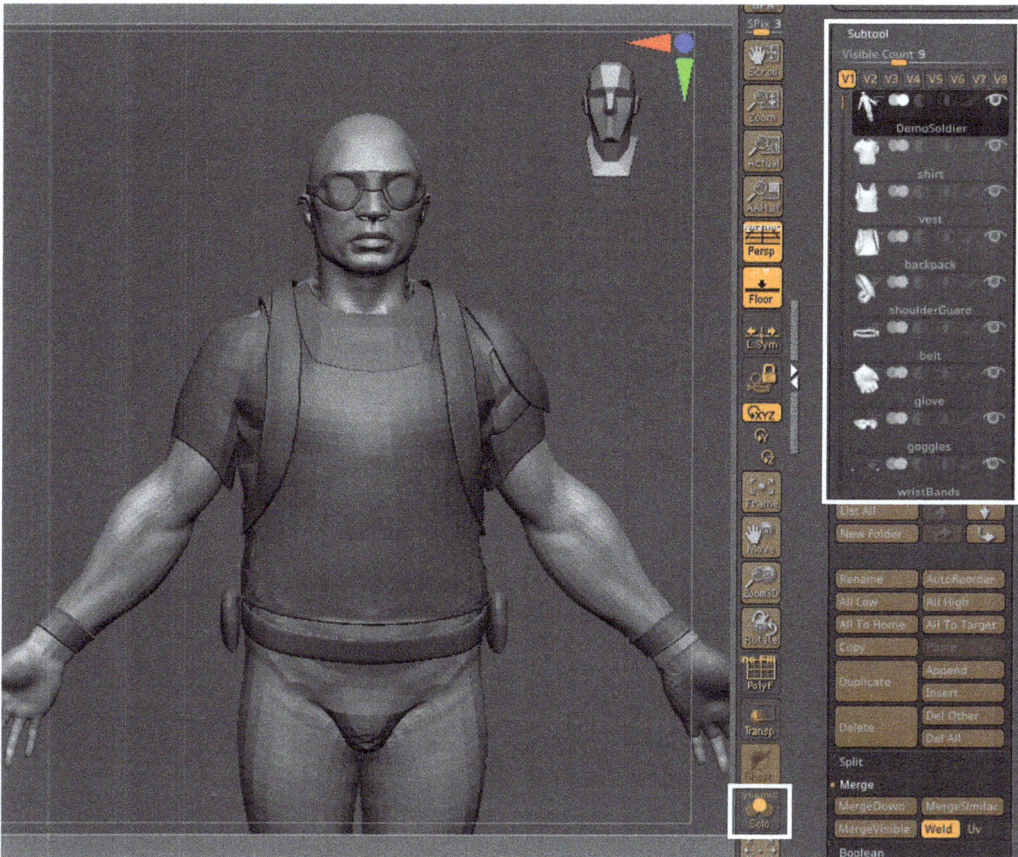

Figure 3.2 – The Subtool list (large box) containing the 3D models
of your scene and the Solo button (small box)

You can isolate subtools with the **Solo** button, located near the bottom of the vertical toolbar on the right side of the screen; this allows for the easier inspection and modification of the isolated subtool.

Each subtool has a couple of icons next to it:

- The eye icon, which turns the visibility of subtools on and off (note that the subtool will always be visible when selected)

- The paint brush icon, which toggles the color of the subtool on and off (if color was applied to the model)

- The three circular icons are related to Boolean operations, an advanced modeling technique that allows for the merging and subtracting of multiple subtools

Below the subtools, we can see some of the most common subtool commands. The following table explains their functions:

Command	Function
1: **New Folder**	Lets you group subtools, to keep the subtool list small and organized.
2: **Select up/down** (**straight arrows**)	Selects the subtool above/underneath the currently selected subtool. You can also press the up and down keys for the same action.
3: **Move up/down** (**curved arrows**)	Moves the currently selected subtool down in the list by one spot.
4: **Rename**	Lets you change the name of the subtool. This becomes more important as the project becomes more complex.
5: **Duplicate**	Lets you duplicate a subtool.
6: **Append/Insert**	Lets you add a tool from the **Simple Shapes** menu or append a subtool from another open ZTool.
7: **Delete**	Deletes the subtool (be aware that you can only recover a deleted subtool by loading a previously saved file; the **Undo** button does not apply here).
8: **MergeDown** (accessible in the **Merge** sub-menu, below the previous commands)	Combines the selected subtool with the subtool underneath to create a single subtool.

Figure 3.3 – The main subtool commands

Now you should have a basic understanding of subtools and know the main commands for modifying and managing them. Next, let's begin to manipulate the model using **Symmetry** mode.

Symmetry mode

Symmetry mode allows you to work on one side of the model and have the changes automatically applied on the other side. This is great for creating objects that are more symmetrical by nature, such as humans, animals, and many hard surface structures. There are a number of sculpts that require symmetry, which is difficult to achieve without this mode, but ultimately, this method will save you more time.

You can activate **Symmetry** mode by pressing X on your keyboard or navigating to **Transform |
Activate Symmetry**. By default, objects will be mirrored on the x axis, but there is also the option to
choose different or multiple axes.

You will know if **Symmetry** mode is enabled because there will be two brush cursors (red dots) visible
on each side of the model:

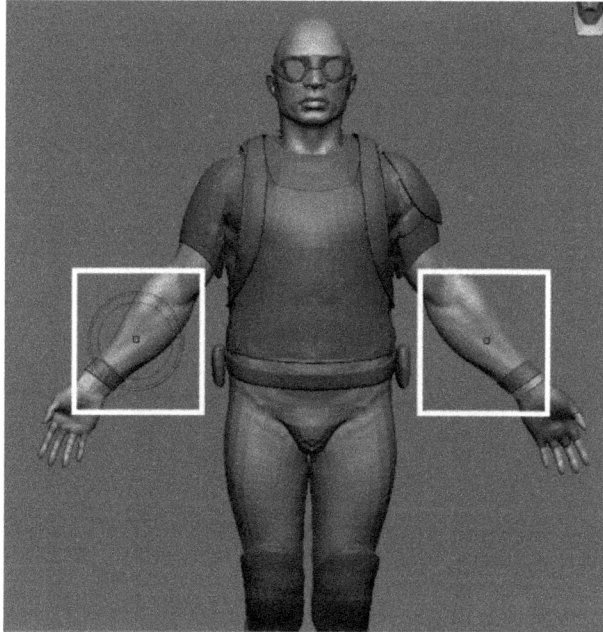

Figure 3.4 – Symmetry mode indicated by a brush cursor on each side of the model (the red dots)

Before using **Symmetry** mode, there are a couple of things you should pay attention to:

- To begin, make sure your mesh is symmetrical. This is because ZBrush will have trouble applying
  mirrored changes if the model is not symmetrical. To do this, go to **Tool | Geometry | Modify
  topology | Mirror and Weld**.

> **Important note**
>
> This operation will change the topology of your model and will not work on a mesh with
> subdivision levels.

- Check whether your model is centered in the scene at the axis you want to work symmetrically
  on. If the model is not centered, **Local Symmetry** provides a solution by assuming the center
  of the model to be the center of symmetry instead of the center of the scene. To use this, go to
  Transform and enable **Local Symmetry**.

Alternatively, you can go to **Tool | Masking | Go to unmasked center**, and then using the Gizmo (explained more in the next section), select the **Mesh to Axis** symbol in order to move the object to the center. Now, your mesh should be in the center of the scene, and you can work in **Symmetry** mode from here.

In the following figure, you can see how local symmetry works on a model that is offset from the center of the scene, as long as the model itself is symmetrical:

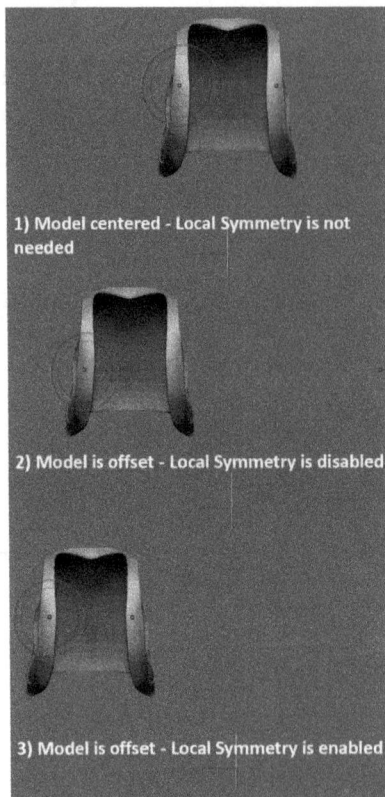

Figure 3.5 – The effects of Local Symmetry mode on a model

- If you have an almost symmetrical mesh, but you moved some parts by accident, making it asymmetrical, you might be able to fix it easily by navigating to **Tool | Deformation** and using the **Smart ReSym** button. This only works if the mesh is centered and has a symmetrical topology to begin with, though.

In this section, we learned how to manage the models that appear in the **Subtool** menu, including covering useful operations such as duplicating and appending subtools. We also looked at **Symmetry** mode and how it can be used even for an offset model. In the next section, we will discuss the Gizmo and how to move objects.

Using the Gizmo to move objects

The **Gizmo** is a tool with multiple functions – though it allows you to complete simple tasks such as translating, rotating, and scaling objects, it also lets you assemble and build up your 3D scene, change proportions, center objects, or apply more sophisticated deformations to models (the last of which we will see more of in *Chapter 9*).

To access the Gizmo, press the **Move**, **Rotate**, or **Scale** button located above the canvas in the default ZBrush UI. Alternatively, you can access them through the **Transform** menu, or by pressing *W* for **Move**, *E* for **Scale**, or *R* for **Rotate** on your keyboard. A colorful icon with two arrows should appear, like so:

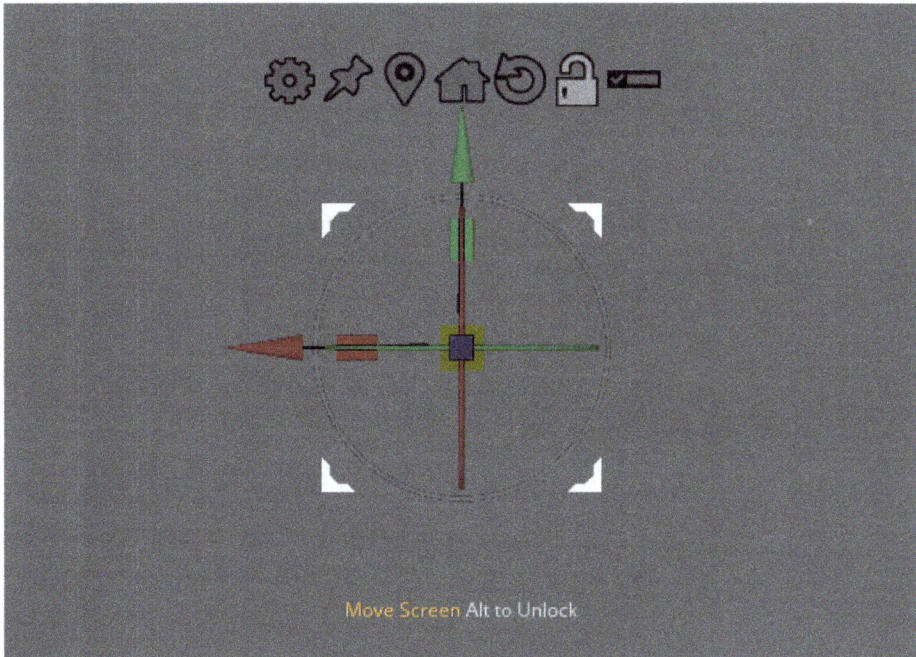

Figure 3.6 – The Gizmo

The Gizmo tool should appear close to the selected subtools; however, in certain scenarios, it can be located somewhere in empty space and hard to locate. In that case, you can select **Tools | Masking | Go To Unmasked Center**.

Once you have the Gizmo visible on screen, you can adjust your model using the following options:

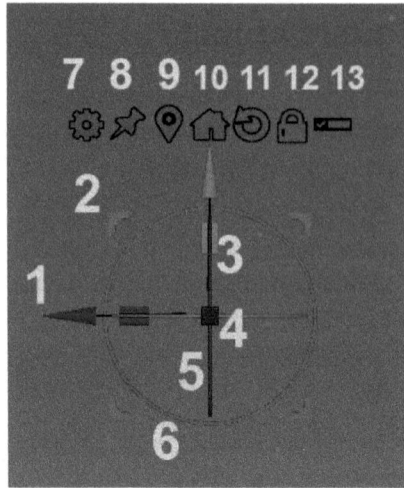

Symbol	Function
1: Red, green, and blue arrow ends	Move the subtool in one dimension.
2: White arrow ends	Move the subtool freely.
3: Red, green, and blue arrow rectangles	Scale the subtool in one dimension.
4: Center yellow square	Scales the subtool uniformly in three dimensions.
5: Red, green, and blue circles	Rotates the subtool around the circle.
6: White circle	Rotates the subtool relative to the orientation toward the Gizmo
7: **Customize**	Lets you turn the active subtool into a simple shape and offers various tools that let you deform your model in different ways.
8: **Sticky** mode	Makes the Gizmo snap back to its original location after moving it.
9: **Go To Unmasked Mesh Center**	Locates the Gizmo at the unmasked center of the selected subtool.
10: **Mesh to Axis**	Moves the subtool to the center of the 3D scene.
11: **Reset Mesh Orientation**	Resets the orientation of the subtool based on a reoriented Gizmo.
12: **Lock/Unlock**	If it is unlocked, all the previous transforms are applied to the Gizmo itself, rather than to the subtool.
13: **Transpose All Selected Subtools**	When enabled, the Gizmo transform will be applied to every visible, unmasked subtool in the scene.

Figure 3.7 – The Gizmo transform symbols and their functions

In order to use some of the transforms that we have discussed, let's make some changes to the soldier. We're going to duplicate the **Wristband** subtool and fit it around the soldier's upper forearm:

1. Select **WristBand** in the **Subtool** menu and use the arrows and circles on the Gizmo to position the wristband on the upper forearms. You should be in **Symmetry** mode by default, because this subtool was last saved with Symmetry mode enabled.

2. Before you scale it, make sure **Local Symmetry** is enabled so the model is scaled based on its own center. You can test the different effects of using **Local Symmetry** and disabling it.

3. Try making the band a bit thinner without distorting the shape by adjusting the Gizmo to be aligned with the scale direction (remember to hold *Alt* or enable **Unlock** to move the Gizmo and not the model).

Don't worry if the shape of the band does not match the forearm perfectly for now, as this is just a simple practice; you will be able to adjust shapes more precisely by using the **Move** brush (we will look at this brush in more detail in the next chapter).

Here is what the soldier model should look like. Note how the Gizmo is perpendicular to the band model, allowing you to scale it down in one dimension:

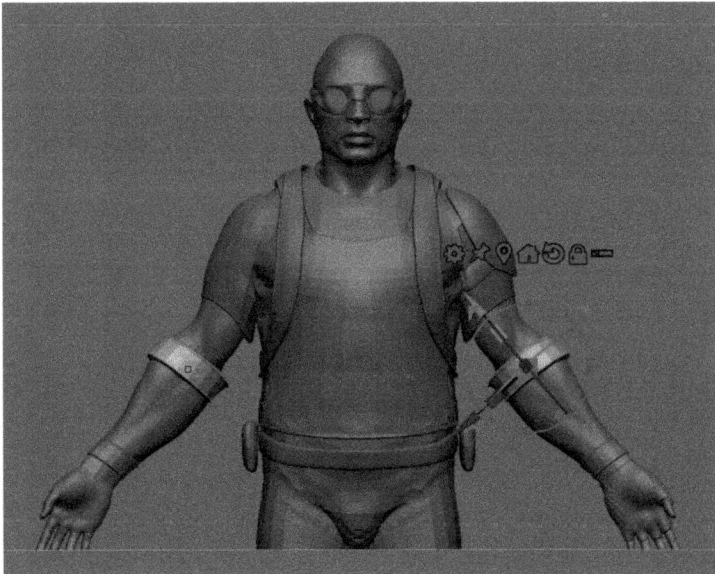

Figure 3.8 – The soldier model with a new band around the forearm

That concludes the section on the Gizmo – we learned how to move, rotate, and scale subtools, as well as translate and orient the Gizmo itself. In the next section, we will take a look at PolyGroups, which can be used to quickly select parts of your models, allowing for greater speed and efficiency in your workflows.

Understanding PolyGroups

PolyGroups are groups of polygons that share a unique color. A PolyGroup can be selected, isolating it from the rest of the mesh (3D model) by hiding the remaining PolyGroups. This makes it easier to modify the model without affecting parts of it accidentally. It is especially helpful when working on specific parts repeatedly, as it saves time.

The division of your mesh into PolyGroups is relevant for a variety of tools and operations. Some of these will be covered in later chapters, but by far the most common use is to hide part of the mesh for a more convenient modeling or sculpting experience.

> **Important note**
> You can see the PolyGroups by pressing *Shift + F* or enabling **Polyframe** by hitting the **PolyF** button in the **Transform** palette.

The following screenshot shows how a complex armor on a 3D character is made up of many PolyGroups, allowing you to isolate parts and work on them conveniently:

Figure 3.9 – The armor is a single subtool, but multiple PolyGroups allow fast
access to individual parts of the subtool for convenient editing

There are several ways to create PolyGroups, but here are the most common ways:

- Apply a mask to part of your model and press *Ctrl + W* – the masked part will be turned into a new PolyGroup. Alternatively, you can achieve this by going to **Tool | PolyGrouping | Group masked**.

- If there is no masked area on your model and you press *Ctrl + W*, the selected (visible) part of your model will be a new PolyGroup. Alternatively, you can perform this operation through **Tool | PolyGrouping | Group visible**. If you hide part of a mesh, only the visible portion is made into a new PolyGroup.

These options let you create PolyGroups based on selection and masking, which I will explain in the next section.

Here are other useful PolyGrouping tools, located under **Tool | PolyGroups | Auto Groups**:

- **Auto Groups**: If your selected subtool consists of multiple meshes, not combined by polys, **Auto Groups** will turn every continuous mesh into an individual PolyGroup. You can try it for the soldier model by selecting the **Boots** subtool and pressing **Auto Groups**. Since the boots are two separate objects they will appear in two different colors, indicating their PolyGroup.

- **Groups By Normals**: In order to understand this option, we must first understand that a normal in 3D represents the orientation of a polygonal surface. You can see the normals of this torus shape indicated by the lines extruding from the polygonal faces:

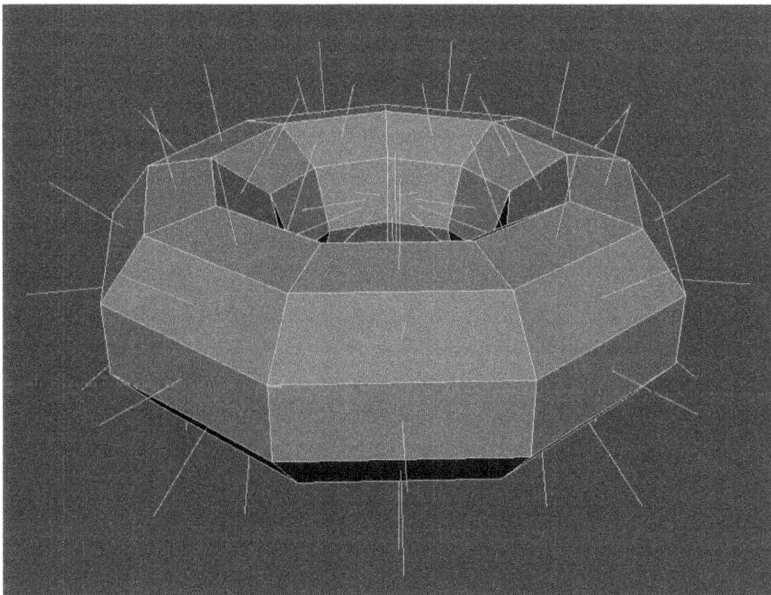

Figure 3.10 – Normals displayed on an object in Maya (3D software)

The **Groups By Normals** function groups polys based on the angle from one normal to the next. You can set the tolerance of this angle based on the desired result.

In the following screenshot, *A* shows an approximately 90-degrees angle between polygons that starts a new PolyGroup along that edge, whereas with *B*, polygons with a normal angle difference of less than 45 degrees are part of one PolyGroup:

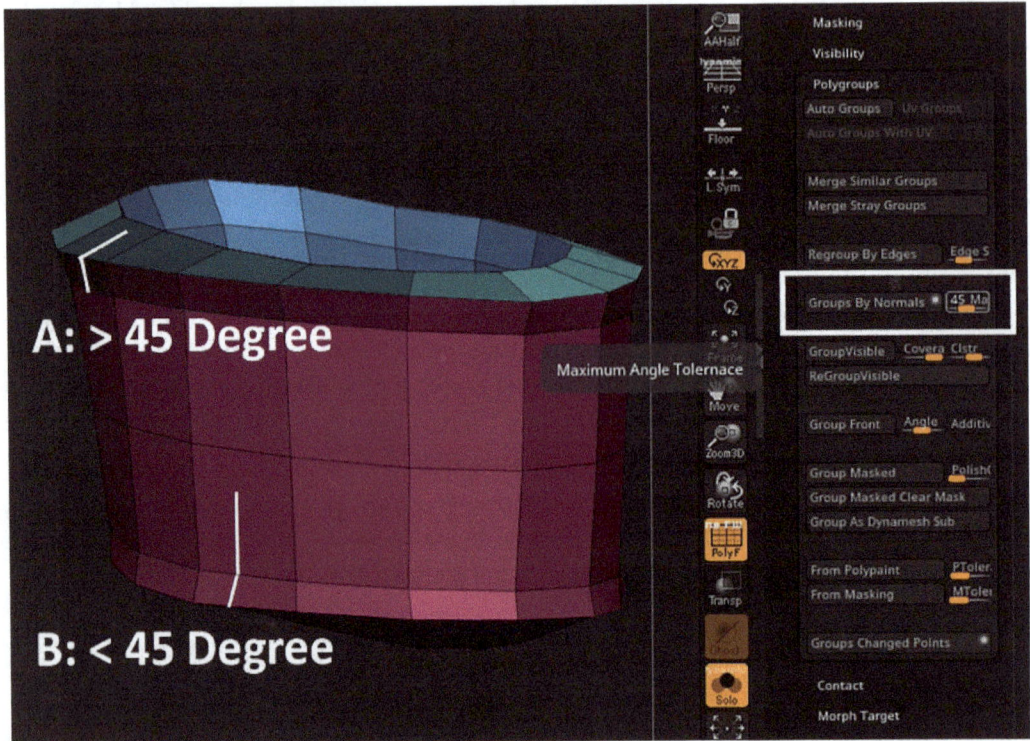

Figure 3.11 – PolyGrouping via the Groups By Normals function

This option is useful for hard surface objects with strong edges and a small amount of large normal changes so the number of generated PolyGroups is small and useful for your personal workflow. The **WristBand** subtools of the soldier model are a good example of that – you can see how there is a sharp edge between the large outward- and inward-facing parts of the wristband and the border side, which determines the wristband's thickness.

To sum up, we can say that the primary purpose of PolyGroups is to allow easy and quick access to parts of a model. Now let's look at the **Masking** and **Selection** tools, which will help you create PolyGroups most effectively.

Using the Selection and Masking tools

In this section, we are taking a look at the **Masking** and **Selection** tools. Both play an important role in organic sculpting by allowing us to focus on a smaller target area in a variety of ways; hard surface and polymodeling would not be possible without them.

Selection

Making selections is important for a range of operations, but the most common use lies in the ability to work on a smaller area of polygons while making sure the rest stays unaffected.

The **Selection** tool becomes active when you hold down *Ctrl + Shift*. By default, you will have **SelectRect** equipped. Once you switch to this selection mode, the brush icon on the left side of the canvas will change from your primary sculpting brush to the icon of the selection brush.

In order to make a selection on your model with **SelectRect**, hold *Ctrl + Shift* and drag the green rectangle over the part of your model that you want to isolate:

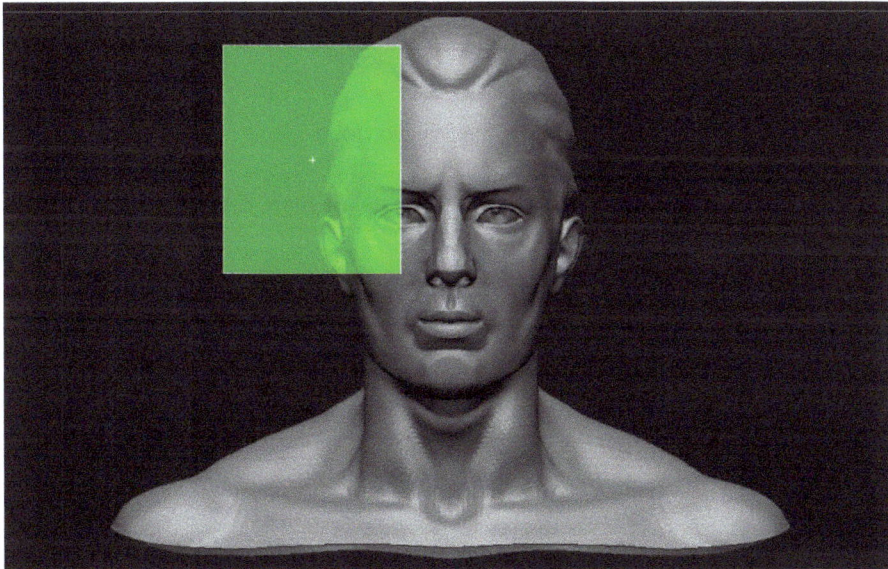

Figure 3.12 – Applying a selection on a 3D model

If you want to change the selection brush, hold *Ctrl + Shift*, then click on **SelectRect** on the **Primary Brush** icon on the left side of the canvas. This will open a menu showing alternative brushes.

Although ZBrush refers to these brushes as "selection" brushes, most of them are not actually used for selection. However, there are two available selection brushes, **SelectRect** (as we have just seen) and **SelectLasso**, which are some of the most used and most important brushes in this menu. Other brushes, such as **TrimCurve** and **SliceCurve**, that are not used to make a selection will be explained later in the book.

You can select those brushes by left-clicking on their icon:

Figure 3.13 – The Selection tools

Now select the **SelectLasso** brush. This tool allows you to make a more detailed selection by drawing a selection area:

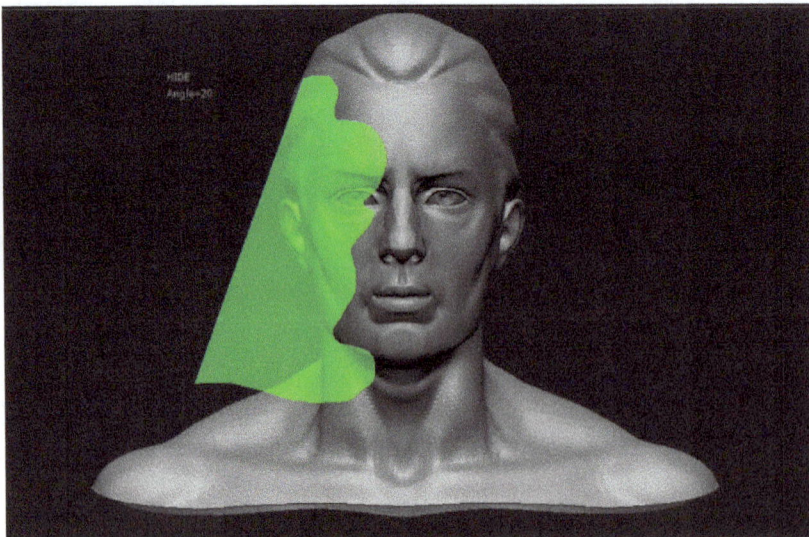

Figure 3.14 – Using SelectLasso

When you select one of the available brushes in the **Select Brush Slot** window, ZBrush will ask you to confirm the brush as your new automatic selection. Then, from that point forward, holding *Ctrl + Shift* together will always equip that brush until you select another one of the brushes.

Now you can select faces (polygons) with any selection brush of your choice, and the rest of the mesh will be hidden. If you make a selection in empty space, it will invert the selection.

Let's return to our soldier example and try to use one of the two selection tools to isolate the pouches attached to the belt and scale them without affecting the belt itself:

1. Start by selecting the **Belt** subtool and switching to **Solo** mode in order to view the model without accessories obstructing the view.

2. Make sure that **Symmetry** mode is enabled, and then use **SelectRect** or **SelectLasso** to select the pouch.

3. Make sure that **Local Symmtetry** is enabled, and then go into **Gizmo** mode, place the Gizmo on the pouches, and scale them up.

4. Bring back the full visibility of the belt subtool by pressing *Ctrl + Shift* and clicking on any empty space.

5. Disable **Solo** mode to observe the full character with the new adjustments.

Now you should have a character with bigger pouches on the belt, looking similar to this:

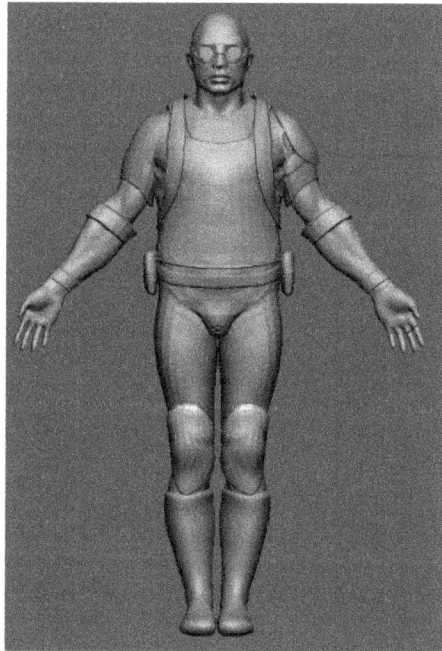

Figure 3.15 – The modified soldier

If you want to delete parts of your model, you simply hide/de-select everything you want to delete, and then go to **Tool | Modify Topology | Delete Hidden**.

Also in the **Tool** menu, under **Visibility** this time, you will find some useful options that help control the visibility of the mesh. This includes **Grow**, which will grow the selection, and **Shrink**, which will shrink the selection.

If you *Ctrl + Shift-click* on a PolyGroup, it will isolate this group. If you click on the isolated PolyGroup again, the selection will be inverted. Now you can keep clicking on other PolyGroups to hide them and continue using hiding and inversion to create the exact selection you need.

In this section, you learned about selections in ZBrush, and the two most important selection tools, which let you isolate parts of your model for further, focused modifications.

Next, you will explore an equally essential tool, which is masking. Here, you will, once again, learn about the most important brushes, and practice on the soldier model example.

Masking

Masking is a vital tool in ZBrush that allows you to freeze polys, making them not react to any sculpting, modeling, or deformation you apply to the mesh. This is a great way to achieve specific results and there is a lot of room for creativity in how you apply different kinds of masks in combination with different kinds of deformations.

To apply a mask to your subtool, hold the *Ctrl* key and left-click/paint with your pen. **MaskPen** is equipped by default and lets you paint a mask like a regular paint or sculpting brush.

Holding the *Ctrl* key will make you switch into **Masking** mode and make the **Masking Brush** icon appear on the primary brush icon spot, similar to how the selection brush appears when applying a selection. Click on it to open the **Masking Brush** menu. There are a couple of other useful masking brushes from this menu to consider:

- **MaskCircle**: Useful to create oval or round masks
- **MaskCurve**: Can be used to create precise, geometric masks with sharp corners or smooth curves
- **MaskLasso**: Great for masking irregular shapes quickly

You can see the effect of all four brushes here:

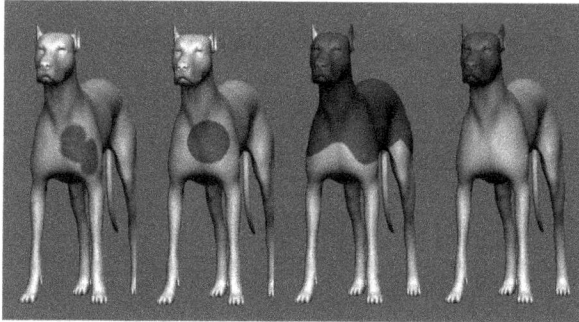

Figure 3.16 – The results of different masking tools: (from left to right)
MaskPen, MaskCircle, MaskCurve, and MaskLasso

When you select one of the **Masking** brushes, ZBrush will ask you to confirm that brush as your new automatic selection. Then, from that point forward, holding *Ctrl* will always equip that masking brush until you select another brush.

When precise, specific results are required and/or the model is complex, there are some more masking functions that can help create a more accurate mask faster. These commands are available under **Tool | Masking**. Here is the effect of these masking functions on a mask:

Name	Function
1: **Clear Mask**	Removes the mask from your model. Do this by dragging a mask in empty space.
2: **Blur Mask**	Fades the mask to allow for a smoother result when working on an area that transitions from masked to unmasked. *Ctrl* + left-click on a masked area to blur the mask.
3: **Sharpen Mask**	This is the opposite of the **Blur Mask** function – it increases the contrast of the mask and will result in a more binary state of the model, where your actions either affect the model 100% or not at all. *Ctrl* + *Alt* + left-click on a masked area to sharpen the mask.
4: **Invert Mask**	As the name suggests, this command will invert the mask, in case you need to apply changes in alternating fashion. *Ctrl* + left-click in empty space to invert the mask.
5: **Shrink Mask** 6: **Grow Mask**	Sometimes growing or shrinking a mask with the pressing of a button can be useful if manual masking with selection brushes proves to be challenging. Click **Navigate to Tool	Masking	Grow Mask/Shrink Mask** to grow or shrink the mask on your model.

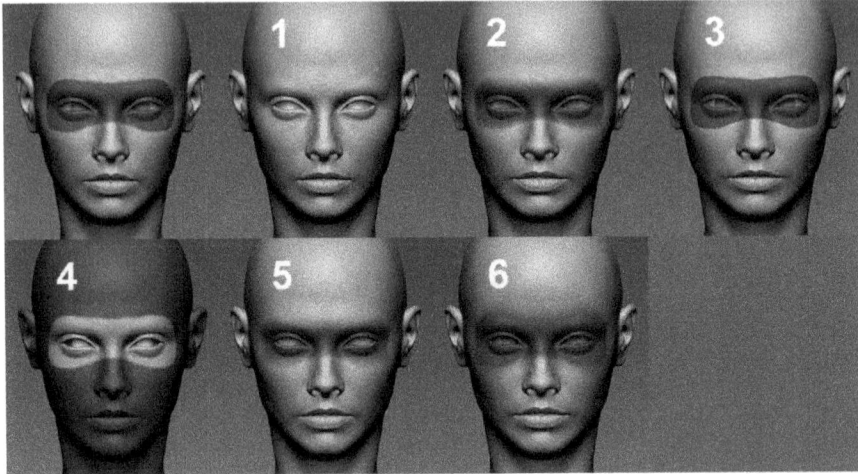

Figure 3.17 – Useful Masking functions

There are more masking techniques for specific cases, which can be very powerful for character creation and digital sculpting, so I will introduce these throughout the book. But for now, let's practice the masking and Gizmo tools on the soldier model.

Posing a character with Masking and Gizmo tools

Now that you understand the Gizmo and the key functions of PolyGrouping, selecting, and Masking, you can use these tools to create a pose for the soldier. Let's rotate the legs, including shoes and armor, placing the soldier's feet farther apart:

1. Start by selecting the body subtool and switching to **Solo** mode in order to view the model without accessories obstructing the view.

2. Then make sure **Symmetry** mode is enabled and mask the legs.

3. Invert and then blur the mask.

4. Go into **Gizmo** mode. Place the Gizmo on the top part of the leg where the legs would naturally pivot.

5. Enable **Transpose all Subtools** in the Gizmo tools, in order to affect all the visible, unmasked subtools with the next transforms.

6. Turn off the visibility of every subtool, other than the body, knee guards, and shoes.

7. Make sure the Gizmo is oriented in a neutral position by *Alt* + left-clicking on **Reset Mesh Orientation** in the Gizmo tools.

8. Rotate the Gizmo to place the feet further apart.

The result should look similar to this:

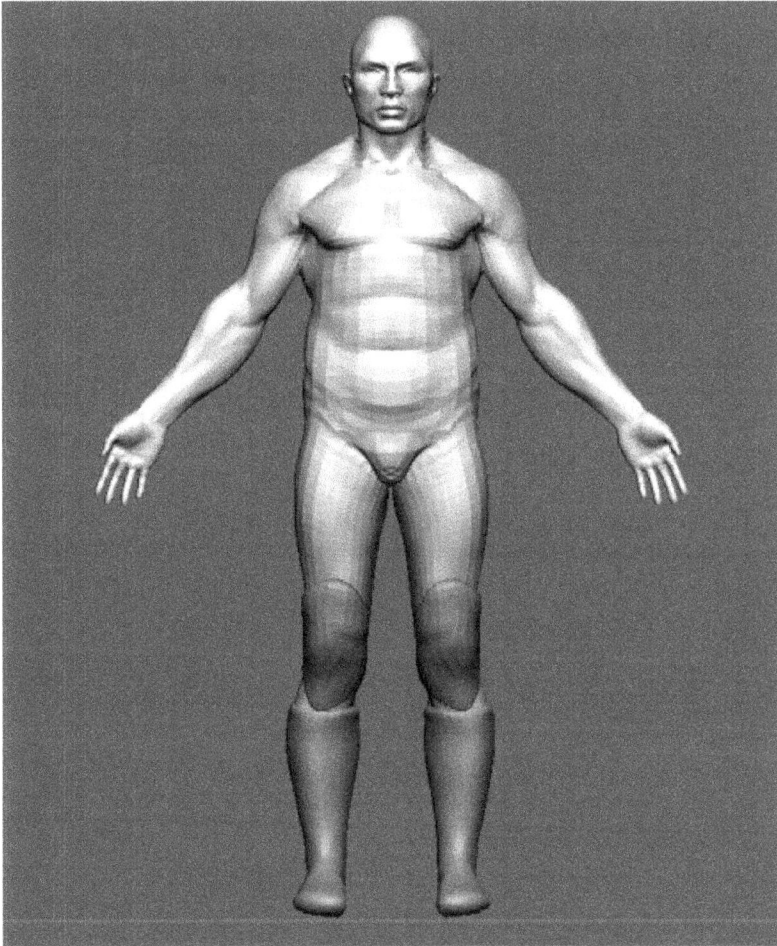

Figure 3.18 – The result of posing multiple subtools using the Gizmo and Masking

By now you have learned the basics about PolyGroups, selecting, and Masking. These tools are versatile and powerful, and you will learn how to use them to their maximum effectiveness as you gain experience with ZBrush.

Summary

Congratulations on completing the third chapter! You covered a lot of ground learning the most essential ZBrush functionality needed for a variety of tasks.

We started this chapter by taking a look at ZBrush's subtools and essential functionality, such as **Symmetry** mode. Then, we were introduced to the Gizmo and showed how to move, rotate, and scale objects, which we also tested with a practical exercise on the soldier model from LightBox.

After that, we explained the basic functions and benefits of PolyGroups, selecting, and Masking, which work in combination to allow for greater efficiency in your modeling workflow and let you deform objects in a variety of ways

The next chapter will explore ZBrush's wide variety of diverse and powerful brushes. We will look at the most important and useful brushes, as well as special kinds that behave differently and come in handy for specific use cases. Plus, we will create and save our own custom brushes, enabling us to be more versatile in creating different surfaces in ZBrush.

4

Exploring Brushes and Alphas

In this chapter, we will take a look at the wealth of brushes that ZBrush offers – this includes standard brushes used for sculpting and detailing, as well as special brushes such as the **InsertMesh** and **Curve** brushes. We will also explore Alphas and how to create our own before using a custom Alpha to create our own detailing brush, which we will test out on the demon bust from the previous chapter.

The reality is, you will most likely use only a few brushes in your sculpting career and there will be only a handful of brushes you use most of the time. However, if you need to create certain models that require very specific brushes, having knowledge of the brushes and brush options available in ZBrush can save you lots of time and give you great results.

By the end of this chapter, you will be able to choose the right brush for any task, and if you can't find the right brush, you'll know how to make one yourself.

We will cover these topics:

- Learning about brushes
- Learning about Alphas
- Creating your own custom brushes

Technical requirements

For the best experience, it is recommended that you have a strong PC that meets the minimum requirements described in the first chapter's *Technical requirements* section. However, you can work on this chapter with just a mouse, a functional PC setup, and a ZBrush license.

Learning about brushes

In this section, we'll look at the various brush modes and properties, as well as brush libraries. The goal is to give you an overview of what brushes are available, so you'll find the right one for any design challenge. Later, we will test and adjust the effect of various brushes until we eventually create our own brushes that fit our needs.

Accessing brushes

There are several places where you can access brushes. In this sub-section, you will learn about these options and their benefits.

3D Sculpting Brushes Menu

The **3D Sculpting Brushes** menu contains the most important and most commonly used brushes that come with ZBrush, and you will be able to create almost anything with them.

You can open the **3D Sculpting Brushes** menu by pressing *B*:

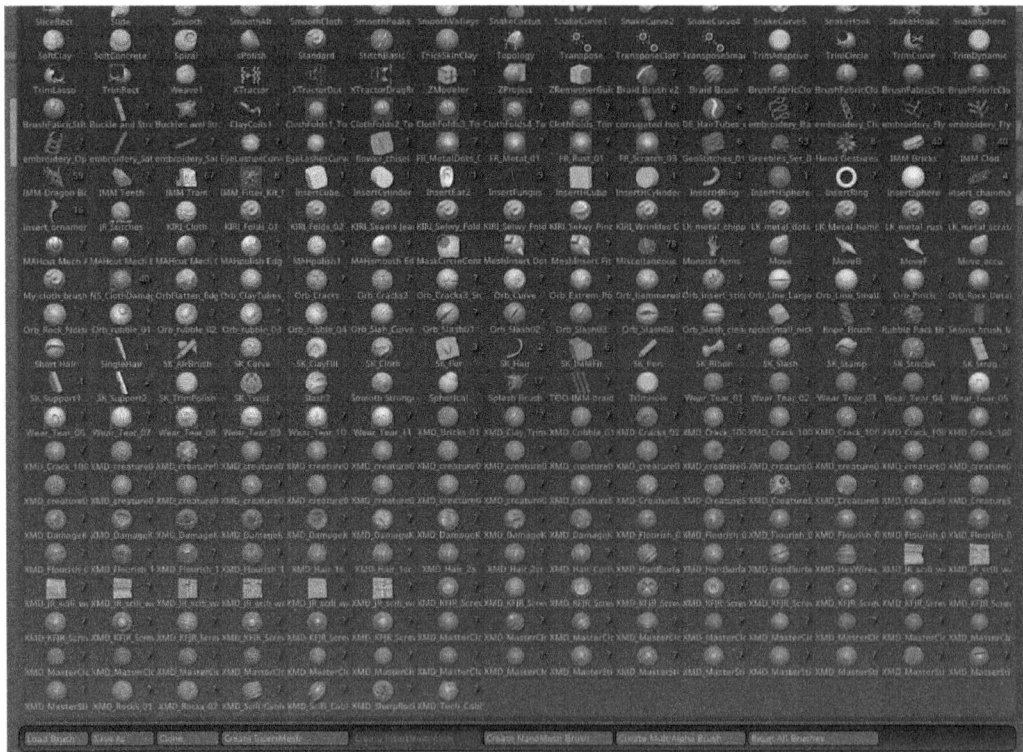

Figure 4.1 – 3D Sculpting Brushes menu

Easily accessible and structured alphabetically, this menu is convenient and lets you browse a relatively large amount of brushes quickly. Then, you can simply select one by left-clicking it.

> **Important note**
>
> The **Brush** menu in *Figure 4.1* contains many external brushes, so your menu may contain fewer brushes in it. If you want to add your own brushes to ZBrush's selection, you can add them to a folder called `start up`. Depending on your installation path, you can find the folder here: `C:\Program Files\Maxon ZBrush 2023\ZStartup\BrushPresets`, or `C:\Program Files\Pixologic\ZBrush 2022\ZStartup\BrushPresets`, if you have an older ZBrush version.

However, the brushes in this menu are not all ZBrush has to offer. There are more brushes in LightBox, which you will see next.

LightBox Brush Library

In ZBrush's LightBox Content Library, you can find brush folders containing a variety of additional brushes that can be loaded.

To access LightBox in ZBrush, press , on your keyboard:

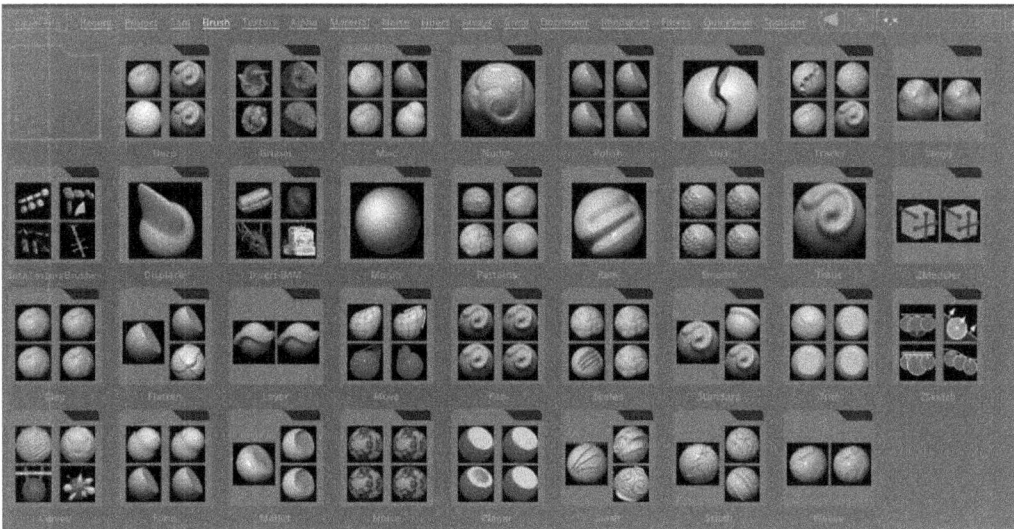

Figure 4.2 – Additional brush folders in LightBox

These folders contain groups of brushes, based on the type of brush or the effect. These brushes might not be needed to the same extent as the brushes from the main brush menu, but it is good to know that these brushes exist so you can take advantage if you need them.

> **Important note**
> Again, these brushes are saved on your computer, and again, depending on your installation path, you can find the folder here: `C:\Program Files\Pixologic\ZBrush 2022\ZBrushes`.

Here are some of the most useful and commonly used brush groups and the characteristics of their brushes:

- **Insert-IMM**: This folder contains various IMM brushes that let you add geometry, from simple primitive shapes to more complex models, such as dragon bones or train parts. Some of these brushes can be an efficient tool.

- **Scales**: These brushes are great for adding scales to animals such as snakes, fish, or even dragons.

- **Smooth**: These brushes can be used to smooth out your mesh in a specific way. These brushes have characteristics that make them only affect parts of the model while keeping others the same. For example, the smooth peaks brush smooths out convex shapes, while keeping crevices sharp, which creates a unique result that can be used to achieve certain surface effects.

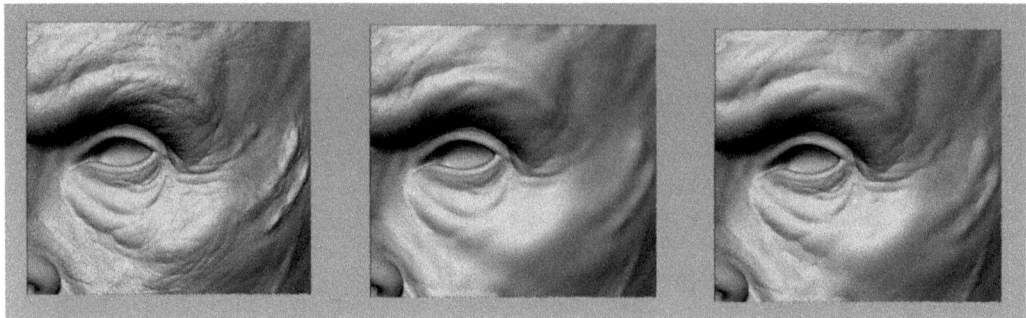

Figure 4.3 - Before (left) and after smoothing with the regular (middle) and smooth peaks (right) brushes

> **Important note**
> If you don't find the brush you are looking for in ZBrush's Brush Library or LightBox, the ArtStation.com marketplace has a great variety of affordable brushes, created by ZBrush enthusiasts: `https://www.artstation.com/marketplace/game-dev`.

Custom user interface brushes

Another place to store brushes is the **user interface** (**UI**). If you've created a custom UI already (covered in *Chapter 1*) it may contain a few frequently used brushes that are now very easy to access, but you personally want to access quicker.

This is what your custom UI with brushes could look like:

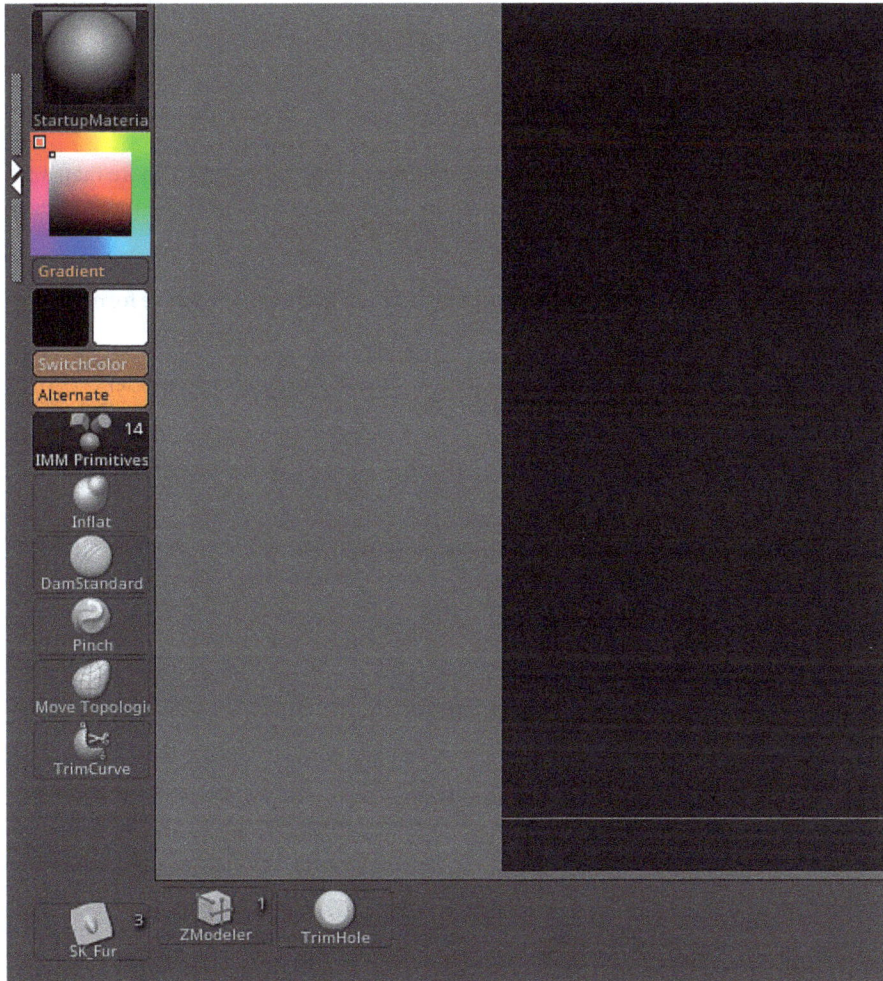

Figure 4.4 – Brushes around the canvas, stored in a custom UI

While adding too many brushes to your custom UI might overcrowd it and defeat its purpose, this is a great way to access brushes even faster than opening the main brush menu.

Hotkeys for brushes

Hotkeys allow you to access brushes as fast as, or faster than, brushes that you store on your custom UI. In order to assign a hotkey to a brush, simply hover over a brush icon, press *Ctrl + Alt*, and press the key that you wish to be the hotkey for the brush. Simply pressing a key is a fast way to access a brush, and it does not even take away space on your UI, so this is a great option for that reason as well.

Now you should be aware of the different places to access brushes, and what could be a suitable place to store a particular brush. In the next sub-section, we will go over some of the most important brushes for the digital sculptor.

Exploring important brushes

ZBrush comes with a large variety of useful brushes, but most users will never try more than a fraction of them. To save you some time with experimentation, I'll introduce some of the most popular and powerful brushes with examples showing their effect:

- **ClayBuildup**: Works great for general sculpting and can build up or remove form quickly. You can go a long way with this brush.
- **CurveMultiTube**: Great for quickly adding tube-shaped structures to a model.
- **CurveStrapSnap**: Ideal for adding belts or similar accessories wrapped around a model.

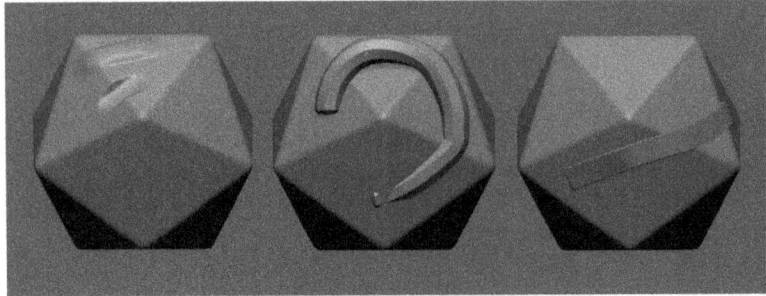

Figure 4.5 – (From left to right) ClayBuildup, CurveMultiTube, and CurveStrapSnap brushes

- **DamStandard**: Make sharp cuts or protrusions (this brush is another sculpting essential).
- **HPolish**: Lets you flatten an uneven surface and can also sharpen up edges.
- **IMM Primitives**: This lets you insert primitive shapes. This is a fast, efficient way to quickly add geometry for various purposes.

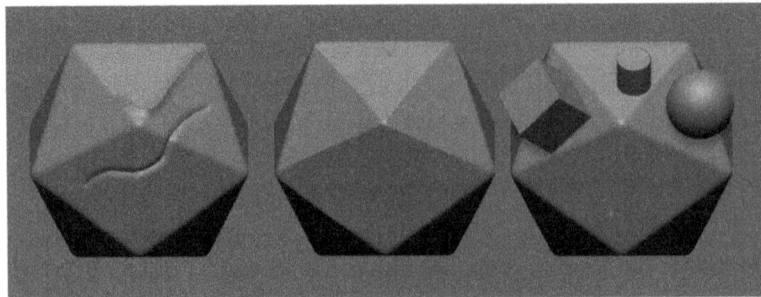

Figure 4.6 – (From left to right) DamStandard, HPolish, and IMM Primitives brushes

- **Inflate**: Useful to expand surfaces and increase the volume of meshes. This brush is also great for fixing thin DynaMesh meshes, as previously discussed.

- **Move**: Great for making large proportion changes, while not affecting the surface detail.

- **Smooth**: One of the most important brushes, letting you smooth out meshes to get rid of detail or unwanted forms.

Figure 4.7 – (From left to right) Inflate, Move, and Smooth brushes

- **Standard**: This brush is great for making more subtle changes and is especially useful for refining forms and enhancing detail.

- **TrimDynamic**: This lets you trim off sharp and protruding forms quickly, which is great for creating objects with multiple flat planes or chipped-off parts, such as rock or damaged metal.

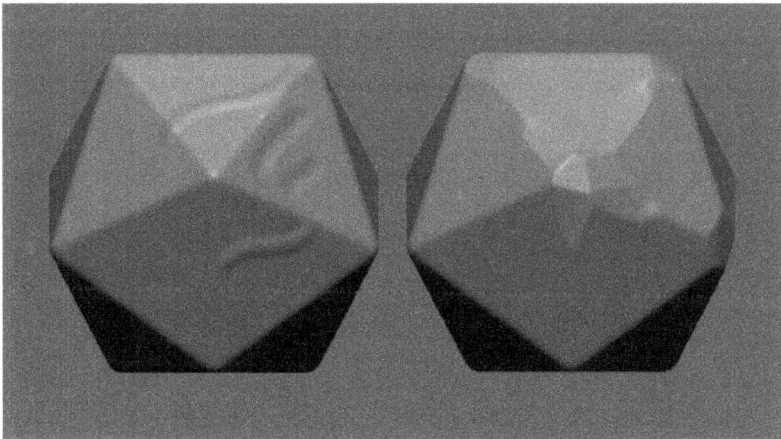

Figure 4.8 – (From left to right) Standard and TrimDynamic brushes

- **MoveTopological**: This moves the surface you apply it to, with the affected area not determined by the proximity to the brush stroke alone. In the following screenshot, *1* shows the starting point of a subtool consisting of 2 spheres; *2* shows the regular **Move** brush, which is applied to the left sphere, but affects the right sphere regardless, due to the large brush radius, and *3* shows the **MoveTopological** brush moving the mesh based on the continuous mesh surface, instead of just the distance from brush stroke to the mesh:

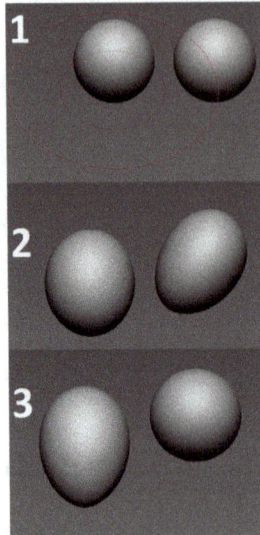

Figure 4.9 – Using the MoveTopological brush

Since the spheres are not connected by any polygons, the sphere on the left, where the brush stroke was applied, is the only affected mesh here. This makes this brush great for moving certain parts, while not affecting close, but separate, meshes.

- **ZModeler**: This brush allows you to make poly modeling operations on your meshes, making it great for non-organic, hard-surface objects, accessories, and costumes.

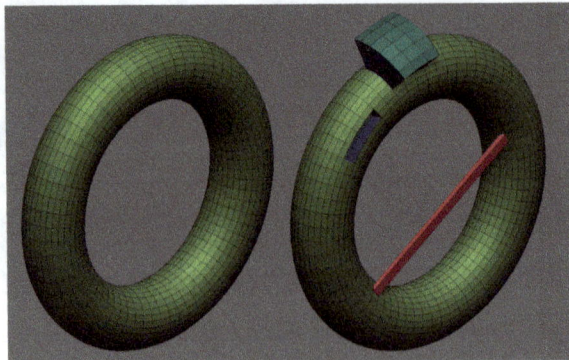

Figure 4.10 – ZModeler brush

Now that we have looked at some of the most important brushes, I hope they will come in handy in your projects and inspire you to experiment with some of the new brushes you've learned about.

Please note that the selection and masking brushes are also very common and useful brushes. You can find a quick overview in *Chapter 3* in the sections about masking and selection.

While you can go a long way just using the brushes available in ZBrush by default, there are many instances in which you want to make changes to brushes in order to achieve a specific effect.

In the next sub-section, you will learn about different brush modes, which are important properties that alter how brushes function in a significant way.

Exploring brush modes

Brush modes determine how the strokes will be applied to the mesh. You can find the icon belonging to this functionality on the left side of the canvas, where the main brush properties are displayed:

Figure 4.11 – Basic brush properties, located on the left side of the canvas

The upper brush icon lets you pick a brush. By clicking on it, you can access the Brush Library.

The second icon sets the brush mode. By default, the **ClayBuildup** brush has **FreeHand** mode selected, but there are more modes to choose from if you click on the icon. Let's take a look at the effect of each brush mode on the same brush (I'll choose the **ClayBuildup** brush for this):

- **Dots**: This mode applies the selected Alpha with a high frequency, repeating the Alpha many times along a brush stroke. For a round Alpha, this results in a continuous line:

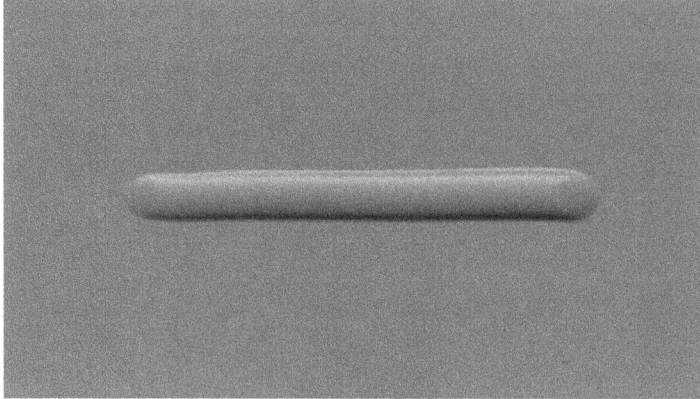

Figure 4.12 – Dots mode

- **DragRect**: This mode enables you to apply the Alpha once, where you click on the mesh, while also being able to control its size.

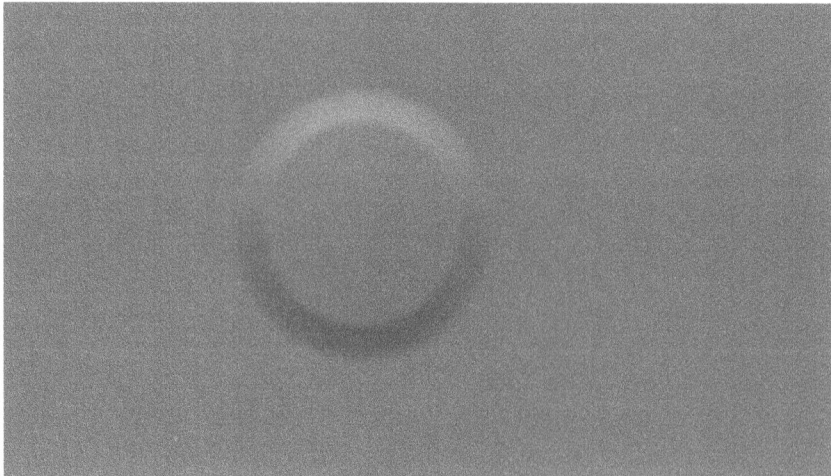

Figure 4.13 – DragRect mode

- **FreeHand**: Similar to the **Dots** mode, this mode applies Alphas at a high frequency, and lets you draw on the surface.

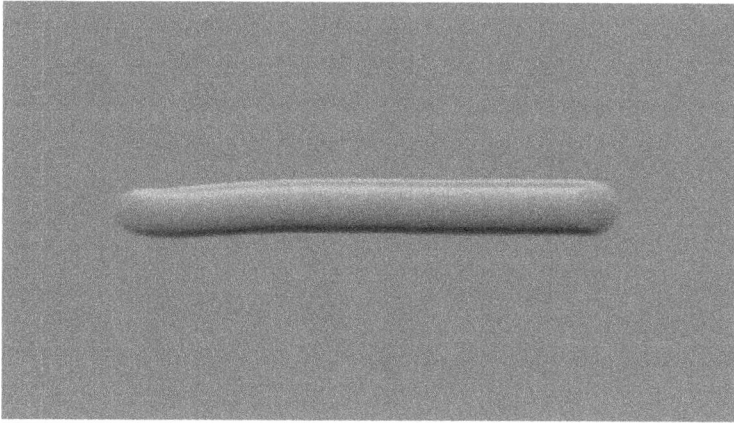

Figure 4.14 – FreeHand mode

- **Spray**: This brush applies the selected Alpha on the mesh with variance in size, placement, and intensity. You can adjust those parameters to control how strong the effect is.

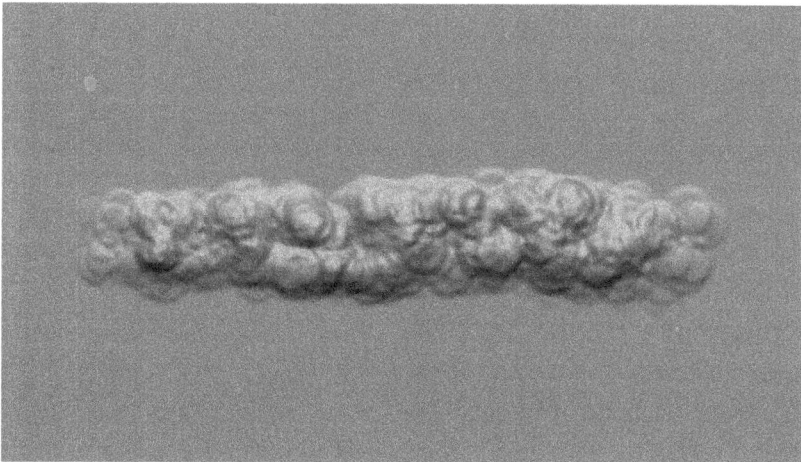

Figure 4.15 – Spray mode

- **DragDot**: Like the **DragRect** mode, this will apply a single instance of the Alpha with each stroke. However, unlike **DragRect**, you can move the Alpha across the mesh and place it wherever you want. The size of the Alpha is based on the draw size and cannot be changed once you start drawing.

Figure 4.16 – DragDot mode

At this point, you can equip a brush and use it with the desired brush mode enabled. Next, we will take a look at some brush properties to give you more options when using them.

Exploring brush properties

Some of the most essential brush properties are located in the **Draw** palette, where you can adjust settings such as brush radius, brush strength, and brush contrast. The **Draw** palette also includes different **Paint** and **Material** modes.

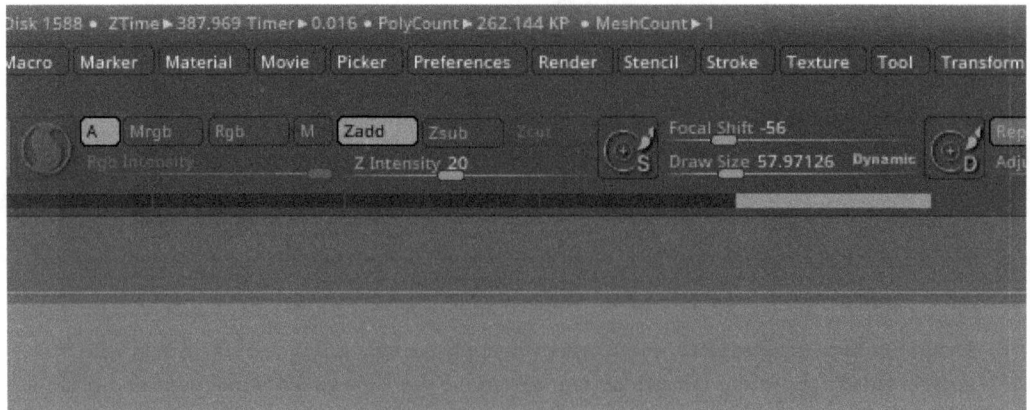

Figure 4.17 – Essential brush modifiers above the canvas

Let's take a look at these essential properties:

- **Draw Size**: This will make your brush radius bigger or smaller.

- **Z Intensity**: This adjusts the strength of your brush.

- **Focal Shift**: This increases the contrast of your brush stroke, resulting in a smaller but more intense stroke at -100, and a larger but less intense stroke at 100.

- **Mrgb**: Your brush will apply the selected color and material. You can find the **Color Picker** and **Material Library** on the left side of the canvas. These will be covered in more detail in *Chapter 6*.

- **Rgb**: Your brush will apply the selected color.

- **Rgb Intensity**: If you want to apply color with your brush, this value adjusts the intensity with which the color is being applied to the mesh.

- **M**: Your brush will apply the selected material.

- **Zadd**: Your brush is in addition mode and will create protruding forms.

- **Zsub**: Your brush is in subtraction mode and will create concave forms.

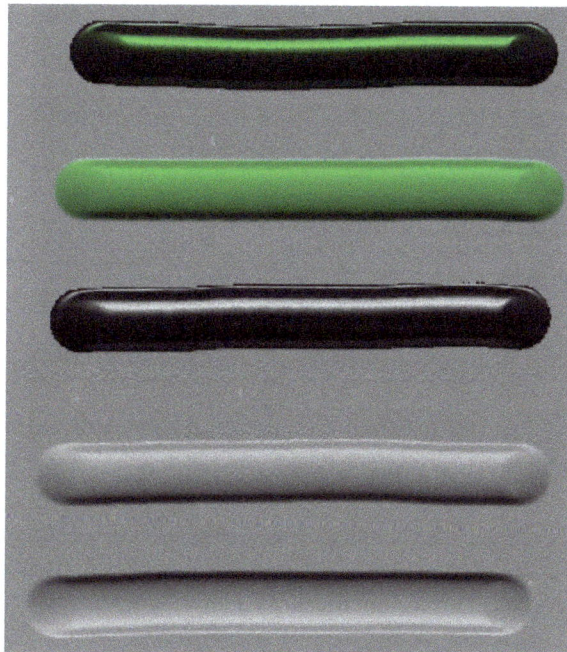

Figure 4.18 – (From top to bottom) Mrgb, Rgb, M, Zadd, and Zsub modes

Now that you are able to apply a wider range of brush strokes, we will take a look at **LazyMouse**, a brush-stabilizing function that helps you create smooth, clean lines.

Making smoother strokes with LazyMouse

Sometimes you need to make precise, clean brush strokes. This is often necessary for detailing, especially if you want to create smooth, flowing lines in ornamental models. **LazyMouse** can help by stabilizing your brush movement and eliminating jagged lines.

You can access this by going to **Stroke | LazyMouse**. Then to get smoother strokes, increase the **LazyRadius** value. This will enable a stabilizing functionality that keeps the brush stroke steady.

If you find that your brush strokes look "dotted," as the Alpha is not applied at high enough frequency, you can adjust another attribute in the **LazyMouse** menu, **LazyStep**. The lower this value, the more times the brush Alpha is applied during a stroke.

The results of **LazyMouse** and setting variations can be seen in the following screenshot:

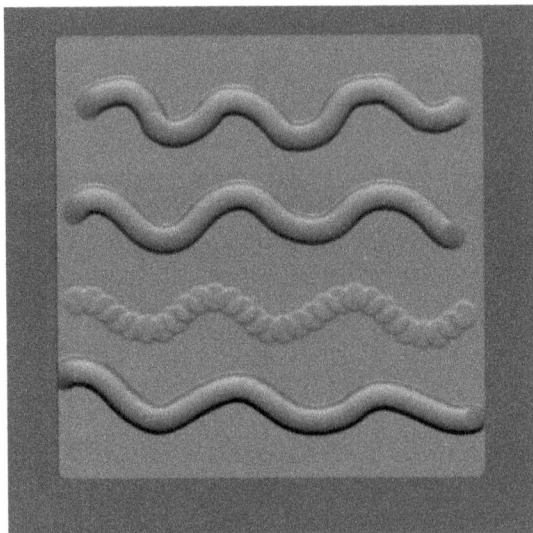

Figure 4.19 – (From top to bottom) no LazyMouse, LazyMouse, LazyMouse
with high LazyStep, and LazyMouse with low LazyStep

So, in this section, we have covered a lot about brushes, including brush settings, modes, and menus, and how to access brushes in different ways. In the next section, I will discuss Alphas and how to create them, which will let you create custom brushes for any of your needs.

Learning about Alphas

Alphas are grayscale images that give height information to brushes, which displace the surface of your sculpts accordingly, creating cavities, protrusions, or a mix of these. Most brushes use Alphas, which makes a good knowledge of Alpha creation an essential part of your ZBrush education.

To access and select Alphas, navigate to the **Alpha** palette or click on the Alpha icon on the left side of the canvas. Then, by left-clicking on one of the Alphas, you equip it to your active brush.

Figure 4.20 – ZBrush's Alpha library

In the following sub-sections, you will learn how to modify Alphas and create custom Alphas.

> **Important note**
>
> The Pixologic website has a great selection of Alphas, from organic patterns to skin and rock detail. There are many categories, so it is worth checking out to see whether some of them can be useful for your project. You can download them for free here: `https://pixologic.com/zbrush/downloadcenter/alpha/`.

Modifying your Alphas

Sometimes the Alphas that are already available are not exactly what you need. ZBrush has a solution for this – it lets you modify Alphas easily through options in the **Alpha** palette located in the top bar:

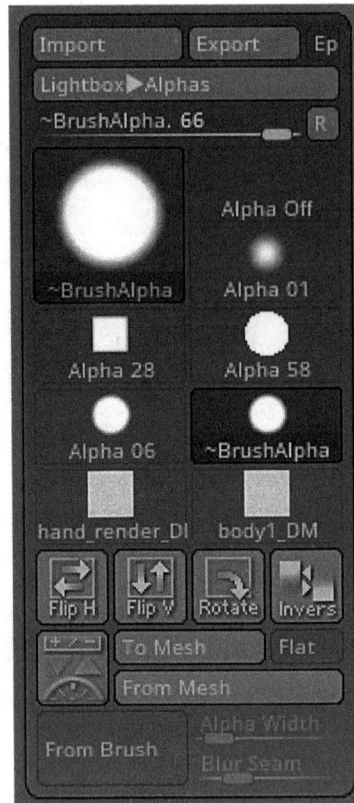

Figure 4.21 – Alpha modification options in the Alpha palette

Here are some of the important buttons in the panel:

- **Flip H/Flip V**: Flips the Alpha horizontally/vertically
- **Rotate**: Rotates the Alpha 90 degrees clockwise
- **Invers**: Inverts the Alpha

Below those buttons are several submenus. Let's take a look at the **Modify** submenu, which is the second submenu in the **Alpha** palette. This menu gives further, more advanced options to modify the Alpha:

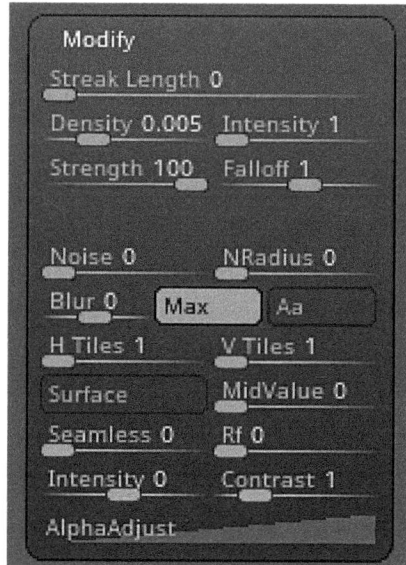

Figure 4.22 – The Modify submenu

Here are some of the more commonly used options:

- **Blur**: This lets you blur the Alpha to achieve a smoother effect.

- **H Tiles/V Tiles**: This lets you tile the Alpha if you want to apply multiple instances with each brush stroke.

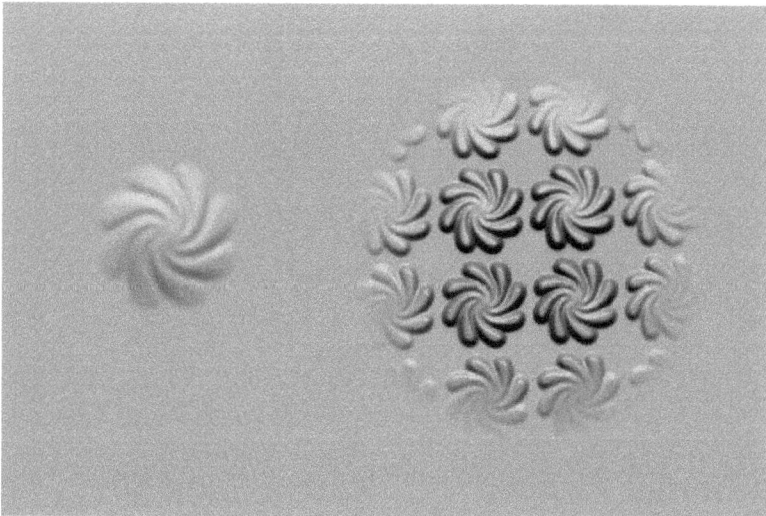

Figure 4.23 – Example of tiling

- **MidValue**: This lets you adjust the depth of the Alpha effect. A value of 0 makes the Alpha only push outward, creating a protruding effect, while a value of 100 pushes inside the model instead. An in-between value will let you create protruding and concave effects at the same time.

Figure 4.24 – MidValue, with values of 0, 50, and 100 (from left to right)

- **Intensity**: This increases the Alpha's intensity, creating a stronger displacement effect in the brush.

- **Contrast**: This increases the contrast, which will create a higher contrast and a more defined result.

- **Rf** (which stands for Radial Fade): This fades the Alpha toward the edge, which can be useful to focus the Alpha toward the center, and to remove fragments that can appear on the edge of the Alphas when it is applied.

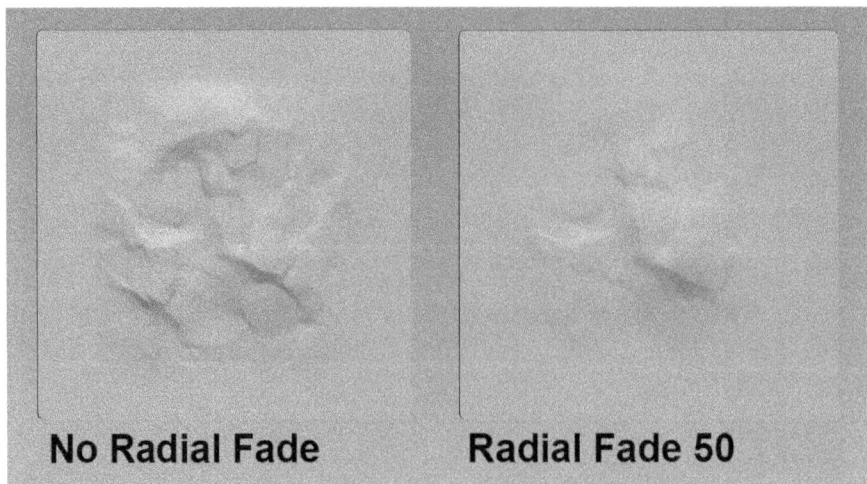

Figure 4.25 - The effect of Rf (Radial Fade)

Being able to modify existing Alphas will give you more possibilities, but often you will need to create your own Alpha to achieve specific effects. We will learn how to do that next.

Creating our own Alpha for the demon bust

Being able to create your own Alpha for a custom brush is essential to be able to be versatile in the detailing of your sculptures. Here we will create our own Alpha, using the demon bust example that we started working on in the last chapter. So, do the following:

1. Go to the **Tool** palette | **Subtool** and append a **Plane3D** model.
2. Navigate to **Tool** | **Geometry** and disable **Smt**.
3. Click **Divide** four to five times in order to increase the polycount of your mesh.
4. Next, use **ClayBuildup**, **DamStandard**, or any other brush you want, and sculpt something on the plane. I created little horns, which I will apply to my demon bust.

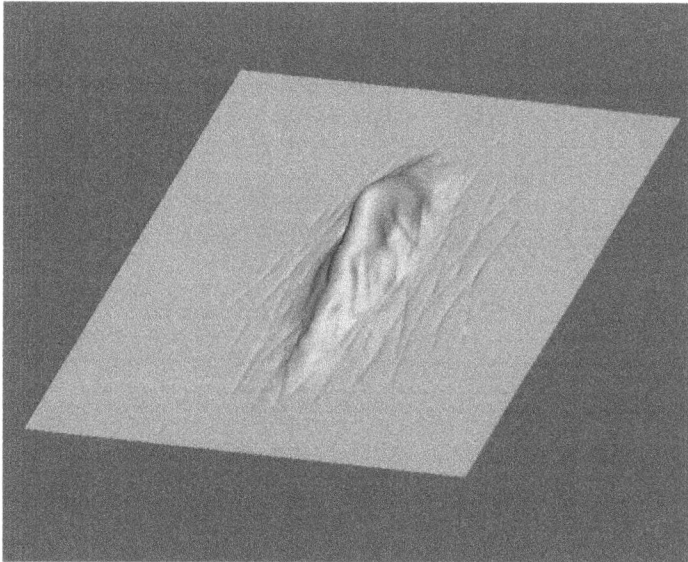

Figure 4.26 – Sculpting a horn detail for the Alpha

5. Open the **Alpha** palette and select **From Mesh**. Now, you can set the resolution of the Alpha with the **Map Size** value and hit **OK**.

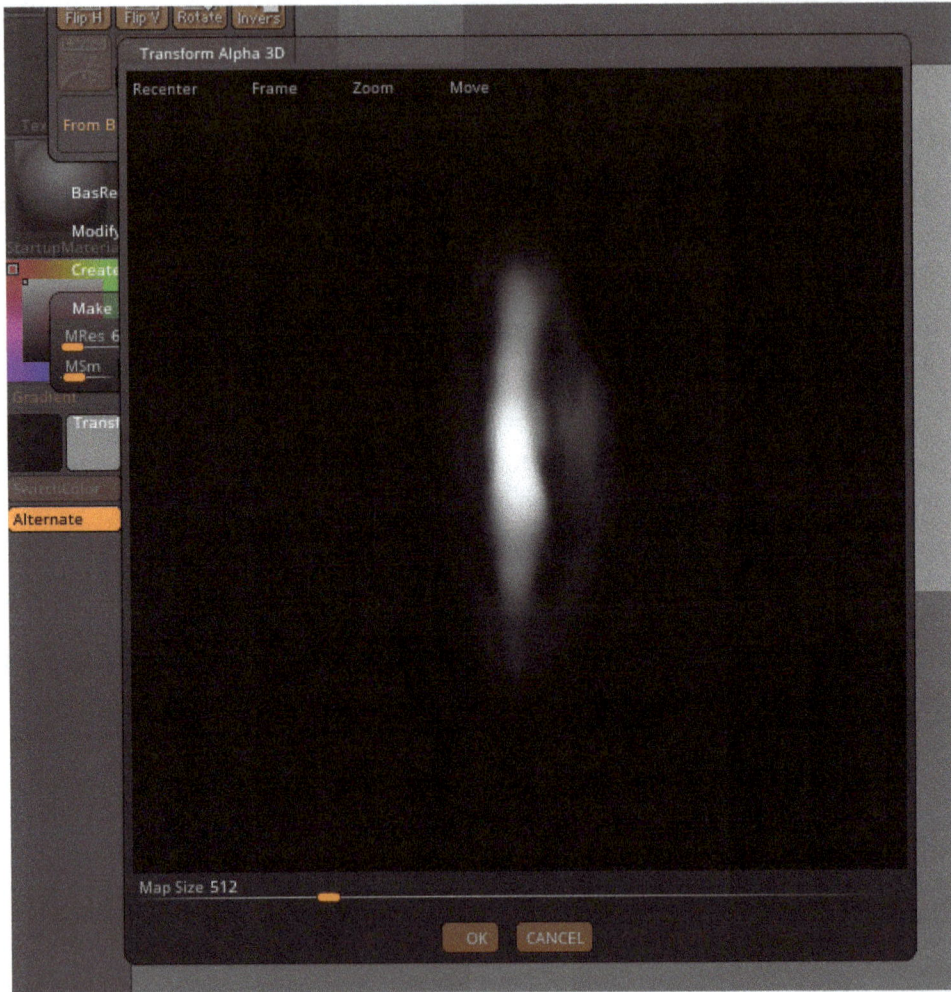

Figure 4.27 – Extracting an Alpha from a mesh

The Alpha is now equipped on the active brush. If you want to export the Alpha, you can do so by clicking on the Alpha icon and selecting **Export**.

This concludes the section on Alphas, in which I introduced the basic Alpha settings and showed how to create an Alpha from a plane. In the next section, I'll show you how to create a custom brush using this Alpha so we can apply it to our demon bust.

Creating your own custom brushes

ZBrush offers a wide variety of useful brushes, but there are many instances in which you want to achieve a specific effect on the sculpture's surface. Custom brushes offer a great solution for this.

We will use an Alpha we created in the previous section and create a custom brush with it. Then we will use that brush to add some detail to our demon bust and save it for later use.

Returning to our demon bust, using the **DragRect** mode makes the most sense because it will allow us to precisely place the horn on our model while scaling it interactively. This type of **DragRect** brush with a custom Alpha equipped is among the most commonly used custom brushes and one of the most simple too, so you can use this blueprint to create a variety of custom brushes for your specific needs.

In order to create this custom brush, complete the following steps:

1. Select the **Standard** brush (since this brush has basic settings, it is a good choice for creating regular sculpting brushes).

2. Press *B* to open the **Brush** menu and click on **Clone**. This will create a clone of the **Standard** brush, so while you create your custom brush, you still keep the original **Standard** brush unchanged.

3. Click on the **Stroke** icon and switch the mode from **Dots** to **DragRect**.

4. Then click on the **Alpha** icon and select the **Horn** Alpha you created.

5. Now you use the brush to add some detail to your demon bust:

Figure 4.28 – Applying horns with our custom brush

6. If you want to save the brush, just press *B* to open the brush menu and select **Save as**.

This concludes this last section of the chapter. You learned how to create a basic **DragRect** mode brush with a custom Alpha. We used that brush to add some detail to our demon bust, saving a lot of time compared to manually sculpting in that detail.

Summary

In this chapter, you learned about ZBrush's most important brushes. You got to know the default brushes from the main brush library, as well as extra brushes in the LightBox menu. We looked at different brush modes such as **FreeHand**, **DragRect**, and **Spray**, and various options and menus that allow us to adjust the brush properties to our liking. From here on out, you'll be able to sculpt on surfaces while applying color and material, as well as use **LazyMouse** to create smooth, flowing strokes.

We also looked at Alphas and modification options and used a **Plane3D** model to create our own custom brush, which we used to apply horns to our demon bust. Finally, I listed some of the most popular sculpting brushes that will be useful for a wide range of tasks.

The next chapter will introduce a workflow using subdivision levels, ZRemesher, and projection tools. We'll follow along with our demon bust by adding more resolution and detail to it – a great point to start texturing it in the following chapter.

5
Creating an Optimized Mesh Using ZRemesher and ZProject

In this chapter, you will learn about ZBrush's powerful retopology tool, **ZRemesher**, and how to use it to create clean and even topology for the demon bust that you have been sculpting so far. This will help prepare the model for further, more detail-oriented sculpting later on.

Then you will get familiar with subdivision levels, which allow meshes to be subdivided many times over, increasing resolution and ultimately allowing for higher levels of detail. Using the **ZProject** tool, we will project the detail from the demon concept sculpt onto the new, cleaner topology that we created using ZRemesher at the beginning of the chapter.

At this point, we will also look at increasing the mesh resolution, allowing you to add extremely fine detail. This will produce a sculpt that can be viewed from a close distance, without loss of resolution, which is often needed in VFX and sometimes games as well.

So, this chapter will cover the following topics:

- Creating new topology with ZRemesher
- Adding subdivision levels and transferring detail using ZProject
- Adding detail to the demon bust

Technical requirements

For the best experience, it is recommended that you have a strong PC that meets the minimum requirements described in the first chapter's *Technical requirements* section. However, you can work on this chapter with just a mouse, a functional PC setup, and a ZBrush license.

More specifically, it is recommended that you have completed *Chapter 2* and have a DynaMesh model available to practice the ZRemesher and ZProject tools. Alternatively, ZBrush's LightBox library contains many models that could be used with the same effect.

Please note that examples in *Chapters 6* and *7* will use the demon bust we started creating in *Chapter 2*. Skipping the retopology step of this chapter may lead to an unoptimized mesh, which cannot be properly painted and textured in subsequent chapters.

Creating new topology with ZRemesher

Just like how DynaMesh produces new topology, so does ZRemesher. However, unlike DynaMesh, ZRemeshing will not close holes in the model or merge close vertices. Rather, it tries to capture an existing shape as closely as possible using a polycount that is determined in its settings.

Before we examine the settings of ZRemesher and what they do, it is useful to know what proper topology looks like so you can judge the results. **Proper topology** refers to a type of topology that allows the animator to create animations and at the same time gives the modeler something optimized to work with. We will look at proper topology in relation to facial animations.

> **Important note**
> Make sure to duplicate your demon sculpt because you will need the original DynaMesh sculpt later to transfer its detail to your duplicated mesh.

Facial topology

Facial topology is the edge flow on the face, which must support the movement of facial muscles and the expressions they create. If the topology does not support this, the animator will have a hard time creating good facial animations.

The following figure illustrates what proper facial topology looks like:

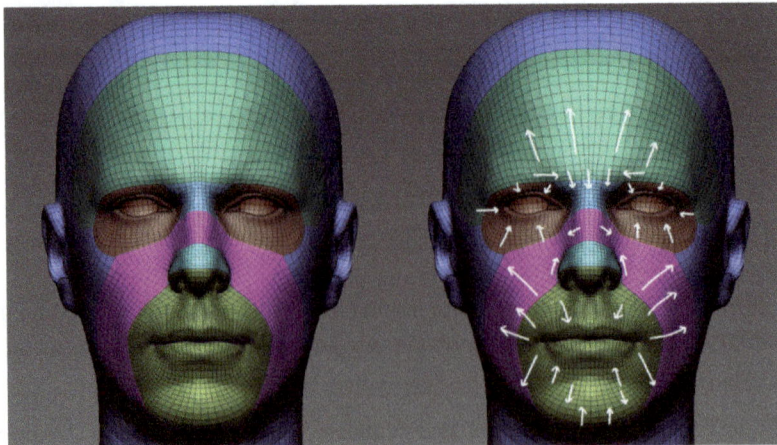

Figure 5.1 – Edge flow supporting facial animations

This demonstrates a good approach to facial topology with edge loops (multiple edges, creating a circular shape) around the eyes and mouth, which support the stretching and compression of different facial muscles, as indicated by the white arrows in *Figure 5.1*.

As a modeler or sculptor, you should follow the shape of wrinkles and facial shapes when creating edge flow because it will make your sculpting experience smoother, as opposed to having to sculpt diagonally across polygons. We can see the difference in the following figure:

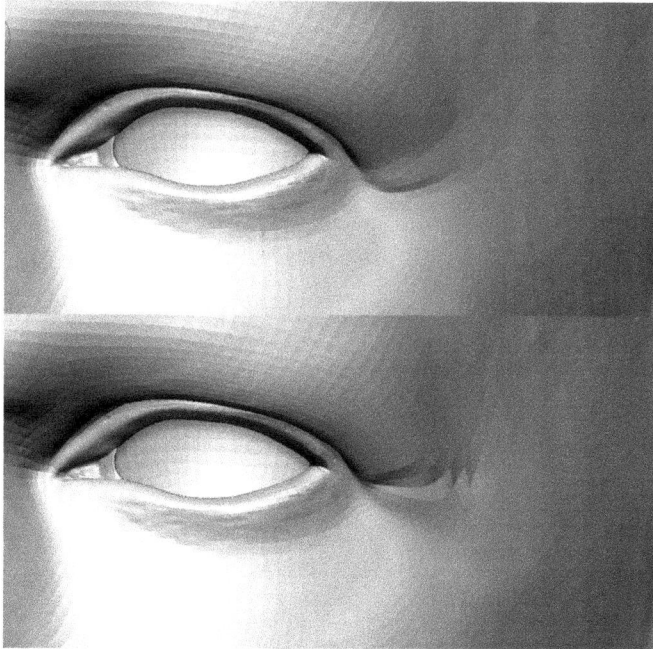

Figure 5.2 – Sculpting with the edge flow (upper) versus sculpting across the edge flow (lower)

A higher resolution will help make up for bad topology, but it will never be easy to sculpt on bad topology, and the results will be inferior compared to sculpting on good topology.

With this information in mind, you can create some new topology with ZRemesher.

Overview of the ZRemesher settings

You can find ZRemesher in the **Tool** palette, under the **Geometry** submenu:

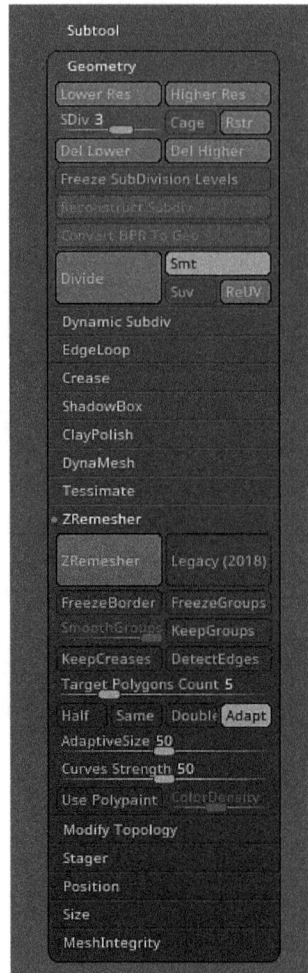

Figure 5.3 – ZRemesher located in the Tool palette

Let's examine the different settings to see the effect on the ZRemesher demon bust.

Legacy (2018)

Legacy (2018) is an old ZRemesher algorithm that can be enabled to produce different results. The new ZRemesher is optimized for hard-surface objects, but Legacy ZRemesher often produces superior results when ZRemesh-ing organic objects.

Let's look at what **Legacy (2018)** looks like with our demon sculpture. Duplicate the model twice by selecting **Duplicate** in the **Tool** palette. Then test the effect of ZRemeshing each duplicate with **Legacy (2018)** turned on and off.

Here we can see our sculpture with **Legacy (2018)** disabled:

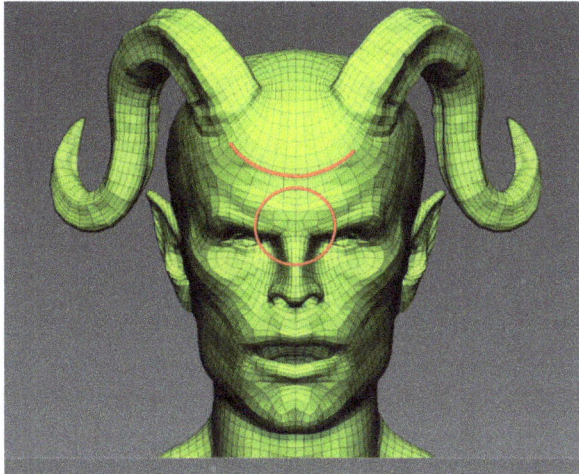

Figure 5.4 – The ZRemesher results with Legacy (2018) disabled

The results in the eye and mouth regions are problematic because there are no edge loops around those areas. The stretched polygons in the middle of the brow area will be uncomfortable to sculpt on, even after increasing the resolution of the mesh.

Overall, there is a big discrepancy between polygon sizes, which means some areas will have insufficient resolution for proper detail, and even if enough subdivision levels are added, it is not efficient and will unnecessarily require more computer performance.

Now, let's look at our sculpture with **Legacy (2018)** enabled:

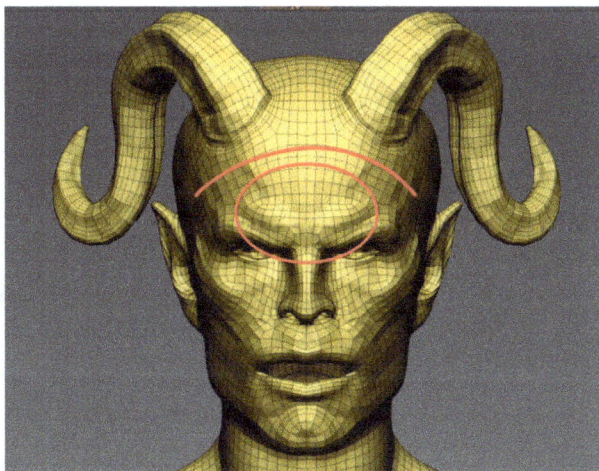

Figure 5.5 – The ZRemesher results with Legacy (2018) enabled

Similar to the previous result, the eye and mouth regions lack edge loops. However, the **Legacy (2018)** ZRemesher provides a superior result in the other areas.

The brow bone shape is captured nicely for the **Legacy (2018)** version of the ZRemesher mesh, while the default version had stretched polys in the middle.

The **Legacy (2018)** version also contains more evenly sized polygons, which will produce better results in the detailing stage.

Please note that the **Legacy (2018)** option is no longer present in ZBrush 2023 and higher. There, you will only have one ZRemesher mode, but the following options still apply.

Next, let's look at the **KeepGroups** option, which gives you more control over the results of ZRemesher.

KeepGroups

KeepGroups allows you to get better results in the eye and mouth areas by creating topology based on PolyGroups that you create. As the name suggests, this option will give a ZRemesher result that keeps PolyGroups where they previously were. This can help create the edge flow you want as you have the freedom to "draw" them with PolyGroups on the mesh.

In this case, you want to create PolyGroups around the eyes and mouth, which flow like the edge flow shown in *Figure 5.1*. To create the PolyGroups, simply mask the area, and hit *Ctrl + W* to create the Polygroup, as shown here:

Figure 5.6 – Using masking to create PolyGroups for the ZRemesher retopology

After creating PolyGroups in this fashion, the result could look like this:

Figure 5.7 – PolyGroups on the demon bust

Once you have created your PolyGroups, it can be useful to smooth the transition between those groups, in order to get better results from ZRemesher.

To do so, go to **Tool** | **Masking** and select **MaskByFeature** with **Groups** turned on. Next, press *Ctrl* and left-click on the canvas to invert the mask. Then go to **Tool** | **Deformation** and use the **Polish By Features** option to smooth out the transition. Finally, you can get rid of the mask by *Ctrl* and dragging a mask into the empty space on the canvas. Now you should have nicely rounded PolyGroups. Here are the steps to achieve this result:

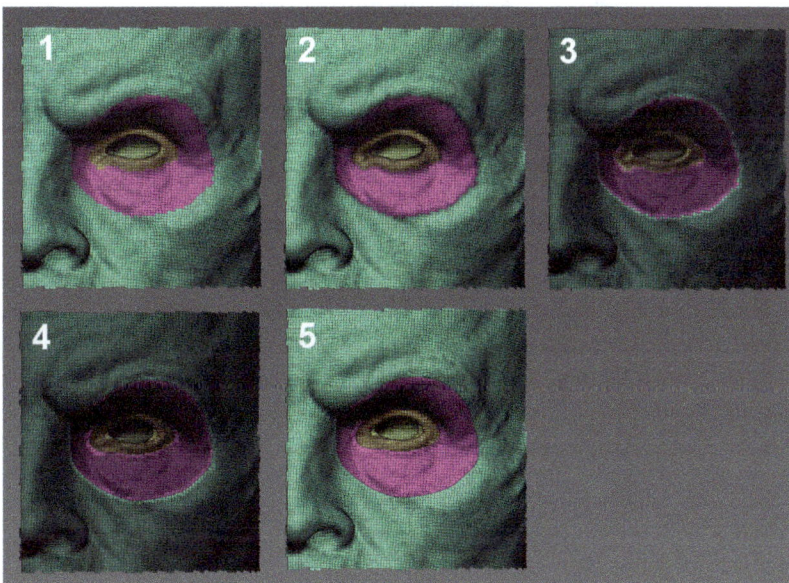

Figure 5.8 – Using the MaskByFeature option with Groups turned on and the
Polish By Features option to smooth the borders between PolyGroups

At this point, we set things up so that ZRemesher will give us the desired topology. With **Legacy (2018)** and **KeepGroups** enabled in the ZRemesher submenu, you hit **ZRemesher** again. My result looks like this:

Figure 5.9 – ZRemeshing with KeepGroups

This topology is still far from perfect, but at least we have some loops around the mouth and eyes. The next two settings can resolve those issues.

Target Polygons Count and AdaptiveSize

The **Target Polygons Count** and **AdaptiveSize** settings determine the polygon count of the ZRemeshed result.

The **Target Polygons Count** number stands for the polycount in thousands, meaning the default value of 5 will produce a 5,000-polygon mesh. However, the actual polycount depends on the **AdaptiveSize** value – higher values give a tolerance, allowing ZRemesher to create more topology in areas where it is needed to preserve the structure of smaller shapes. This makes it a possible solution for inconsistent results around the eyes of the demon model.

Here you can see how an **AdaptiveSize** value of 0 produces very evenly sized polygons, while a value of 100 creates dense topology around the edge of the beveled part of the axe, preserving its shape more accurately:

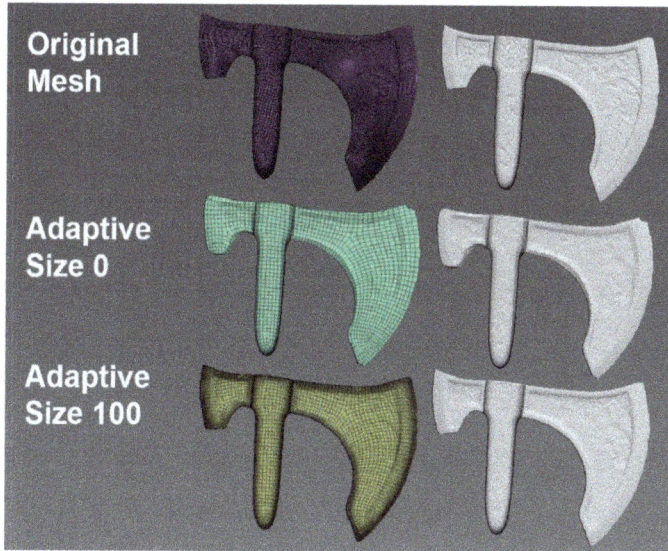

Figure 5.10 – Results of ZRemeshing with AdaptiveSize 0 and 100

Here I worked on the demon again with a **Target Polygon Count** value of 10 and an **AdaptiveSize** value of 80:

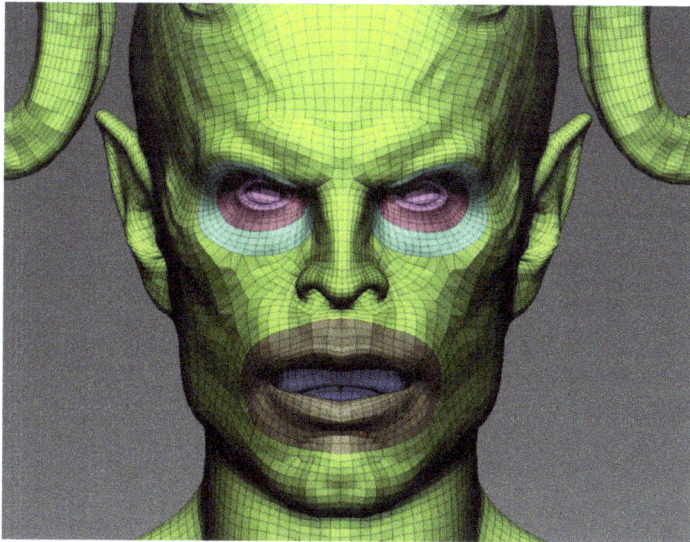

Figure 5.11 – ZRemesher with increased polycount and AdaptiveSize

We now have almost perfect edge loops around the eyes and mouth, but other areas such as the cheekbones still need refinement. Refining these areas requires some experimentation with different PolyGroups and ZRemesher settings to get the best possible results. You should take the time you need to create topology you feel good about as this will support the sculpting process later on.

> **Important note**
>
> It is important to note that when creating a model for facial animation, it is always necessary to do some manual retopology in ZBrush or another 3D modeling program. This is because animation requires very precise and specific results that are unlikely to be achieved with ZRemesher alone. However, when creating a model for 3D printing or concept visualization purposes, ZRemesher is an excellent solution.

This concludes the section on ZRemesher. You learned about the essential settings so you can adjust them according to the specific meshes and get good results.

In the next section, you will learn about ZTools' subdivision structure and how to use ZProject to transfer detail between meshes, which can be useful when you need to update topology on a high-resolution mesh.

Since we are creating topology with proper edge flow for the demon bust, but want to keep the sculpted detail from DynaMesh, ZProject is the solution here as well.

Adding subdivision levels and transferring detail using ZProject

In this section, you'll use the new mesh that we just created with ZRemesher and start increasing its resolution so that you can add a lot of detail in *Chapter 6*. Then we will use the ZProject functionality to transfer detail from the original DynaMesh sculpt to the new mesh.

Subdividing meshes

Subdividing a mesh is an operation that creates four new polygons from each four-sided polygon, and three new polygons from each three-sided polygon, while also creating a new subdivision level.

Here is what subdividing a mesh looks like:

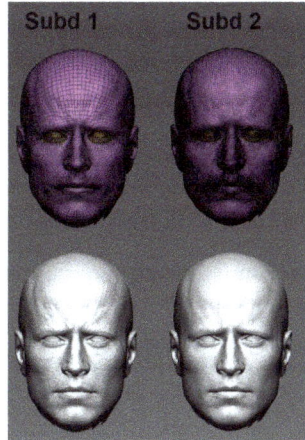

Figure 5.12 – The effect of subdividing a model

You can find the **Subdivide** option in **Tool | Geometry,** where you will find the **Divide** button. By default, **Smt** (smooth) will be turned on – this means that each time you subdivide your model, all of its polygons will be smoothed out; this is especially useful when working on organic models because it helps avoid visible edges. *Ctrl + D* is the keyboard shortcut for dividing your mesh.

Figure 5.13 – Subdivision levels indicated in the Tool palette

Once you have subdivided your mesh, you can toggle between the different subdivision levels using the **SDiv** (subdivision) slider. Alternatively, you can hit **Lower Res** or **Higher Res** or use the corresponding keyboard shortcuts, *D* to increase the subdivision level or *Shift + D* to decrease the subdivision level.

This allows you to make bigger changes to the mesh while working on a low subdivision level, and preserve the detail from the higher subdivision levels, which is something that could not be done with DynaMesh, where you can only have one state of the model at a time.

In the following figure, the facial expression is adjusted on subdivision level 1 (middle head). After switching back to the highest subdivision level (right head), the detail is fully restored.

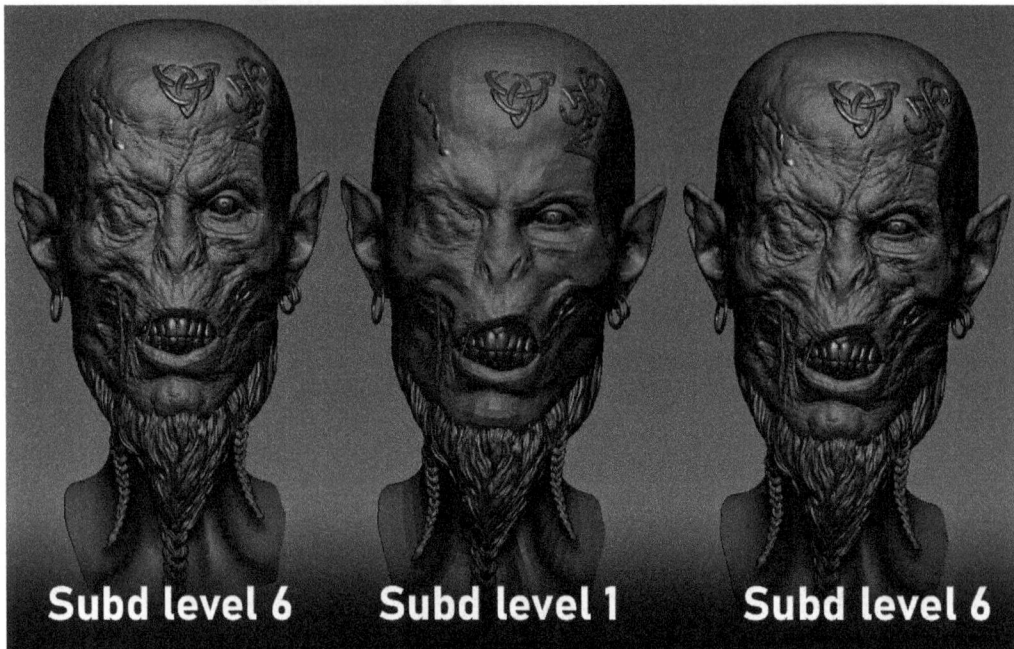

Figure 5.14 – Making changes on a subdivision mesh

Important note

Note that once you subdivide a mesh, many tools and operations can no longer be performed on it. These include all of the operations that change the topology of a mesh (e.g., **DynaMesh** or **Mirror/Weld**). This means that you are less flexible in what you can do after committing to subdivision levels. If you find that you need to change the topology of a mesh with subdivision levels, you will have to create a new mesh using ZRemesher and reproject the detail you have stored in the subdivision level model.

ZBrush allows you to divide your meshes until you have enough resolution for the level of detail you want, or until you reach 100 million polygons per mesh, which is the ultimate limit in ZBrush.

By default, this limit may be set lower than 100 million polygons per mesh; however, to increase this limit, go to **Preferences** | **Mem** (memory) and increase the value of **MaxPolyPerMesh** to your liking and according to what your computer's hardware can handle.

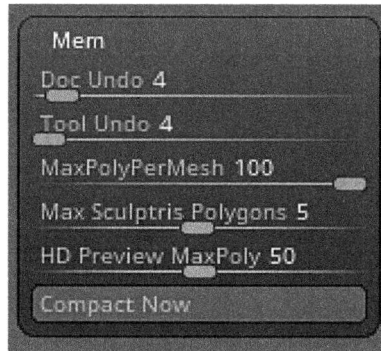

Figure 5.15 – Setting the maximum polycount per mesh

 A large amount of RAM will help keep performance smooth while working on meshes with a high polycount.

For now, only subdivide the demon character once; however, you will add more subdivision levels shortly as part of the ZProject process.

Transferring detail between meshes with ZProject

Depending on the ZRemesher settings you chose earlier, the newly created mesh has a lower polycount than the DynaMesh model you sculpted initially. This means that you lost some of the detail you sculpted. However, there is a great tool for bringing back that detail: ZProject.

Let's look at how to use this tool:

1. Turn off the visibility of all subtools except your original DynaMesh demon sculpt and the demon sculpt that you just retopologized with ZRemesher. Both should be in the same position, with the higher resolution of the DynaMesh subtool being visible, like so:

Figure 5.16 – The original demon sculpt and the new low-poly mesh

2. With the low-resolution (ZRemesher) mesh selected, navigate to **Tool | Subtool | Project** and click **ProjectAll**. Now we can observe that our retopologized mesh matches the original DynaMesh sculpt more closely:

Figure 5.17 – Result of the first ZProject operation

3. Navigate to **Tool | Geometry** and select **Divide** (or hit *Ctrl + D*).

4. Now repeat *Steps 3* and *4* until the new mesh has all the details of the original concept sculpt.

The result should look exactly like the prior sculpt:

Figure 5.18 – The result of ProjectAll, restoring the detail on the new low-poly model

The projection process can be visualized as follows:

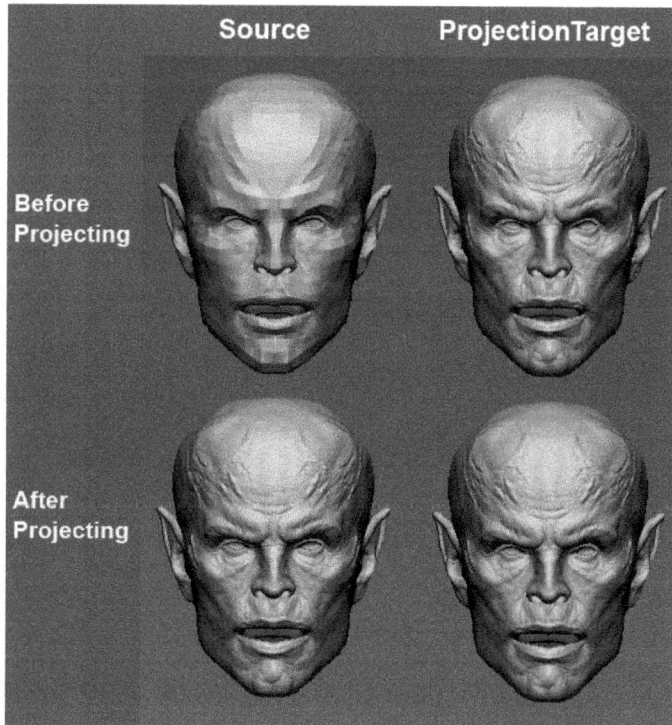

Figure 5.19 – Matching the new low-poly model to the projection target

Here are the stages of projection from the low-poly model from subdivision levels 1 to 4, increasing detail on every level:

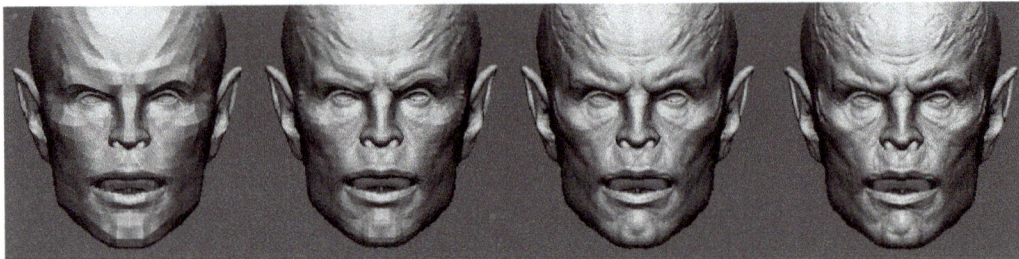

Figure 5.20 – Adding subdivision levels and projecting detail

Once you complete this project, you might find that ZProject created some **artifacts**, which are polygons that get distorted and displaced. The distortions occur especially in tight places such as mouth corners or eyelids, similar to this:

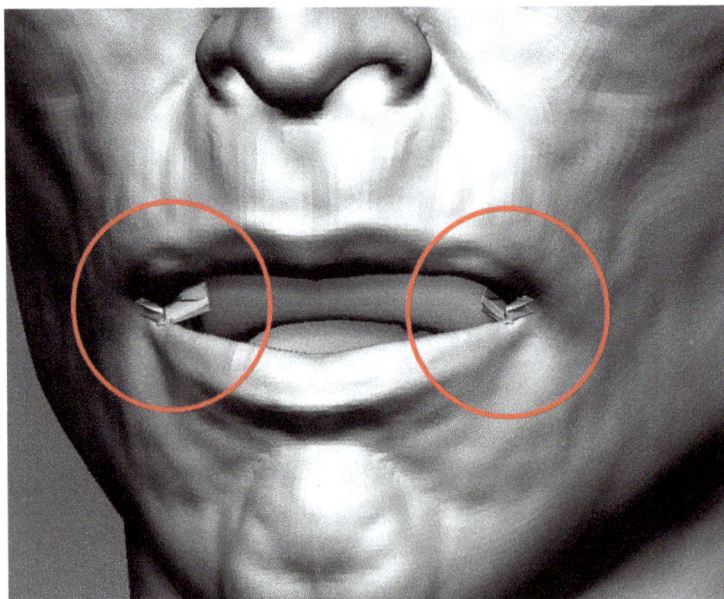

Figure 5.21 – Projection errors

There are several ways to prevent or fix this:

- Lower the **Dist** value in **Tool | Subtool | Project**. This will reduce the degree of distortion in the projection by lowering the distance vertices get attracted to the projection source. In the following figure, you can see the difference between a very big and low distance, with the big distance creating a much more fragmented look.

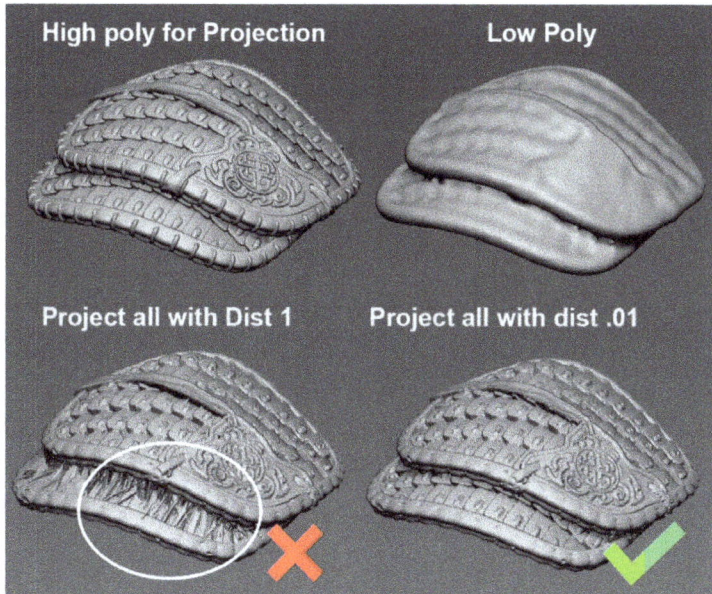

Figure 5.22 – Projection distance

- Mask the area. Masked areas will not be affected by the projection. This could be a good solution in areas where no detail transfer is necessary.

- Clean up after using the **Smooth** brush. If the distortion is not that bad, this can be the quickest and easiest solution.

It is important to note that a clean and effective projection requires some iteration, practice, and experience, so you will need a bit of patience to optimize the results of the projection.

Once you've created a clean, detailed mesh from your DynaMesh concept, you can subdivide the mesh even more so you can focus on adding more detail to the highest subdivision level, or you can decide to switch to a lower subdivision level in order to work more on proportions and larger shapes. The ability to switch between those stages makes the workflow smooth and seamless, and this is one of the big strengths of the subdivision structure.

Now let's continue sculpting the demon bust, adding fine detail, and finalizing the model so that we can add color to it in the next chapter.

Adding detail to the demon bust

Now that you have a proper mesh with several subdivision levels, you can continue sculpting the demon until you are satisfied with the model. There are a variety of ways to add detail in ZBrush. Here are some of the options available for detailing different areas of our demon bust.

Eyes

The eyes are often the focal point of portraits and busts, so it's important to pay attention to this area. You can use the **DamStandard** brush to sculpt small cuts and wrinkles, and the **Standard** brush to give the small shapes between wrinkles some volume.

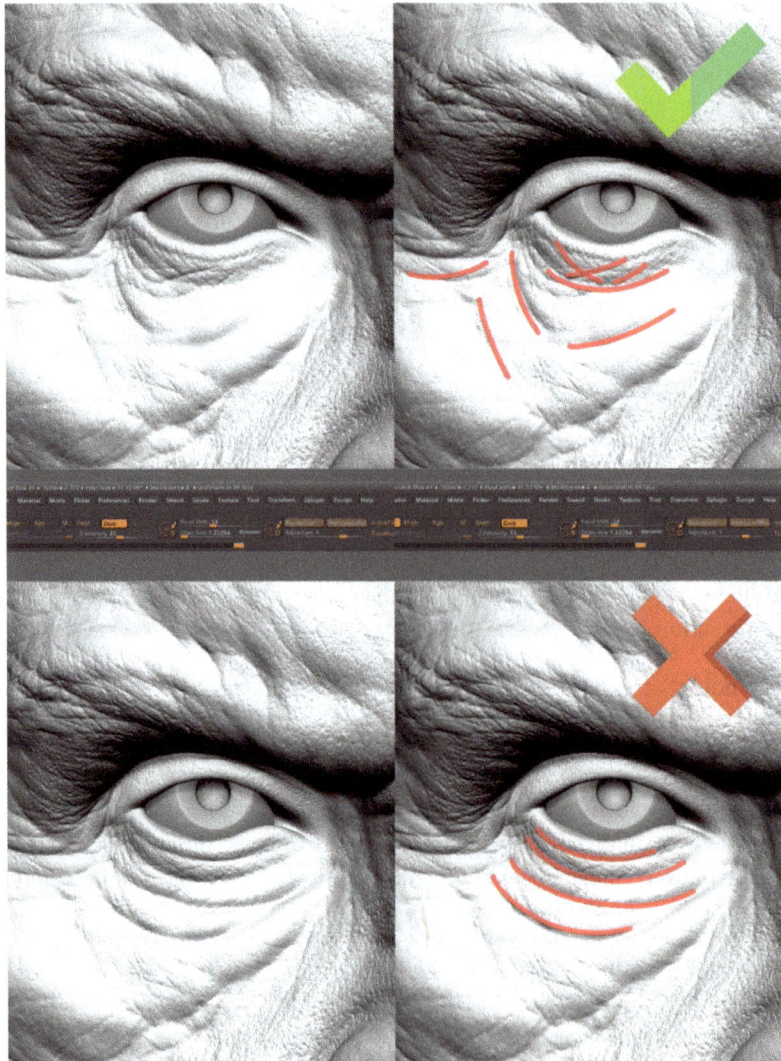

Figure 5.23 – Variation in the direction of wrinkles (upper) and lack thereof (lower)

Reference images are important for this area, as you will see there is a lot of variation in the shape and direction of wrinkles around the eyes. Avoid making them all uniform in size and direction like on the lower sculpt in the figure.

Forehead

The same brushes, **DamStandard** and **Standard**, can be used for wrinkles on the forehead. Plus, the same rule about wrinkles applies to this area: try breaking up the wrinkles and giving them variation in height and angle.

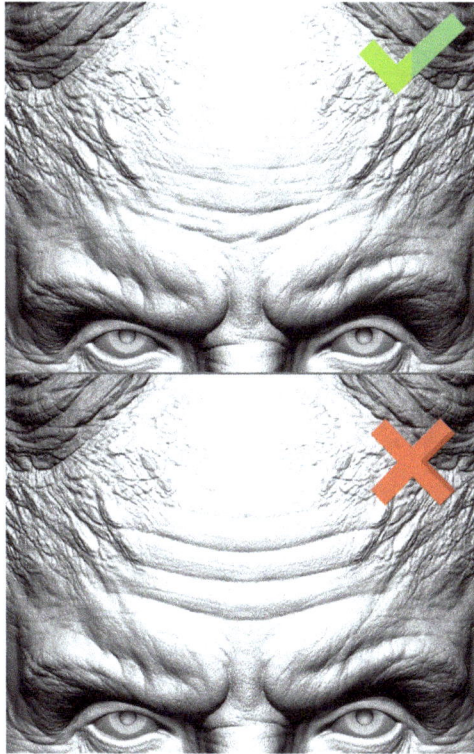

Figure 5.24 – Forehead wrinkles done right (upper) and wrong (lower)

As you can see, variation in intensity, size, and angles will produce a more natural-looking result.

Lips

Lips are another example of an area that requires variation in the direction of wrinkles and cuts. Limiting yourself to simple vertical detailing of the lips would not reflect how lips actually look. Instead, it is worth looking at high-resolution reference images – the **DamStandard** brush is once again a great choice for this kind of detailing.

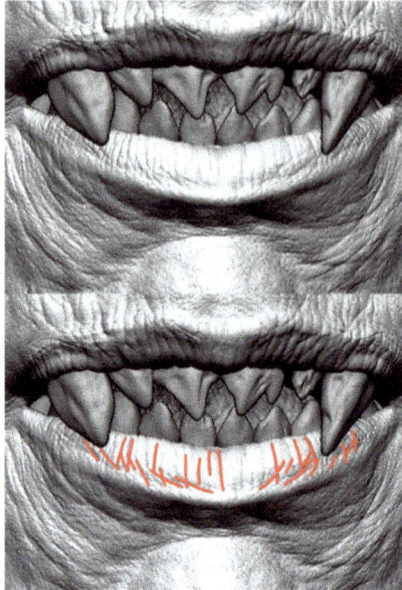

Figure 5.25 – Adding variation to the lip detailing

Horns

The following figure shows the kind of horn structure that can be created using the **Standard** and **ClayBuildup** brush, with **DamStandard** giving the area some cuts and skin damage.

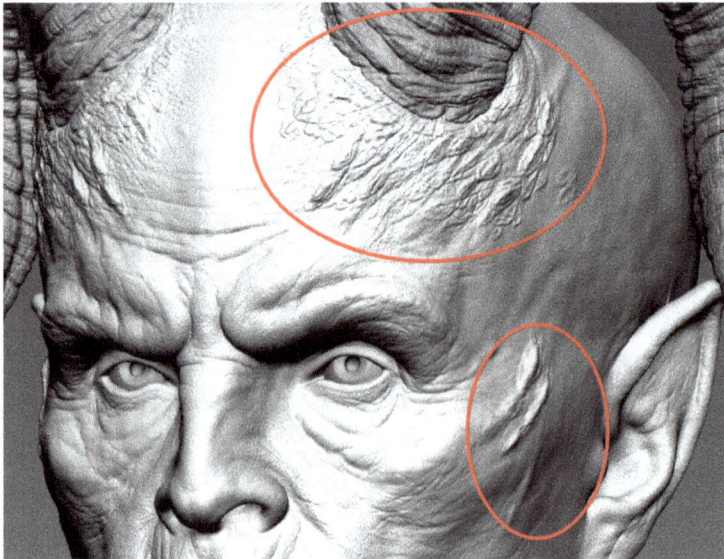

Figure 5.26 – Horn parts

The purpose of these areas is to provide a contrast between the soft skin and the harder horn. This can be supported further by the color that we can add on.

Another reason behind adding this part was to create a transition between the face and the two large horns, making them feel more naturally integrated into the head.

Teeth

The teeth can be sculpted like any other organic shape, using the **Move**, **DamStandard**, and **Standard** brushes. The design of the teeth depends on your concept and preference, but adding variation and small damage will make them visually interesting. Since the teeth are a focal point of this bust, it is a good idea to spend some more time on them – you can try to make them look unique instead of flawless and smooth.

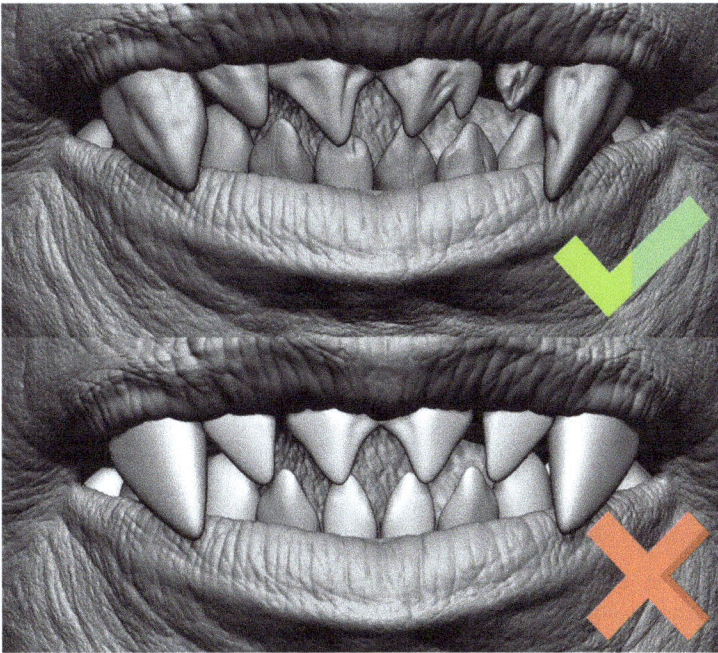

Figure 5.27 – Asymmetrical and detailed teeth (upper) versus symmetrical, smooth teeth

Scar

Scars are a very simple way to add backstory to the sculpture, indicating that the character has been in some fights. Since scars are a heavily used element in character design, make sure to not overdo it and try to do it tastefully.

Beyond the storytelling aspects, scars can be used to create focal points or even guiding lines. The **DamStandard** and **Standard** brushes are ideal for sculpting believable scars.

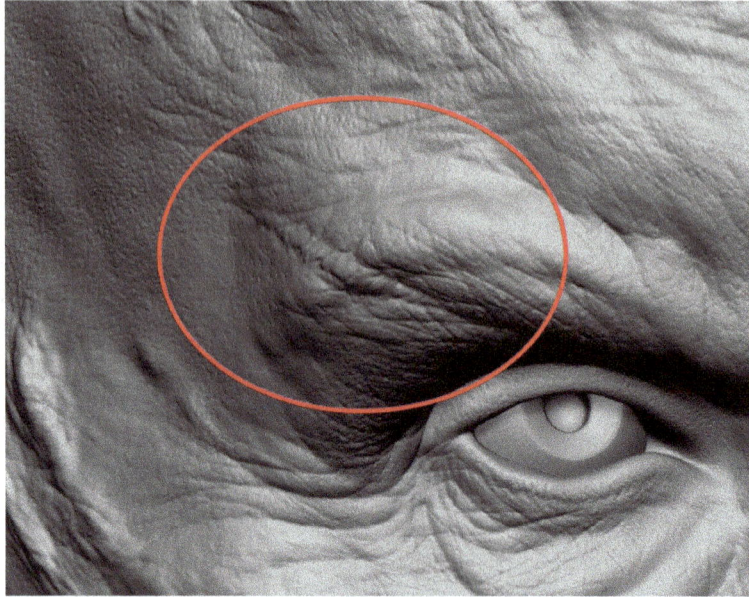

Figure 5.28 – Scars as a way to add storytelling to the character design

All of these points should give you some inspiration to push the detailing on your demon sculpture.

Just note that the fine skin detail that covers the whole bust will be applied in the next chapter after UV creation is covered. For that, we will use NoiseMaker, which is a great way to add detail to large areas of the mesh.

> **Important note**
>
> If you want to get more in-depth surfacing and detailing techniques, you can take a look at the YouTube channel of J Hill for tutorials, or the channel of Ran Manolov for insightful sculpting timelapses:
>
> - J Hill: `https://www.youtube.com/@artofjhill/videos`
> - Ran Manolov: `https://www.youtube.com/@ranmanolov6832`

Summary

In this chapter, we covered ZBrush's powerful retopology tool, ZRemesher, which you used to create proper topology for the demon bust. Here, you learned how subdivision levels allow you to increase the polycount and create meshes with a high level of detail.

You also got to try out the ZProject tool, which lets you transfer detail from a high-resolution concept sculpt onto a newly created mesh with clean topology.

Finally, you explored different ways to add detail to the demon's face using basic brushes and some knowledge of design principles.

In the next chapter, we'll focus on UV creation and materials. With this information and the knowledge you gained from previous chapters, you can finalize your sculpture so you can create some pretty renders to share and present your sculpture in a professional way.

6

Texturing Your Sculpture with Materials, Polypaint, and UVs

In this chapter, you will explore colors and materials in ZBrush, which will let you take your sculpture to the next level and present it in the best possible way. This will set you up for the next chapter, in which you will create beautiful renders and videos of your demon bust.

In the first section of this chapter, you will learn about two types of ZBrush materials – Standard Materials and MatCap Materials – and understand their strengths and weaknesses so that you can implement them into your workflow accordingly. Furthermore, you will be able to create and save custom materials.

Next, you will learn about Polypaint, ZBrush's color system, which will enable you to take your sculptures from good to great by giving them a finished look. We will start with some basic color theory tips to help you effectively apply the coloring tools, and then go through a variety of powerful polypainting techniques demonstrated on the demon bust we started creating in *Chapter 2*.

Finally, we will cover UV creation for ZBrush using the **UV Master** plugin. Once you've created proper UVs and checked them for errors, you can export color maps for your model using the **Multi Map Exporter** plugin.

So, this chapter will cover the following topics:

- Exploring ZBrush's Materials
- Adding color to your model – Polypaint
- Creating UVs and exporting textures

Technical requirements

For the best experience, it is recommended that you have a strong PC that meets the minimum requirements described in the *Technical requirements* section of *Chapter 1*. However, you can work on this chapter with just a mouse, a functional PC setup, and a ZBrush license.

To follow along, you will need the demon sculpture we started creating in *Chapter 2* or a suitable model of your choice. It is recommended that you have a low-polygon model so that UV creation will work well.

You can find suitable models in LightBox, which you can use to practice UV creation.

Exploring ZBrush's Materials

In this section, you will learn about ZBrush's Materials and see how different materials can affect the look of your 3D models. We will start by discussing the two types of available materials: **Standard Materials** and **MatCap Materials**. Both types of materials have unique properties and solve different goals.

You'll also learn how to apply materials to your models, as well as how to modify them and save those custom materials for later use. By the end of this section, you should have a good understanding of the available materials and how to use them in your workflow.

Standard Materials versus MatCap Materials

There are two types of materials available in ZBrush: Standard Materials and MatCap Materials.

The main difference is that MatCap Materials have **Light** and **Specularity** information baked in. This means that objects with MatCap Materials applied do not react to lighting in your scene and always look the same, except for the difference in shading, which is caused by looking at the model from different angles.

Standard Materials, on the other hand, will react to your light setup, which makes them more realistic and allows them to give a better representation of what the model will look like when exported to a different 3D software, or even when 3D printed.

You can see the effect, and lack thereof, on Standard and MatCap Materials in the following figure:

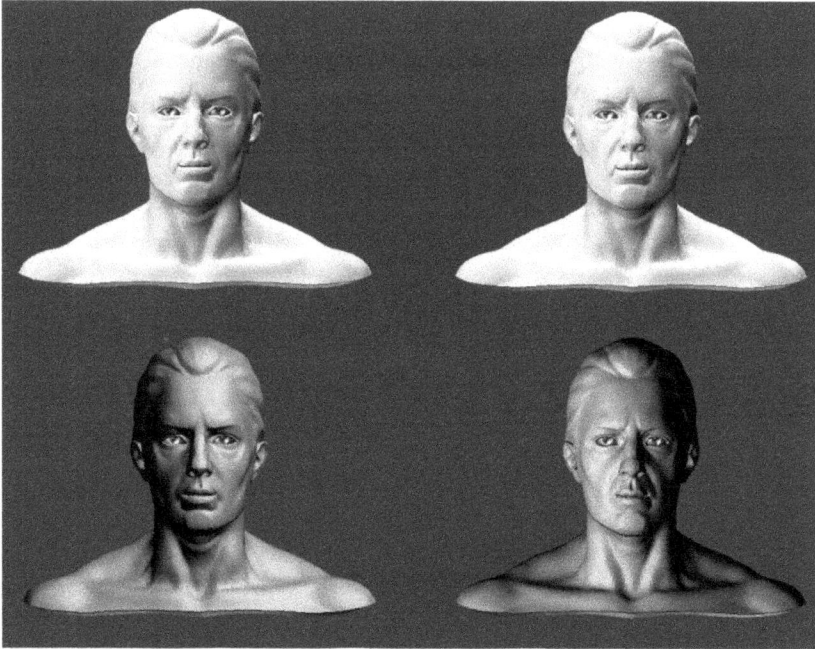

Figure 6.1 – The effect of changing light direction on a MatCap
Material (top) and a Standard Material (bottom)

The ability of Standard Materials to display accurate lighting information is a significant advantage because it allows you to have a better idea of what the mesh looks like so that you can avoid unexpected, often negative, results. This applies especially to human portraits and likeness sculptures, where a high level of accuracy is required.

MatCap Materials such as the MatCap Gray Material can be great for sculpting as they highlight the forms of digital clay in a way that makes it easy to visualize and understand the forms. The material also has fewer dark shadows, which makes it easier to visualize parts of the model that would normally have stronger shadows cast on them.

So, while the lighting is not realistic, for easy and effective sculpting, many sculptors like using MatCap Materials. These are the main differences between the two types of material. Of course, you can always easily switch between materials, using their properties to focus on different aspects of the creation process.

With this information, you can now go ahead and apply a material to your mesh.

Selecting and applying materials

To apply a material to your sculpture, you need to access the Material Library. The Material Library can be found in the **Material** palette in the top bar of the window, though you can access it faster by clicking on the **Current Material** icon on the left-hand side of the canvas:

Figure 6.2 – ZBrush's Material Library

In this window, you will be able to see the materials grouped into **MatCap Materials** at the top and **Basic Materials** at the bottom.

Selecting a material

Clicking on any material will select it, and every subtool in your ZBrush session will be displayed in that material. However, this is only the case if the models have not been assigned a material yet, or if their paint icon in the subtool list is disabled (indicated in the following figure):

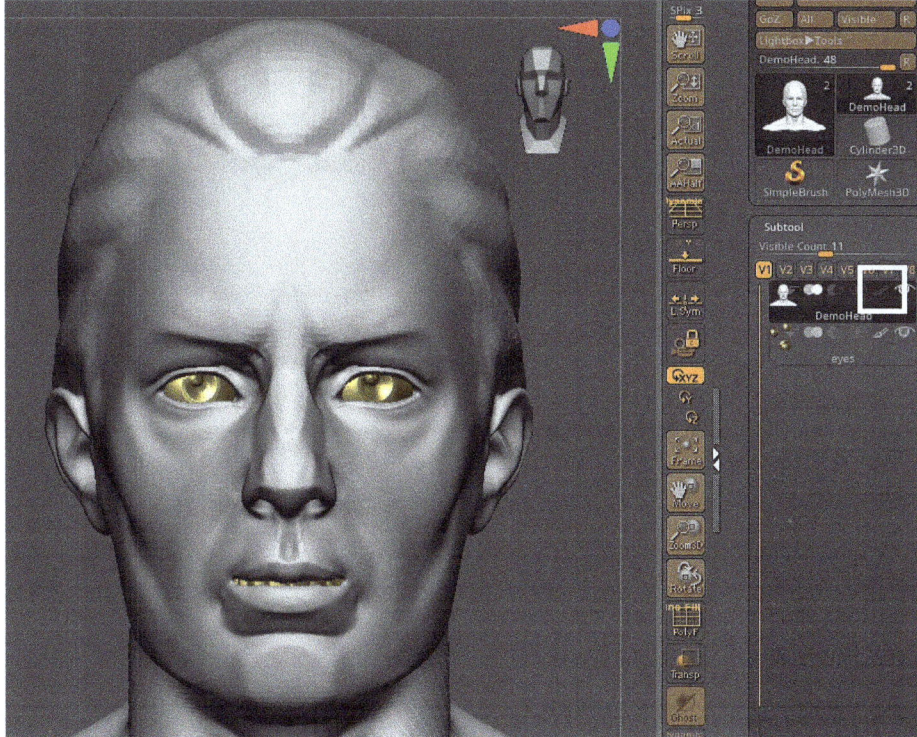

Figure 6.3 – Subtool paint mode disabled and enabled

Otherwise, if material information is stored on a subtool and the paint icon is enabled, such as on the eye subtool here, its material information will override the material you currently have selected.

You can toggle the visibility of the color/material information of a subtool by clicking on the paintbrush icon on the right-hand side of the subtool list.

If you have the color information of your model turned off, you will be able to see the effect the material has on your model and decide if you would like to apply it to your subtool(s).

Applying a material

With a material selected, make sure **M** or **Mrgb** mode is enabled for your current brush – these modes are displayed above the canvas or in the **Draw** palette:

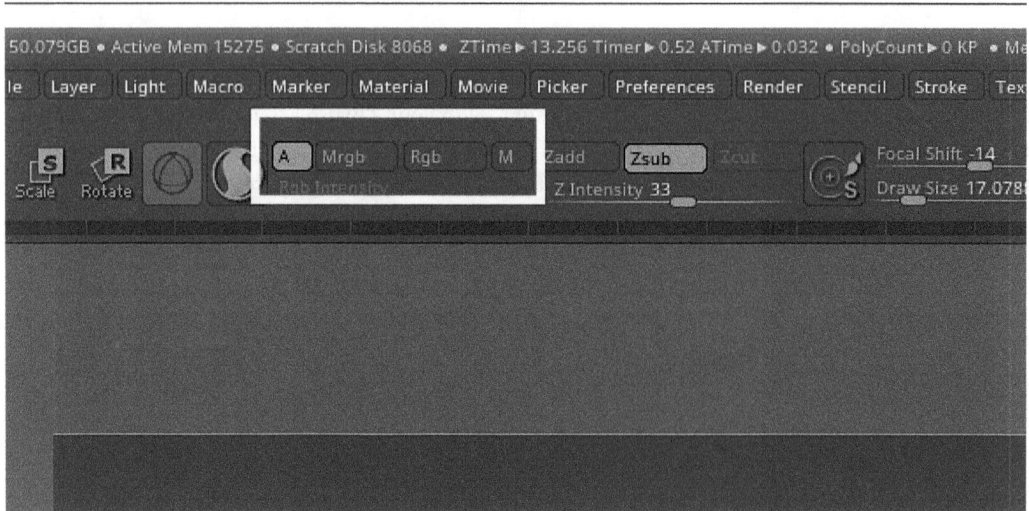

Figure 6.4 – Material and color mode buttons above the canvas

Then, you can open the **Color** palette and click on **FillObject**. If you have **M** mode enabled, this will apply the currently selected material to the object, whereas if you have **Mrgb** mode enabled, it will apply both the material and color.

Once you've applied a material and/or color to a subtool, paint mode will be turned on for that subtool, and selecting a different material in the material library will not be reflected in the active subtool.

Adjusting material modifiers

ZBrush offers a variety of materials, ranging from different metals to plastic, skin, or various clay materials. While you have a lot of materials to choose from, you may need a very specific material that you cannot find. For example, this could be a material that has a certain tint, or a certain specularity profile to it.

In this case, you can create a custom material by adjusting the properties of an existing material. Since MatCap Materials can't be adjusted in the same way as Standard Materials, let's learn how to make a custom Standard Material:

1. Open the Material Library and select **BasicMaterial**.
2. Select **CopyMat**.
3. Click on any other material (you will replace this later) and click **PasteMat** (ZBrush only lets you paste on existing materials).
4. Navigate to **Material | Modifiers**.

This is what the model with **BasicMaterial** looks like by default:

Figure 6.5 – Model with BasicMaterial selected and applied

Let's look at some modifiers while using this model as a guide.

Ambient

The **Ambient** modifier is similar to the **Exposure** setting in image editing software – it brightens up the whole model and illuminates shadow areas, which makes it great for sculpting on areas concealed by shadows:

Figure 6.6 – Effect of the Ambient modifier

When you are polypainting your model, this modifier is great because it gives a low-contrast look that reveals colors more. At the same time, however, it is also harder to visualize forms in brighter areas of the model due to the lack of contrast and depth created by this modifier.

Diffuse

Diffuse will increase the light intensity, but unlike **Ambient**, this will not brighten up the shadow to the same extent, leading to a higher contrast look:

Figure 6.7 – Effect of the Diffuse modifier

Specular

You can adjust the specularity of the surface with the **Specular** modifier – higher values increase the brightness of the brightest spots on the model:

Figure 6.8 – Effect of the Specular modifier

Specular Curve

The **Specular Curve** modifier lets you adjust the shape of the specular highlight. You can make it more diffused on rough surfaces, for example, or narrow and tight on metal or oily surfaces, for example. The following figure shows the effect of different **Specular Curve** profiles on the specular highlight:

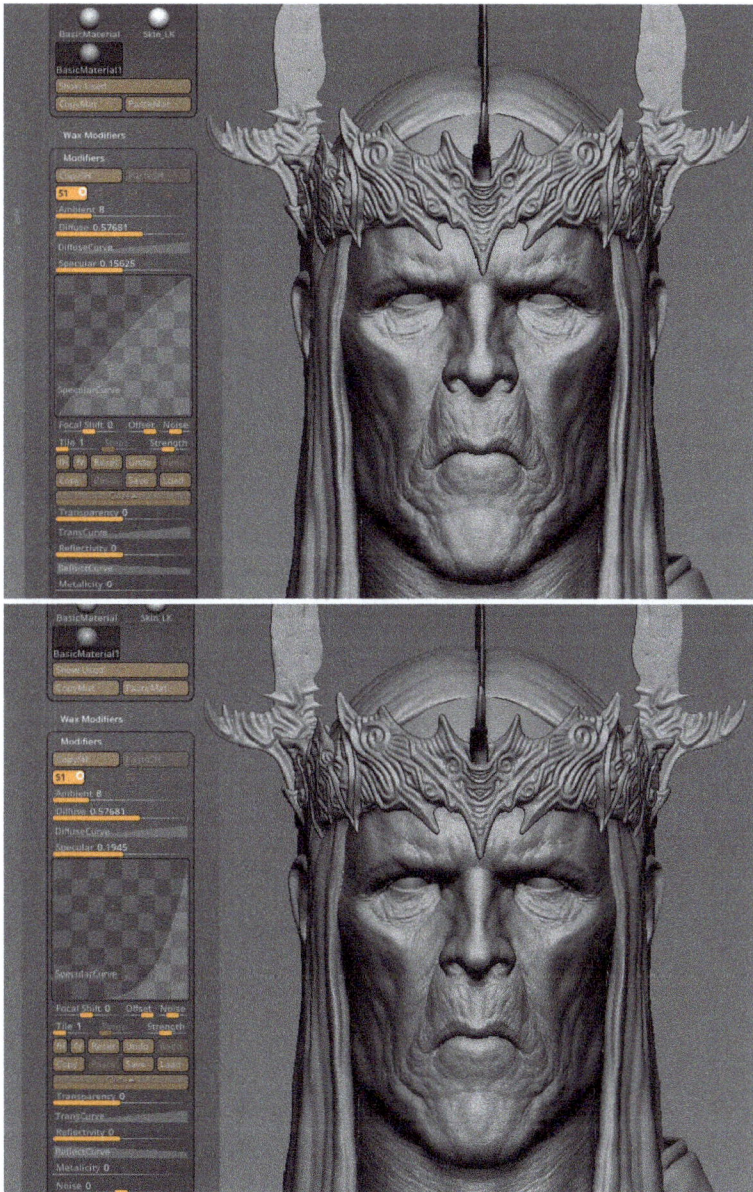

Figure 6.9 – Effect of the Specular Curve modifier with a high contrast curve

Compared to the first image, the specular highlight is slightly smaller and more intense in the second image, creating a bit more contrast overall.

Colorize Diffuse and Dif

If you increase the **Colorize Diffuse** value, you can pick a color for the **Dif** modifier and it will take on that color:

Figure 6.10 – Effect of the Colorize Diffuse and Dif modifiers with green color input

Saving and exporting materials

To save a material for later use, simply open the Material Library, with your material selected, and click **Save**. There is also the option to click **Save as Startup Material**, which will make this the default material that ZBrush launches with:

Figure 6.11 – Save options for materials

This concludes this subsection on important material modifiers. Most sculptors will find materials they like in the Material Library, but knowledge of these modifiers can be helpful when you decide to enhance materials. For example, you might want to add some specularity or reduce the shadow a bit for color painting. Although the use of advanced and special materials in ZBrush is rather rare, especially for commercial work, it's good to know that these modifiers exist.

In the next section, you'll learn about Polypaint, ZBrush's color system, and apply color to our demon bust.

Adding color to your model with Polypaint

Besides its vast sculpting and modeling functionality, ZBrush also gives you the ability to color your models using an array of different tools and techniques. In this section, you'll learn about ZBrush's Polypaint feature, which lets you add color to your demon bust.

What is Polypaint?

Polypaint is a tool for painting on the surface of a mesh, storing color information in the vertices. It lets you paint on any mesh without the need for UVs, which makes it very flexible and great for concept sculpting:

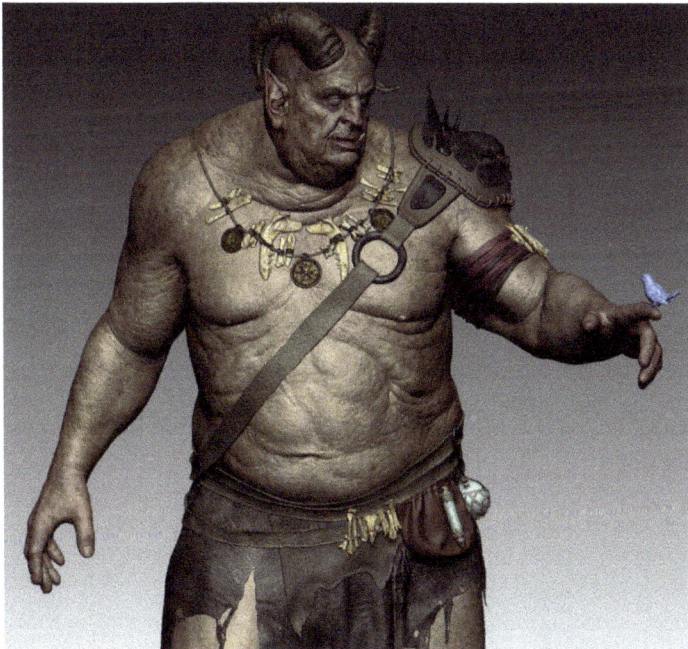

Figure 6.12 – A sculpture created with Polypaint in ZBrush

However, adding detailed color information requires a high-resolution mesh with a large polycount since the number of vertices determines the amount of color information that can be stored. If very sharp color textures are needed, Polypaint might not be the best choice – the maximum polycount of a mesh is limited and your PC's performance is affected when you're working on extremely high poly meshes.

To avoid this, it is always possible to split the model into multiple parts, but this will create other potential issues in a production pipeline.

However, for a concept sculpture such as our demon bust, Polypaint is a great choice!

Exploring Polypaint techniques and tips

In this section, you'll learn about techniques and workflows for adding color to your model. You can follow along by painting on the demon bust, or any other model of your choice.

Before you start implementing those tools, though, it makes sense to take a brief look at some basic color theory so that you know how to use them in a way that produces visually appealing results.

Basic color design tips

First, it's important to note that color will only play a minor role in the quality of your finished model. Great color won't turn a poor sculpture into a good-looking one, but a masterfully crafted sculpture with mediocre or boring color will still look great in most cases.

This means you shouldn't rush through the sculpting phase and time spent detailing will make texturing easier. Color can still enhance the design and tell a story (and a lot of the design principles discussed in *Chapter 2* also apply to color). Let's look at some tips you should keep in mind while you're painting your model.

Color schemes

Not all colors mix well together. However, certain color schemes have proven to work well:

- **Monochromatic:** This color scheme is based on a single color hue, only using variation through saturation and value:

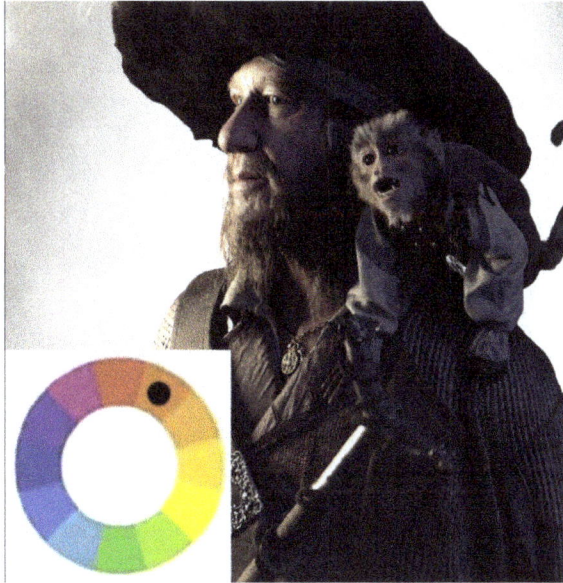

Figure 6.13 – Monochromatic color scheme

- **Analogous**: This color scheme uses one main color and two secondary colors that are right next to it on the color wheel. The secondary colors should only make up a smaller part of the design, while the main color makes up most of it:

Figure 6.14 – Analogous color scheme

- **Complementary**: This color scheme is made up of two colors that are opposite each other on the color wheel. One color should be dominant, while the other should be used as an accent:

Figure 6.15 – Complementary color scheme

- **Split-complementary**: This color scheme uses one base color and two secondary colors that are placed symmetrically around it on the color wheel:

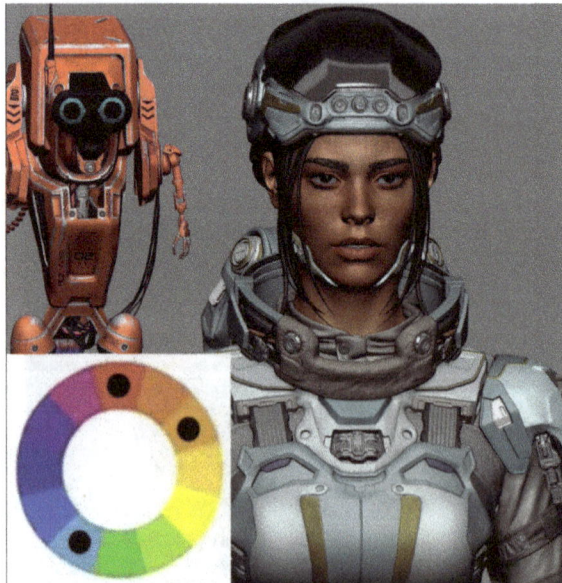

Figure 6.16 – Split-complementary color scheme

Note that these color schemes are not absolute rules, but rather guidelines. They are particularly useful for beginners as it is hard to go wrong with them.

Color meanings

Color can be used to emphasize certain themes. This is often used in business, marketing, and design of any kind, so it's something to consider when you're designing your characters. Let's take a look at what colors can be indicative of:

- **Red**: Evil, passionate, and energetic
- **Blue**: Trustworthy and confident
- **Yellow**: Happy and energetic
- **Orange**: Joyful and enthusiastic
- **Black**: Evil, luxurious, elegant
- **White**: Innocent, wise, divine
- **Green**: Alien, natural, growing
- **Purple**: Sexy, elite, luxurious

Rule of thirds

Color schemes already include a balance of dominant and accent colors. The rule of thirds builds on this idea, suggesting that you should have an asymmetrical distribution of color so that they do not compete with each other, and focal points can be created:

Figure 6.17 – A balance of darker/more saturated color (red) and brighter/less saturated areas (blue)

This rule also applies to other factors, including level of detail. When adding patterns or details in color, it's important to balance them with some simple areas that have little detail. This creates so-called areas of rest, which make the model more pleasing to look at:

Figure 6.18 – A balance of highly detailed areas (red) and less detailed areas (blue)

Color zones of the face

Since portraits are a popular topic and painting skin is a common task, it makes sense to look into practices for painting the skin as well.

When you paint the skin color of your model, it is important to not just use a flat "skin color" tone, but instead to add a variety of colors throughout the head. A useful concept to follow when you're adding this variation is the color zones of the face, as described by illustrator James Gurney.

Used in some form by many famous painters, the color breakup goes as follows:

1. **Upper third/forehead**: An overall yellowish tone comes from the absence of red tones, largely caused by blood cells and muscles.

2. **Middle third/center of the face**: This area has more prominent red tones due to a larger amount of capillaries and the blood flow closer to the surface.

3. **Lower third/chin**: Here, you can observe a more blueish tone. For shaved men, this comes from microscopic hair, and overall, the flow of deoxygenated blood around the lips adds to that effect.

The following figure shows the difference between a flat skin color and a skin color with some color variation:

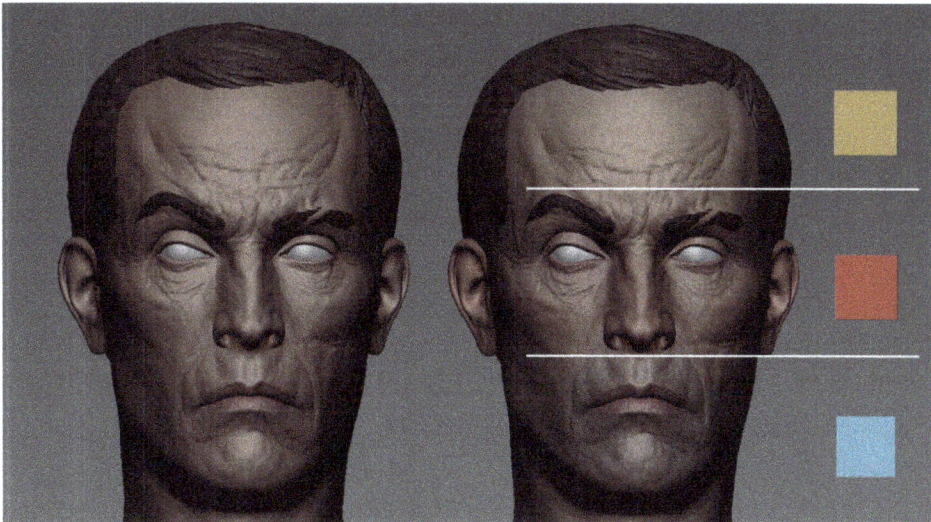

Figure 6.19 – Adding color variation based on the color zones of the face

Of course, these colors have to be used in a very subtle way and appropriately depending on the subject, but they are a great and simple way to make the face feel more "fleshy" and interesting.

These color theory concepts will be useful as a foundation for the following section, which will teach you painting techniques, and allow you to finish with a nicely painted sculpture that is ready to be rendered.

Starting your polypainting

To start adding color, there are a few things to ensure:

- Select the **Paint** brush, which is one of ZBrush's default brushes in the main brush menu.

- Go to the **Stroke** palette | **Lazy Mouse** and disable **LazyMouse**. While this is useful to stabilize the brush stroke while sculpting, it can be hard to paint with.

- If you use a sculpting brush, make sure to enable **Rgb** mode and disable **ZAdd/ZSub** for the selected brush by clicking on the relevant buttons above the canvas or going into the **Draw** palette. Also, make sure **Rgb Intensity** is at a value above **0**; otherwise, there will be no effect (a value of **100** lets you paint with 100% intensity).

- Ensure you have a high enough polycount. This depends on your preference, but if there are large visible edges and polys, the polycount may be too low for proper polypainting:

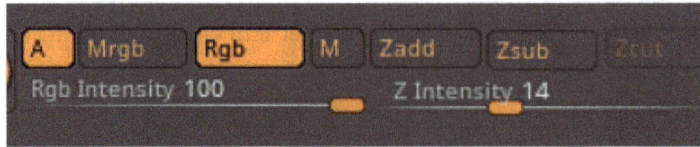

Figure 6.20 – Basic settings for polypainting

At this point, you can pick a color on the color picker on the left-hand side of your canvas and start painting your model (Using your pen/left-clicking).

Polypainting the demon's head

Besides using a brush to paint on your demon sculpture, there are several ways to apply color, creating interesting effects. Here, you will learn about some of the most useful ways to polypaint your model; you can follow along as you paint your demon sculpture.

Before starting the painting process, you have to decide which material to pick. A good choice is **BasicMaterial**, with an increased **Ambient** value. Go to **Material | Modifiers** and increase **Ambient** until the shadow areas of your model are illuminated a bit.

Since the material is gray by default, colors on your screen will appear darker than they are. This can be reduced with a brighter material so that you have a more accurate representation of what the color would look like once it's been exported:

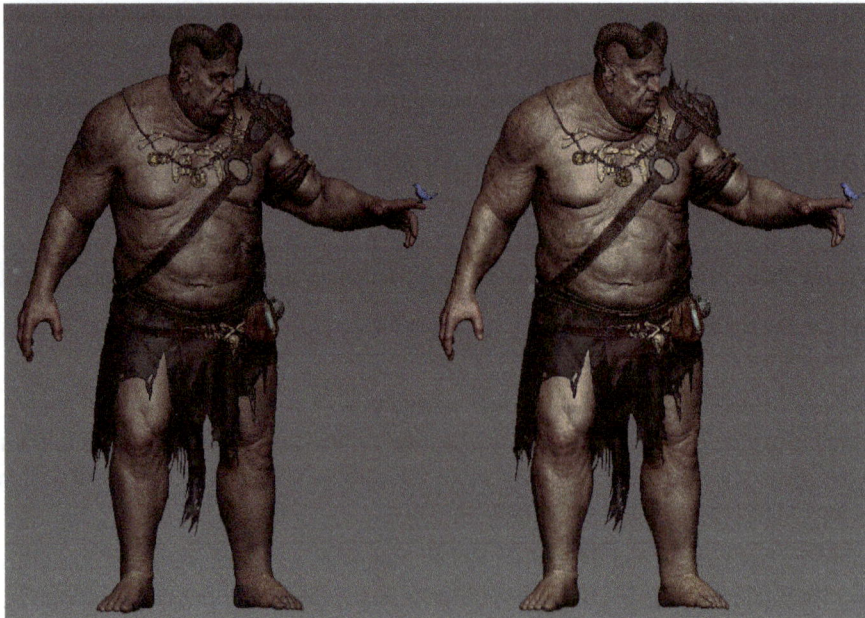

Figure 6.21 – The difference between the default BasicMaterial (left) and a brightened-up material (right)

Color fill

As a starting point, you can begin by giving your model a main color. With a color picked from the color picker and your model selected, navigate to the **Color** palette and click **FillObject**. You can repeat the same for eyes, teeth, and horns (if they are separate).

Now, you should have a very basic, colored sculpture:

Figure 6.22 – The demon sculpture is filled with simple colors

Color Spray

Now, choose your standard painting brush and select **Color Spray** mode on the **Brush** mode icon from the left-hand side of the canvas. Next, click on the **Alpha** icon, and select **Alpha07**. This will give a nice spray-paint effect.

This will let you spray-paint on the mesh with the selected color, as well as some other random colors. The amount of color variation is controlled by a value inside the brush modes menu: the **Color Intensity Variance** option. With a value of **0**, there is no color variation at all; at **1**, it's a rainbow spray paint:

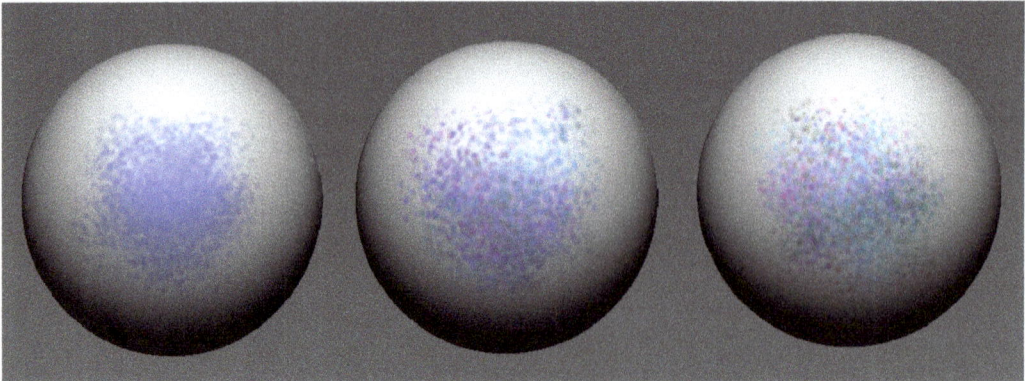

Figure 6.23 – Color intensity variances of 0 (left), 0.5 (middle), and 1 (right)

For the demon bust, you can pick a low **Color Intensity Variance** like **0.1** or **0.2**. Then you can pick a dark color, and start painting some darker areas on the demon. To create a nice contrast to the eyes, paint the eye area darker. You can also paint the horn/scale areas so that they are recognizable as different surface textures:

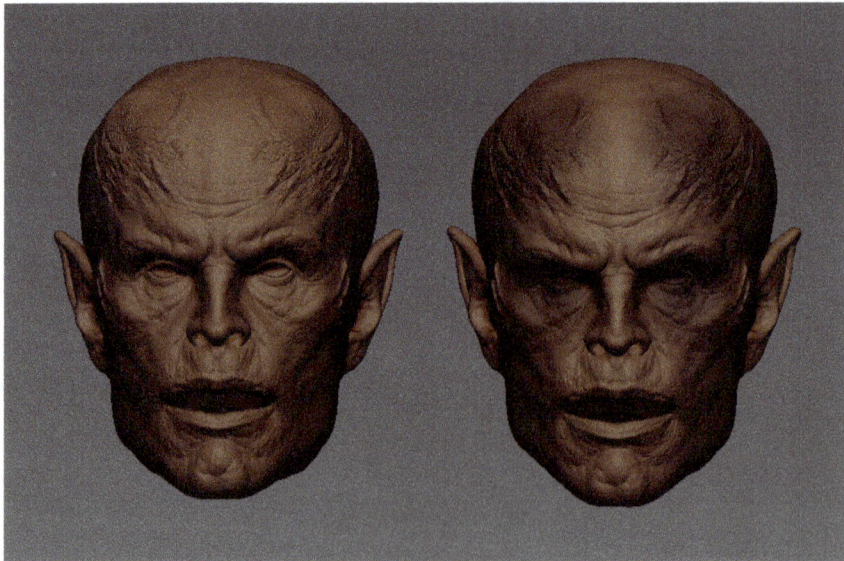

Figure 6.24 – Before and after painting darker areas on the head

> **Important note**
>
> To pick a color from the canvas, hover over the color and press *C*. This is especially useful if you want to keep using the same color. Once you have applied that color somewhere on the mesh, you can keep accessing it. You can even add a new subtool, just to add color to it, turning it into your own color library for your project.

Color variation

For the horns, you can try to create a bit of a gradient, with darker colors at the root and brighter colors at the tip. A smooth transition between dark and light tones can be observed and natural, which makes this a good technique for achieving natural-looking results.

At the root of the horns, make sure that the skin blends nicely with them by applying darker colors there as well.

Around the eyes and for the lips, you can go with desaturated, purple tones:

Figure 6.25 – Applying basic color variation

Adding detail

If you want to add more detail and make the texture more complex, there is a useful Alpha that is great for detailing: **Alpha 60**. You can equip it to your spray brush and use it with a lighter color tone to suggest a fibrous structure, such as wood grain:

Figure 6.26 – Adding color with Color Spray mode and Alpha 60

Adding brighter tones

After adding darker tones, you can also add brighter color variations. For example, this could work well on the corner parts of the horns and on the part of the face where the skin is close to the bone, such as cheekbones.

Here is what we have so far:

Figure 6.27 – Refining the color of the sculpture

DragRect

When you choose the **DragRect** brush mode, you can drag an Alpha on your mesh and scale it interactively. **Alpha 08** could be a good Alpha to create a freckle-like texture, but you could also choose one of the round Alphas to create moles or freckles individually:

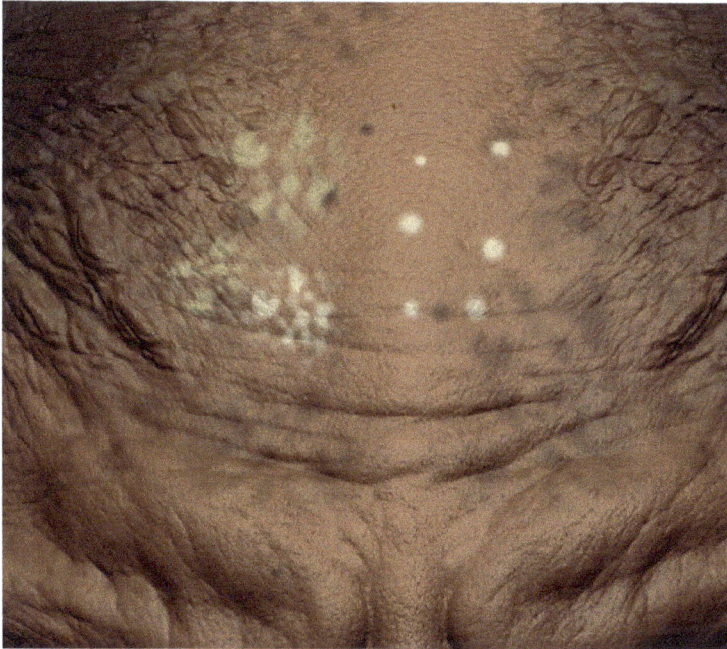

Figure 6.28 – Freckle-like color applied with the DragRect brush mode

This brush mode does not react to pressure sensitivity, so it might be too strong when you use an **Rgb Intensity** value of **100**. If the Alpha won't blend in nicely with its surroundings, you might want to reduce its intensity.

Instead of using the **dragRect** brush with an Alpha, you can use it with a texture. To pick a texture to use, click on the **Texture** slot below the **Current Alpha** icon and select one of the textures in the library.

Texture 22 has an interesting, organic look and a lot of variation that can help add more complexity and detail to your model. You may use this texture with a low **Rgb Intensity** value to make the texture more detailed:

Figure 6.29 – Adding detail with a dragRect brush

Painting using a cavity mask

Detailed models are great for texturing as they give you options for polypainting techniques. If your model has cavities – wrinkles or little holes, for example – you can mask them off. Then, you can either paint on the surrounding area or invert the mask and paint in the cavities directly.

Cavity masks are great for adding dirt or dark colors, and for the demon bust, we can use a cavity mask to add dirt to the teeth. To do this, go to **Tool | Masking | Mask by Cavity** and adjust the **Intensity** value, as well as **Cavity Profile**. You can experiment to see what gives the best results – I use a regular linear curve and an **Intensity** value of **10**.

Then, click **Mask by Cavity**, and then hold *Ctrl* and left-click on the canvas to invert the mask. Now, you can paint with a dark color and you will precisely fill the cavities with that color:

> **Important note**
> You can disable the visibility of the mask so that it does not distract you while you're polypainting the model. To do so, go to **Tool | Masking** and disable **View Mask**.

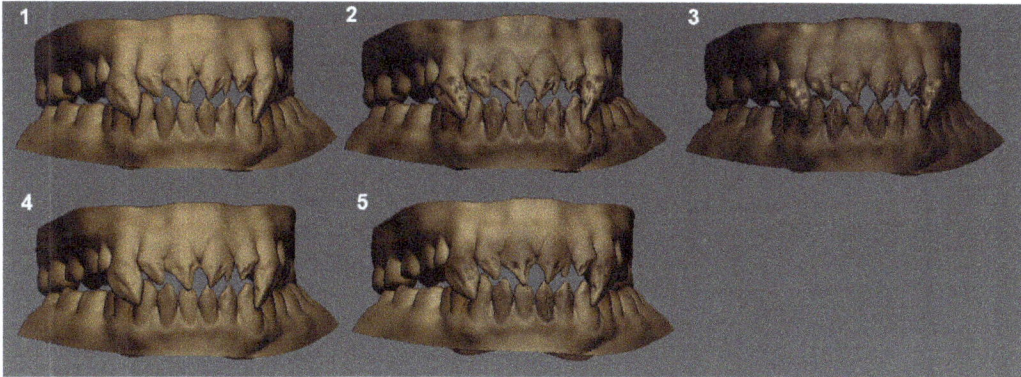

Figure 6.30 – Starting point (1), masking by cavity (2), inverting the mask (3),
disabling the visibility of the mask (4), and adding color to the cavities (5)

You can also use the cavity mask without inverting it and use another color on the protruding/highlighted areas. For example, we can paint on the horns and choose a light beige color to add highlights to the outer forms of the horns, which are unmasked. This could suggest some sort of effect from the weather or minor damage that brightened up the horns:

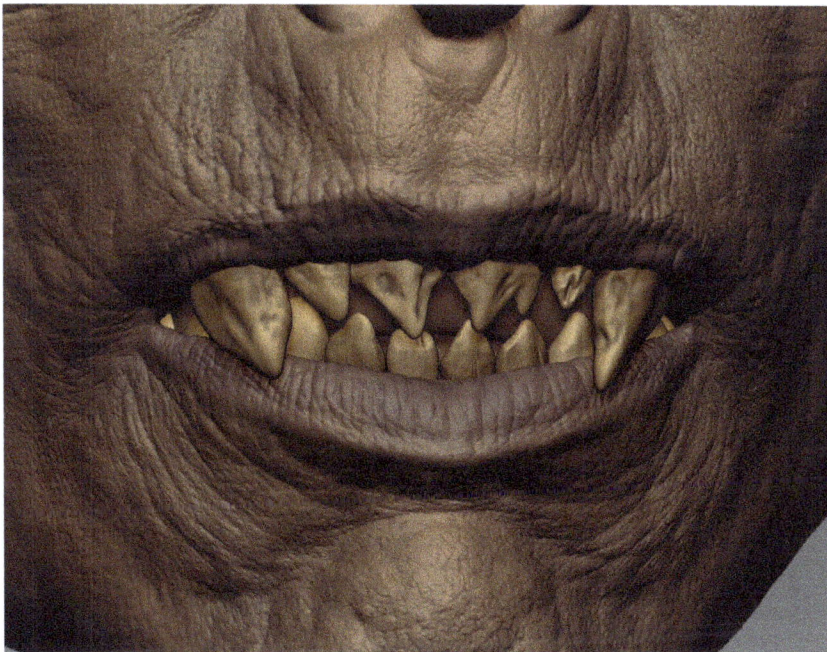

Figure 6.31 – Using Mask By Cavity to add paint to cavities on the model

Painting using a PeaksAndValleys mask

Another useful masking technique for polypainting is the **PeaksAndValleys** mask. As its name suggests, this function will mask polygons that are locally the highest or lowest points in high-detail areas of the mesh. Here is an example that shows how this mask affects detailed areas, while smoother areas stay unmasked:

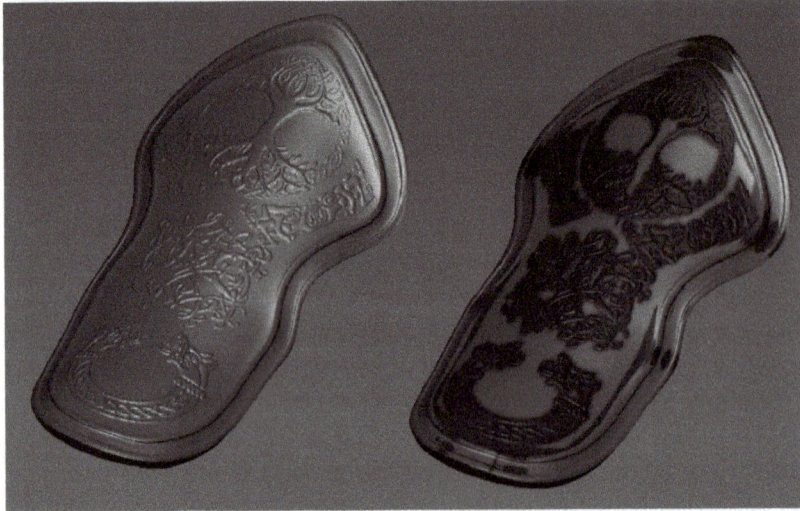

Figure 6.32 – Applying a PeaksAndValleys mask

For the model, it can be used on the horns to add more dark and bright color variations:

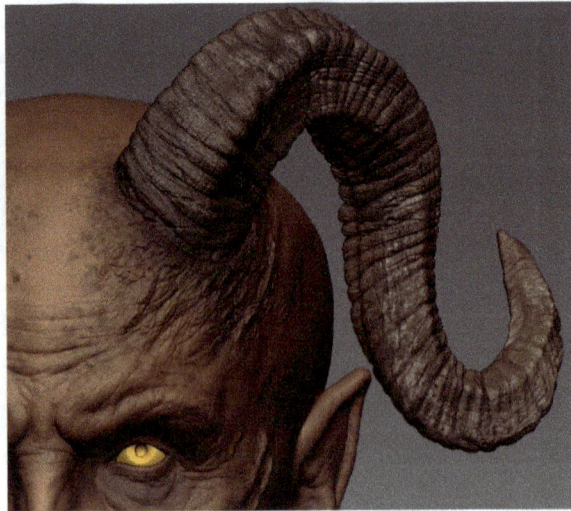

Figure 6.33 – Adding color masking PeaksAndValleys

There are more masking techniques that can potentially be useful – often, you must try out different techniques and see what gives the best results for the given mesh!

Repeating the process for the demon's upper body

After texturing the head, you can texture the body, using the same color and techniques to ensure it matches the head (if you choose to have the head and body as a single mesh, you can ignore this). Here are the steps for adding Polypaint to the body:

1. You can pick the basic color from the head by pressing *C* while hovering with your mouse/ brush over the color on the head that you want to copy. The color wheel will have this color active now. Then, go to the color palette and select **FillObject**. Make sure you have **RGB** mode enabled for this:

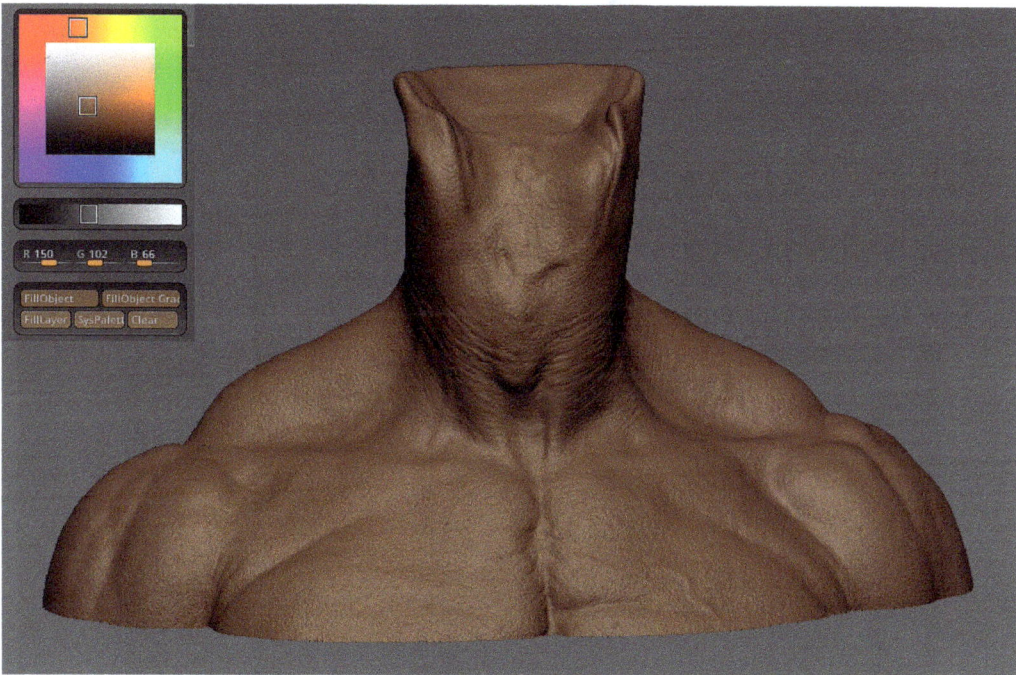

Figure 6.34 – Applying basic color

2. Next, you can pick a dark color to add some simple color variation.

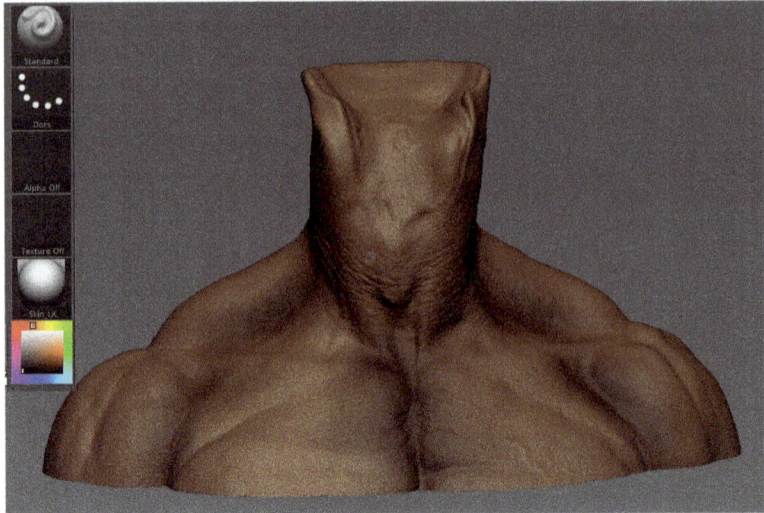

Figure 6.35 – Dark color added to the body

3. Add a lighter color with **Color Spray** mode and **Alpha 60** to give an interesting detail to the body, and increase the color detail. Following the direction of the muscle shapes will make this color detail more believable:

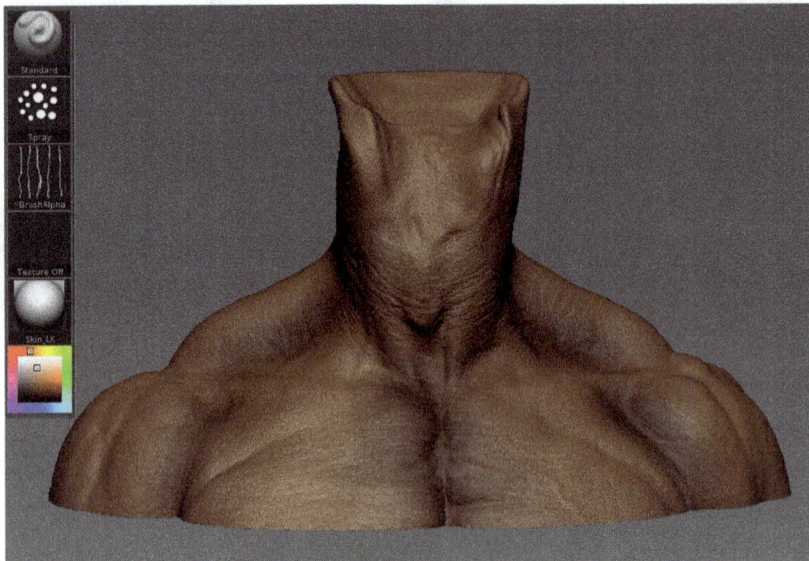

Figure 6.36 – Adding color detail with Color Spray mode

4. Switch the brush mode to **DragRect** and load **Texture 22** to apply some more detail to the body. You might want to decrease the RGB intensity if the effect is too strong:

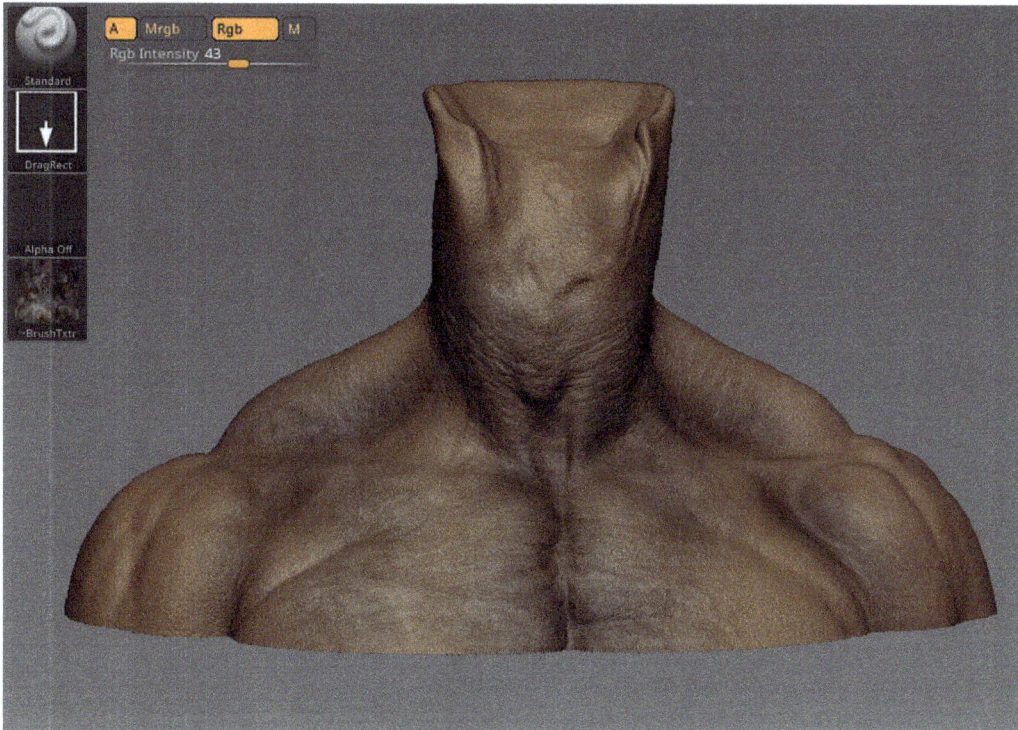

Figure 6.37 – Applying a color texture

5. Finally, you can use the base color again to paint over the color detail you have added. This can reduce the intensity and contrast if you would like to have a more subtle color, which can look a bit more natural and believable. This is a design choice that is up to you:

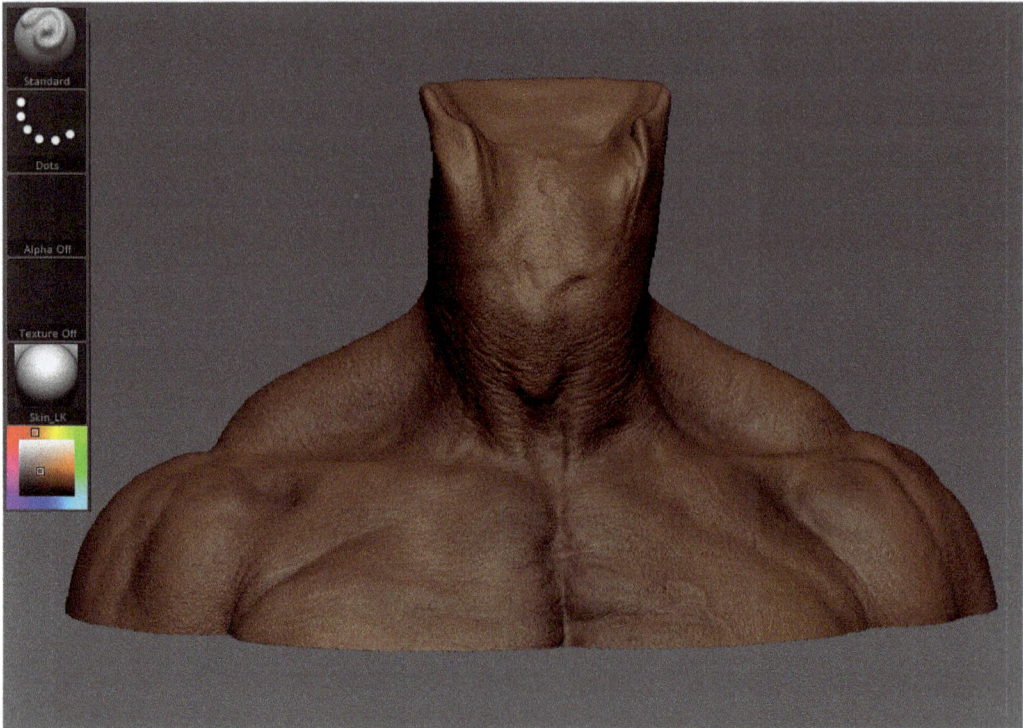

Figure 6.38 – Finalizing the color

That concludes polypainting the demon bust. Here, you learned about some simple tools and techniques that can be used on a variety of objects. Applying them effectively and making them visually appealing will probably require some repetition and additional education on the theory and methods of painting. A fantastic and free resource is the video tutorials of artist Bobby Chiu. You can find his videos here: https://www.youtube.com/@BobbyChiu/videos.

In the last section of this chapter, we'll learn how to create UVs inside ZBrush using the **UV Master** plugin. Having a model with UVs will allow you to export textures, use your model in other programs, and have it contain all the color and sculpting information you created in ZBrush.

Creating UVs and exporting textures

In this section, you will learn how to create UVs using the **UV Master** plugin. First, we will take a look at what UVs are and why they are important. Then, we will take a look at the **UV Master** plugin, and its strengths and weaknesses compared to other UV tools. By the end of this section, you'll be able to export texture files for the demon character using the colors you painted in the previous section.

What are UVs and why do they matter?

UVs are markers that link the vertex information of a mesh to the pixels of a two-dimensional texture file. This allows you to apply textures to your model and get different effects, such as displaying color or adding detail in the form of normal or displacement maps.

UV mapping is especially useful for game characters or animated meshes, which need low-poly models. UV mapping enables these low-poly meshes to display height and color information that makes them appear as detailed and complex as any high-poly model created in ZBrush while having a much better performance and other production-related benefits.

UVs make it look like the model was cut up and flattened on a square image, as shown in the figure image. You can see multiple pieces, called UV shells, assembled in a square space:

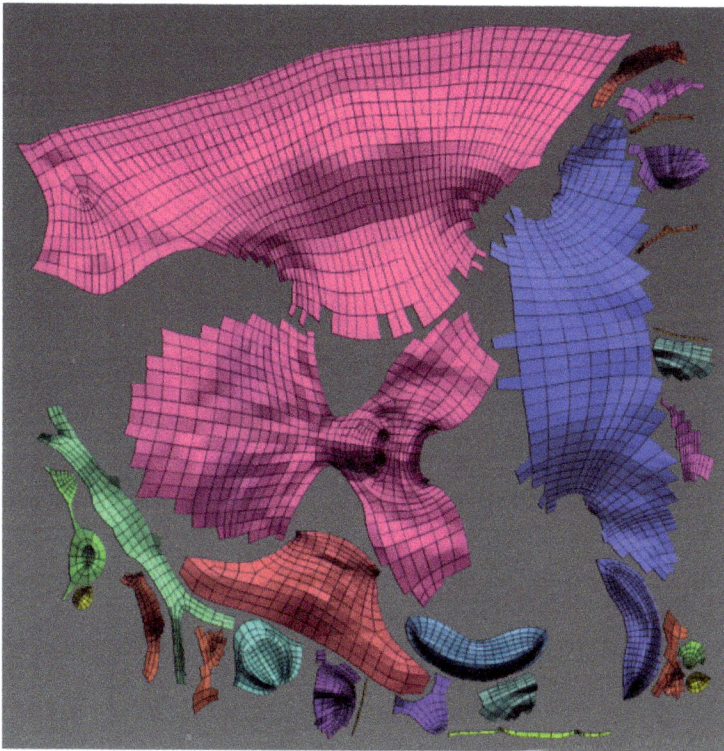

Figure 6.39 – Flattened UVs in ZBrush

UVs that are automatically created in ZBrush will look unorganized and random. In a professional production environment, such as in VFX or video games, it is required to have the UVs organized, rotated, and positioned more consistently. However, it is important to note that at the time of writing, ZBrush does not have the functionality to do this properly, which is why 3D artists mostly use external tools to create UVs.

Creating UVs with UV Master

ZBrush's plugin for UV mapping, **UV Master**, can be found in the **ZPlugin** palette. To create UVs, all you have to do is click **Unwrap** with your model selected. Here is what the UVs of some simple shapes will look like:

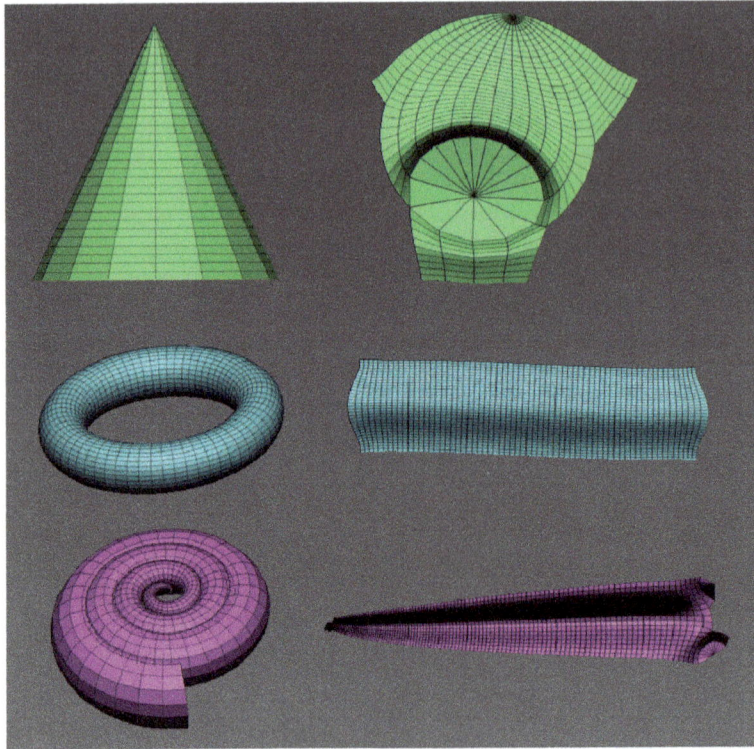

Figure 6.40 – A simple UV unwrap of primitive shapes

While these are functional UVs, they are not ideal in many regards. This can be improved easily by preparing the mesh. Especially when you are working on more complex assets, it is important to follow some good practices to make sure the UV unwrapping process is smooth and has useful results. A badly prepared mesh will often cause ZBrush to freeze or crash!

Preparing the mesh for easy unwrapping

Here are some of the things you should ensure:

- Make sure the model you want to unwrap does not have a high polycount. **UV Master** can create UVs for high-resolution meshes, but beyond 100k polygons, things will get slow and the risk of ZBrush freezing or crashing increases dramatically.

- To help **UV Master** cut and unwrap your mesh, create polygroups. The more complex the topology of your mesh is, the more polygroups you will need for **UV Master** to work properly. To have **UV Master** consider the polygroups in its algorithm, navigate to **Zplugins | UV Master** and enable **Polygroups**.

- Make sure you avoid polygroups that are very large in one dimension, which might lead to the square space of the UVs having lots of wasted space. Since UVs are used to apply texture maps to them and have limited resolution, it is important to maximize the space, and multiple, smaller shells can maximize the space more than fewer, larger shells. This can be observed in the following figure, which compares a mesh with one polygroup and the same mesh with four polygroups:

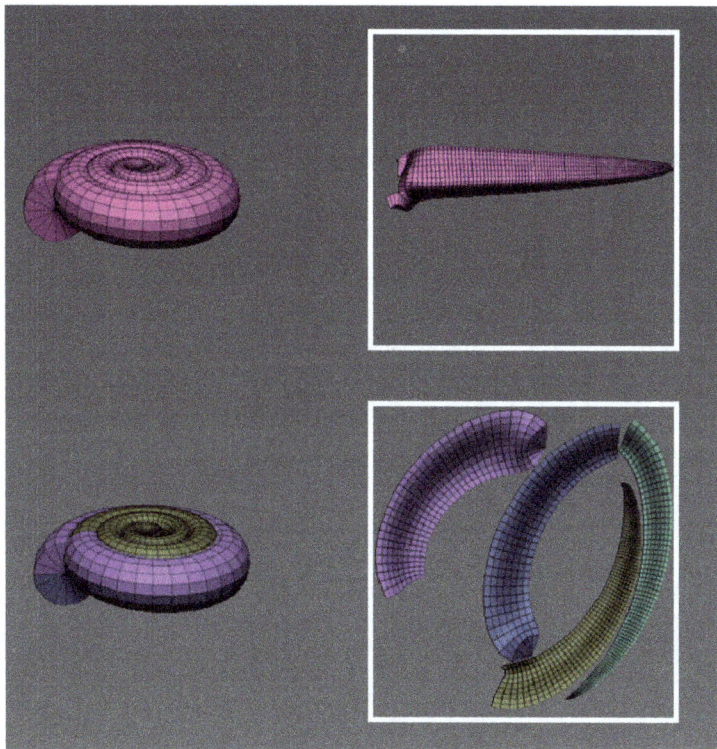

Figure 6.41 – Difference in unused space in the square UV space

- If you have a symmetric mesh, enable **Symmetry** in the **UV Master** menu so that the plugin will create the UVs in a symmetric layout. This allows for more convenient texturing:

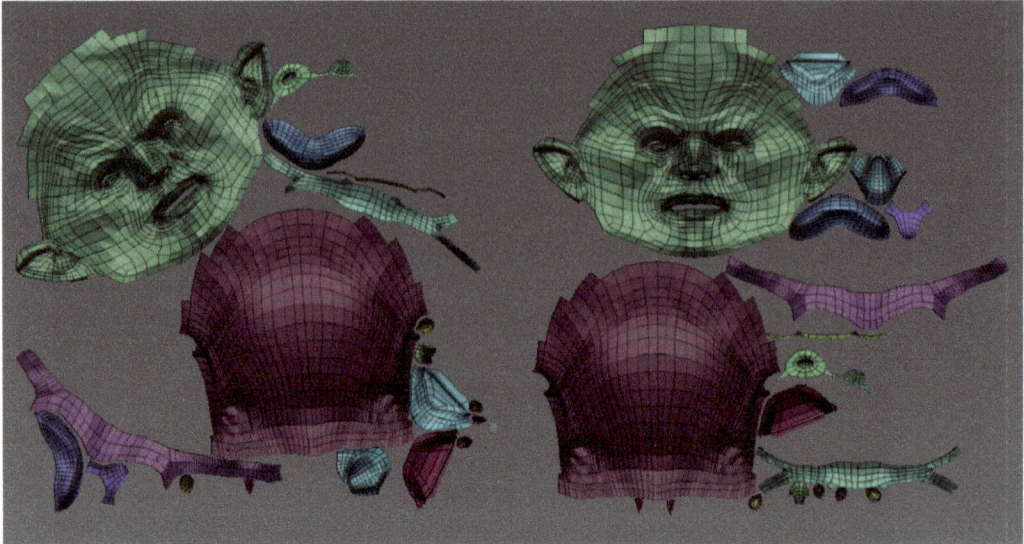

Figure 6.42 – UV unwrapping without Symmetry enabled (left) and with Symmetry enabled (right)

- When creating polygroups, try to hide the seams in less obvious areas. For our demon model, we do not want multiple groups in the middle of the face; the reason for that is that tileable textures applied to the mesh will have a visible seam displayed, where it is very noticeable. When applying a tileable skin color map, there is a potential issue where you may see the UV border creating an unpleasant look:

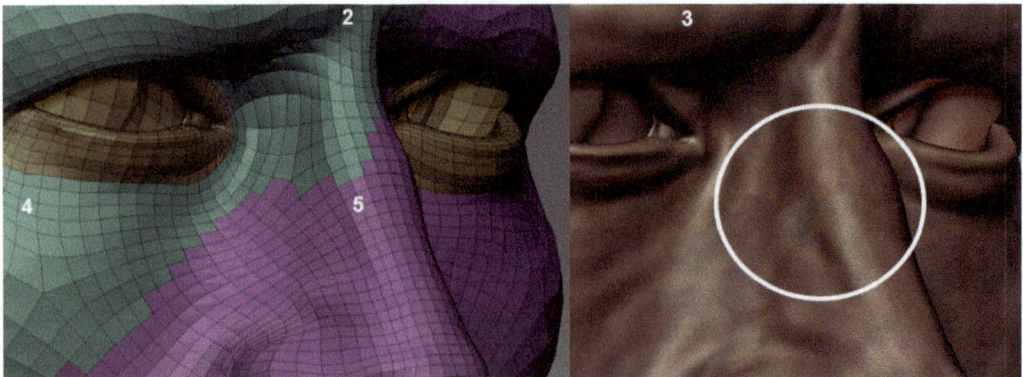

Figure 6.43 – A bad place to create a UV seam, resulting in the tileable skin texture working poorly

With these points in mind, create polygroups for your model by using the masking and selection tools of your choice. That is what polygroups for the head could look like:

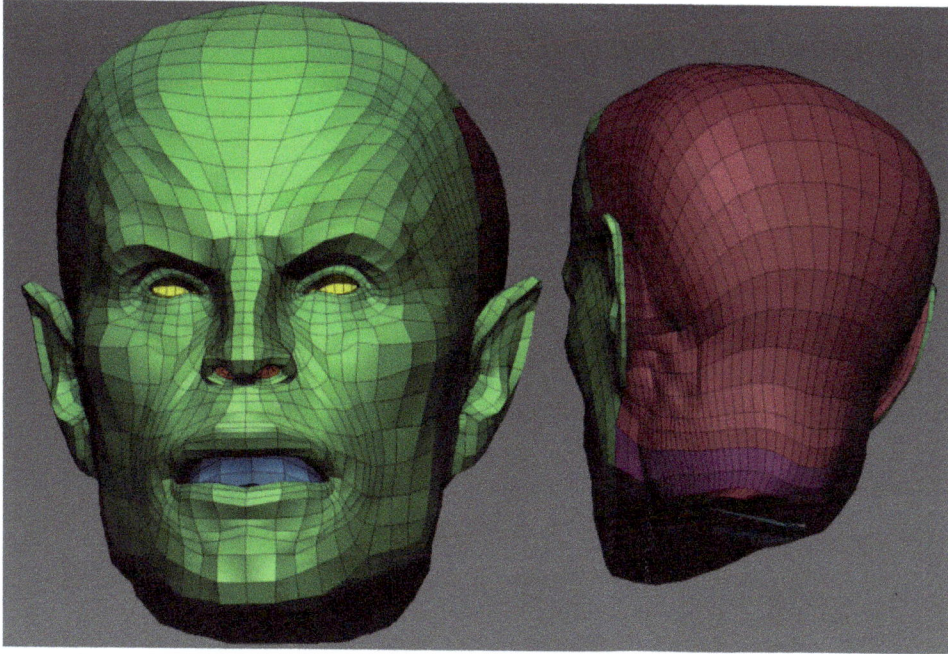

Figure 6.44 – Polygroups to prepare UV unwrapping

As you can see, the front of the face is one large polygroup – this will prevent visible seams from appearing in this area, which can be caused by texturing.

The eye cavities, nose holes, and mouth interior will get additional polygroups because unwrapping a single UV shell including large cavities would result in a lot of stretched and compressed polygons around those areas. This would create issues once you apply tileable textures of any kind.

Then, there are a couple more polygroups on the back and bottom, which will ensure correct unwrapping without you having to stretch their bordering polygroups.

UV unwrapping your mesh

If your mesh has good polygroups and is suited for unwrapping, go to **Zplugin** | **UV Master** and enable **Polygroups** and **Symmetry**. Then, hit **Unwrap**. Depending on the polycount and how many useful polygroups you have created, this process will be quick or take some time.

Evaluating the unwrapping results

To see the UVs, go to **Tool** | **UV Map** and select **Morph UV**. This will morph your mesh into the UV shells, laid out on a flat space, like so:

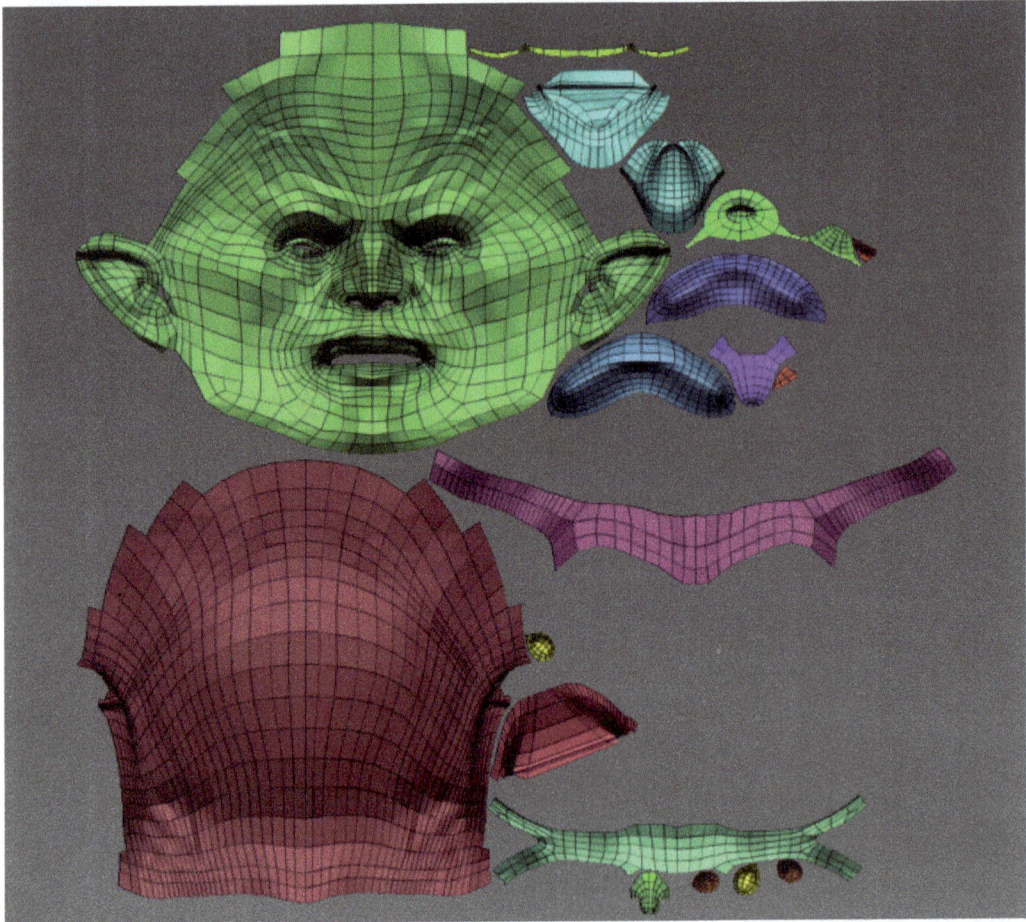

Figure 6.45 – Flattened UV view

The results look clean and symmetrical for the most part. This is a useful UV set, but it is important to note that it will not be optimized enough for games or animation purposes. Professional artists will have to optimize the UV to a degree that is possible only with manual UV unwrapping, which ZBrush does not support. They would do this in 3D software packages such as Maya, 3DS Max, or Blender.

However, UVs are not only needed for rendering but are often essential for modeling and sculpting too. ZBrush is a great tool for this, and the speed and efficiency of **UV Master** are appreciated in any scenario where UVs don't need to be very optimized.

Looking for unwrapping errors

Although the UVs may seem to have unwrapped properly, there may be significant stretching and compressing of UVs present. To check the results, you can apply a checker texture to the mesh, which will let you see whether the checker pattern shows distortion.

To do this, go to **Tool | Texture Map**, click on the image, and select **Texture 03**:

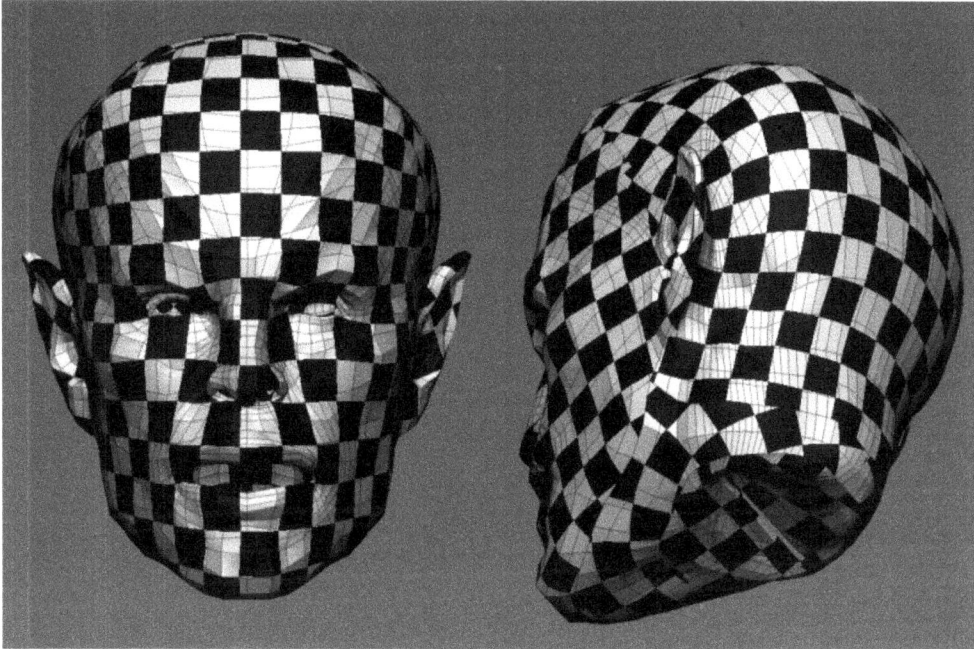

Figure 6.46 – Checking UVs using a checker pattern texture

This is a very good result for automatically created UVs. While there's some minor stretching on the chin, it's nothing too dramatic – you could resolve those issues by adding additional polygroups in areas where stretching/compressing is happening; however, that will come at the expense of additional seams.

If you need a more precise result, you can export your model and tweak the UVs in another 3D program. So, let's learn how to export color information as a texture file.

Exporting textures using the Multi Map Exporter plugin

Once you've created UVs for your model, you can export color and normal maps from ZBrush. These textures will allow you to implement your model in game engines or 3D software, and showcase the full detail and color information from ZBrush.

To export the polypaint from your model as a texture, go to **Zplugin | Multi Map Exporter** and enable **Texture from Polypaint**. Next, you can adjust the **Map Size** slider to determine the size of your textures.

An important setting to enable is **FlipV**. When ZBrush processes textures, it flips them vertically, which would be incompatible with most other software that does not flip the textures. To make the textures apply correctly to your model outside of Zbrush, it is important to flip the texture when exporting by using this setting:

Figure 6.47 – Multi Map Exporter settings

Then, you can click on **Export Options** to open a submenu. Here, you can click on **File names** and set the file type you want to export, as well as the naming you would like the texture files to have.

Once everything has been set up to your liking, you can click **Create All Maps**, at which point Zbrush will export every type of texture that you've enabled.

Exporting different textures

In addition to polypaint textures, there are several other useful types of textures you can export. Here is an overview of the available textures:

- **Displacement**: This map exports grayscale textures that include height information for your mesh. It allows you to export a low subdivision level mesh, but with a displacement map displacing the low poly, you can render the mesh with its full detail from the highest subdivision level. To use this, make sure you go to **Export Options** | **Displacement Map** and set **SubDiv level** according to the subdivision level of the mesh you export (other displacement settings will depend on the software you plan to render with; make sure you look up documentation for whatever software you plan to use):

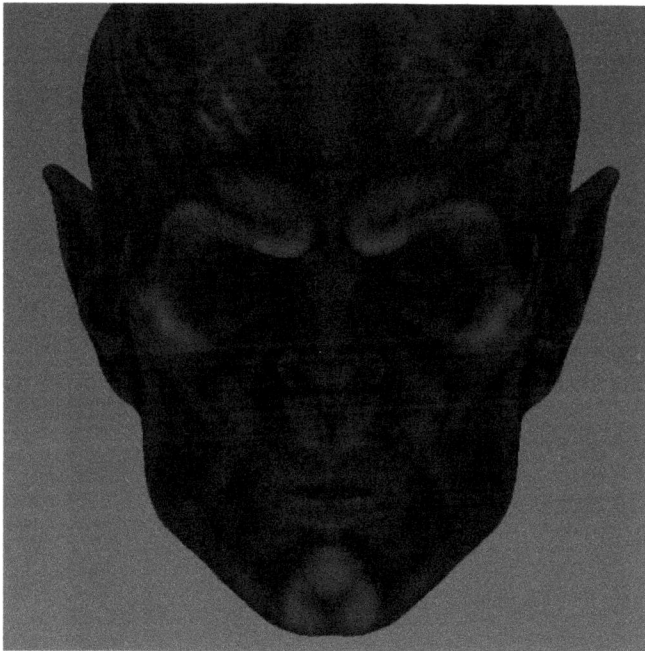

Figure 6.48 – Displacement map

- **Vector Displacement**: This map combines the benefits of both normal and displacement maps. It includes color information, which allows you to displace a model in all directions instead of only perpendicular to its surface. This is great for models with overhanging structures:

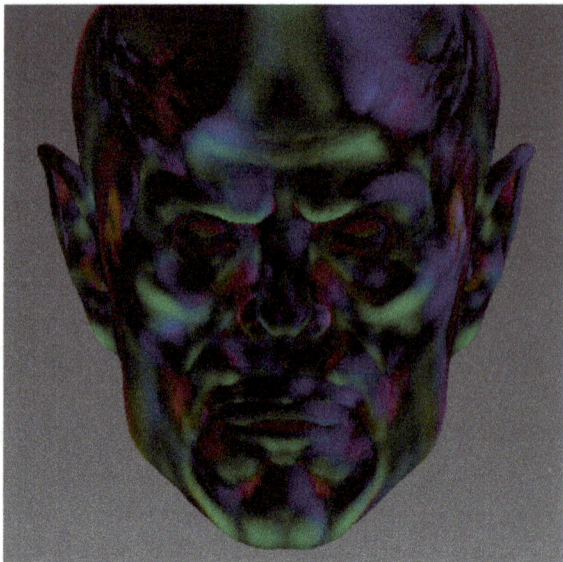

Figure 6.49 – Vector Displacement map

- **Normal**: Normal maps simulate detail on the surface of a model that is visible in the render engine. They render faster than displacements because they do not displace the geometry; instead, they just "fake" surface detail. However, that means that they cannot affect the silhouette of the model like displacement maps can. Therefore, you have to be mindful when deciding whether a normal map is sufficient or whether a displacement map can give better results:

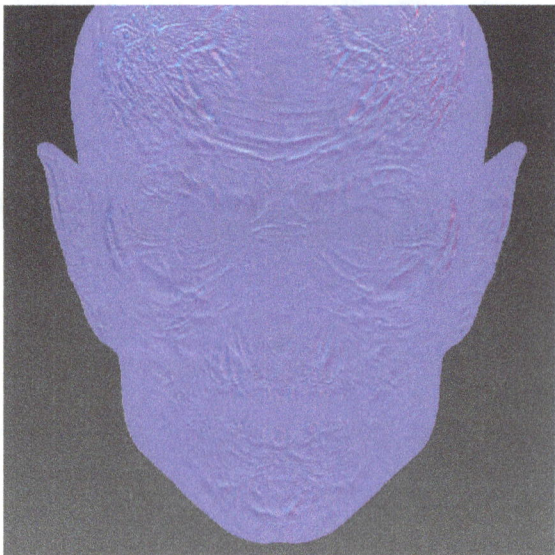

Figure 6.50 – Normal map

- **Texture from Polypaint**: This texture stores the color information from the polypaint that's been applied to the mesh:

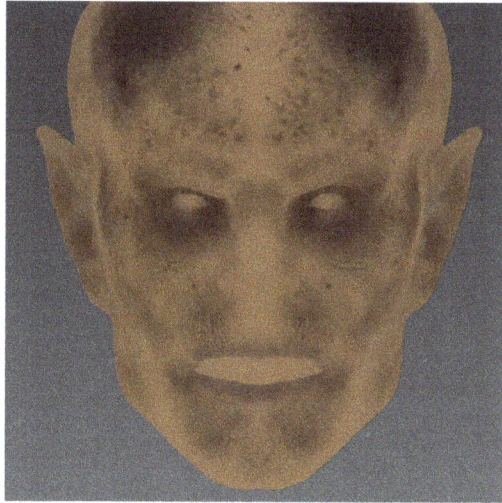

Figure 6.51 – Texture from Polypaint

- **Ambient Occlusion**: This grayscale texture simulates the self-shadowing of objects. Crevices and tight spaces that naturally produce shadows will be captured by this map, which helps fake shadows in models. Since the computation of shadows in CG takes up significant render time, these fake shadows can help add detail and realism without increasing render times, making them useful for game and real-time implementation:

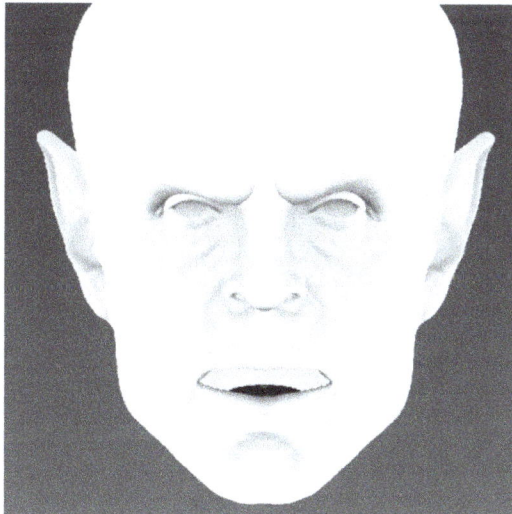

Figure 6.52 – Ambient Occlusion

- **Cavity**: This is a map that shows the cavities of your model in grayscale. It can be used in the texturing workflow or when rendering to emphasize the cavity detail of your model. This map contains a lot of detail, and it is very useful for adding detail to your color textures:

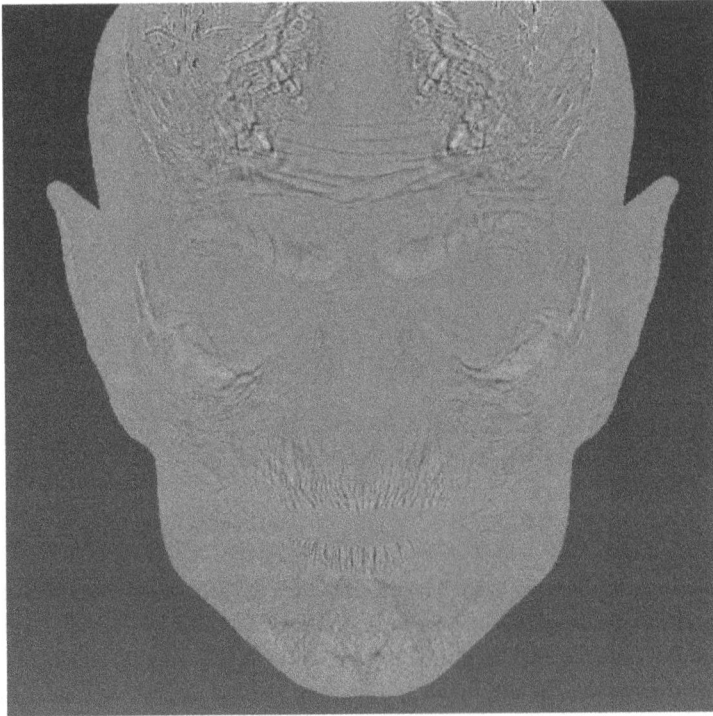

Figure 6.53 – Cavity Map

- **Export Mesh**: This setting does not export your texture, but it will export your model so that when you create all the texture maps, it will export the model along with them.

This concludes our discussion of UV creation and texture export. You've learned about two essential ZBrush plugins, **UV Master** and **Multi Map Exporter**, which will help you prepare your mesh for implementation in game engines or 3D software.

Summary

In this chapter, we covered material, color, and UV creation. We also saw how to export textures from ZBrush. First, we introduced materials in ZBrush and the differences between the two available types of materials: Standard Materials and MatCap Materials. You learned how to modify them and create custom materials.

Next, you explored Polypaint, a tool that allows you to apply color to your sculptures in a variety of ways. The section provided some color theory tips and showcased the various texturing techniques on the demon model, allowing you to follow along with your model.

Finally, you learned how to create UVs using the **UV Master** plugin and polygroups to optimize the UV unwrapping results. After testing the results of the unwrapping process using a checker texture, you were introduced to the texture export setting, as well as a variety of useful maps you can export.

Now that your demon model is complete, the next chapter will show you how to light and render it. By the end of the next chapter, you will have rendered and exported high-quality images and video of your demon bust. By presenting your digital sculpting in the best light, you can share the rendered material with the world!

7

Lighting and Rendering Your Model

In this chapter, you will learn about different lighting techniques from cinematography, and how to build these light setups inside of ZBrush.

You will also get familiar with rendering basics, using the BPR functionality to create renders that include simulated shadows, anti-aliasing, ambient occlusion, and more. Then we will go over the available render passes and export renders of the demon bust that we have been working on across the previous chapters.

Last but not least, you will learn how to create a turntable video in ZBrush, which will let you make impressive personal projects featuring images and video material that showcase your sculpture from every angle.

It is important to note here that in a professional environment, lighting and rendering are usually done outside of ZBrush, but when it comes to preparing a simple presentation of your work, some of these tips can come in handy.

This chapter will cover the following topics:

- Managing lights in ZBrush
- Rendering in ZBrush
- Creating turntable videos

Technical requirements

For the best experience, it is recommended that you have a strong PC that meets the minimum requirements described in the first chapter's *Technical requirements* section. However, you can work on this chapter with just a mouse, a functional PC setup, and a ZBrush license.

To follow along, it would be useful to have the demon sculpture we started creating in *Chapter 2* or a suitable model of your choice.

You can find some good models in LightBox that you could use to experiment with the lighting techniques.

Managing lights in ZBrush

Lighting is an important part of the modeling process, as the way your model is lit can affect its appearance. In this section, you'll learn about the lighting terminology and options in ZBrush, and how to set up the kind of lights that bring out the best in your model.

The effect the lighting has on the model can be seen in the following figure:

Figure 7.1 – The effect of the light setup on the appearance of the model

Even though the model is the exact same, the lighting on the left side makes the character appear more sinister and fierce, while the model on the right looks a bit more peaceful and gentle. This is the reason why it is worth spending time to experiment with lighting, to find the setup that supports your vision for your sculpture the most.

Adjusting the lighting in ZBrush

To start this section, we want to dock the **Light** palette on the left side of the canvas. Since you will be using these options quite a bit, it is more convenient to keep the palette accessible this way; especially when experimenting with the angle of your light sources, it makes sense to keep these menus open.

So, open the **Light** palette and left-click and drag the small circle icon on the upper-left corner of the palette to drag it to the left side of the screen. The palette will now be docked and stay open:

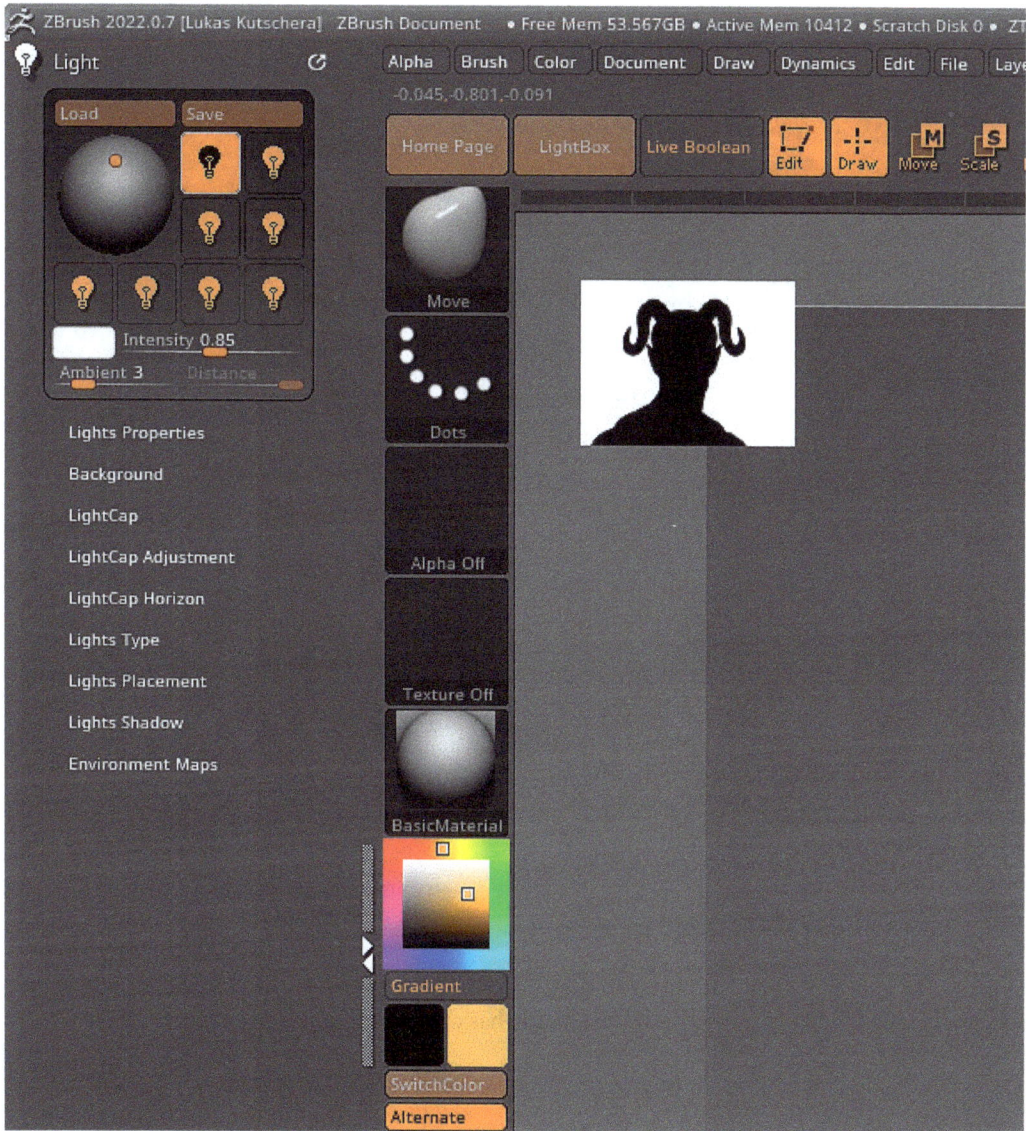

Figure 7.2 – Docking the Light palette to the canvas

Now, on the top of the **Light** palette, you can see a sphere with an orange dot on it – this dot indicates the light direction. By default, the light comes from the front and a high angle, as we can see here:

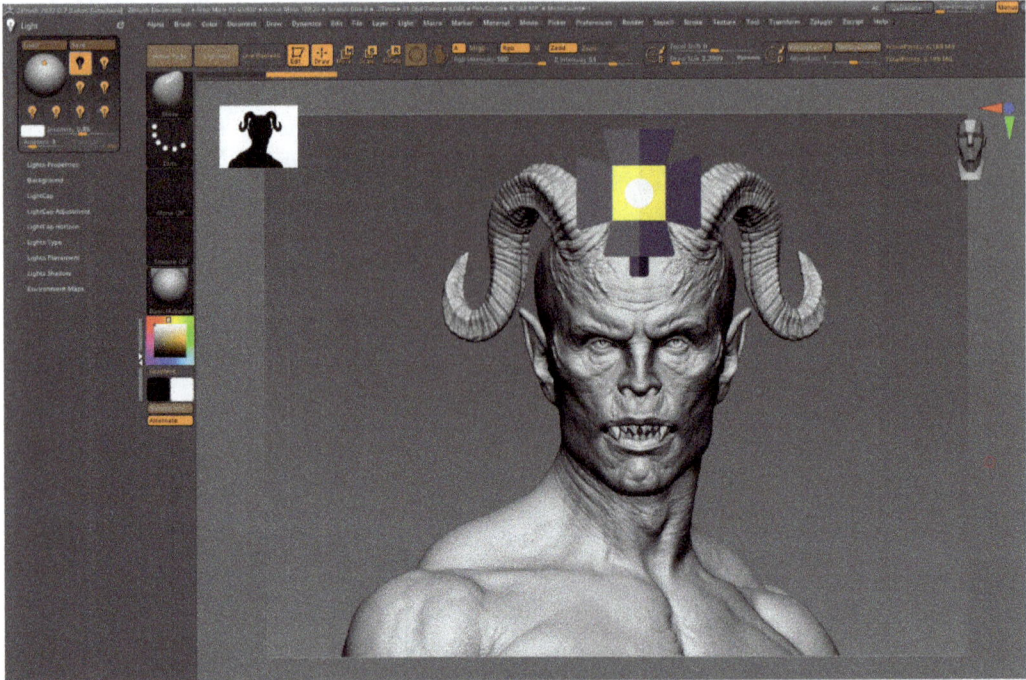

Figure 7.3 – Default light in ZBrush

This high angle highlights the facial features of portraits nicely and creates a high-contrast look.

But let's look at how to adjust this default lighting.

Changing the light direction

To change the light direction, left-click and drag the dot on the sphere to the place where you want the light to come from. You can see how the lighting has changed in the following screenshot:

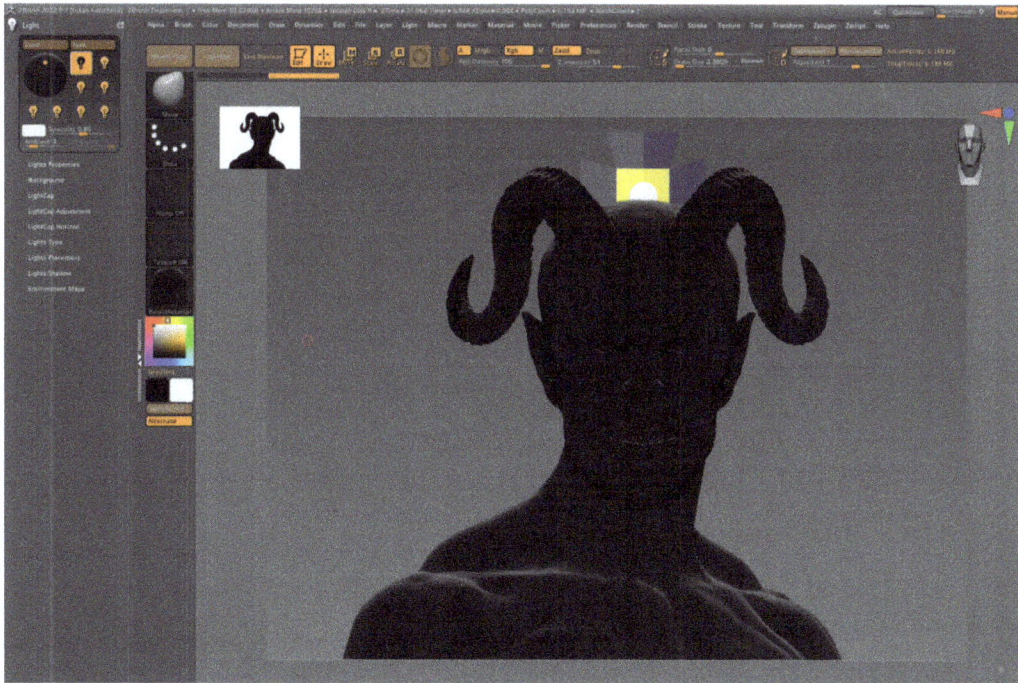

Figure 7.4 – Switching a light to be on the back side

If you left-click on the sphere once, the dot will switch to the back side, lighting your model from behind. Clicking again will bring the dot back to the front.

Adding multiple light sources

Next to the sphere, you'll see multiple lightbulb icons. The first orange lightbulb icon indicates that this light is currently enabled. But if you click on another lightbulb icon, you can adjust the direction and settings of that light.

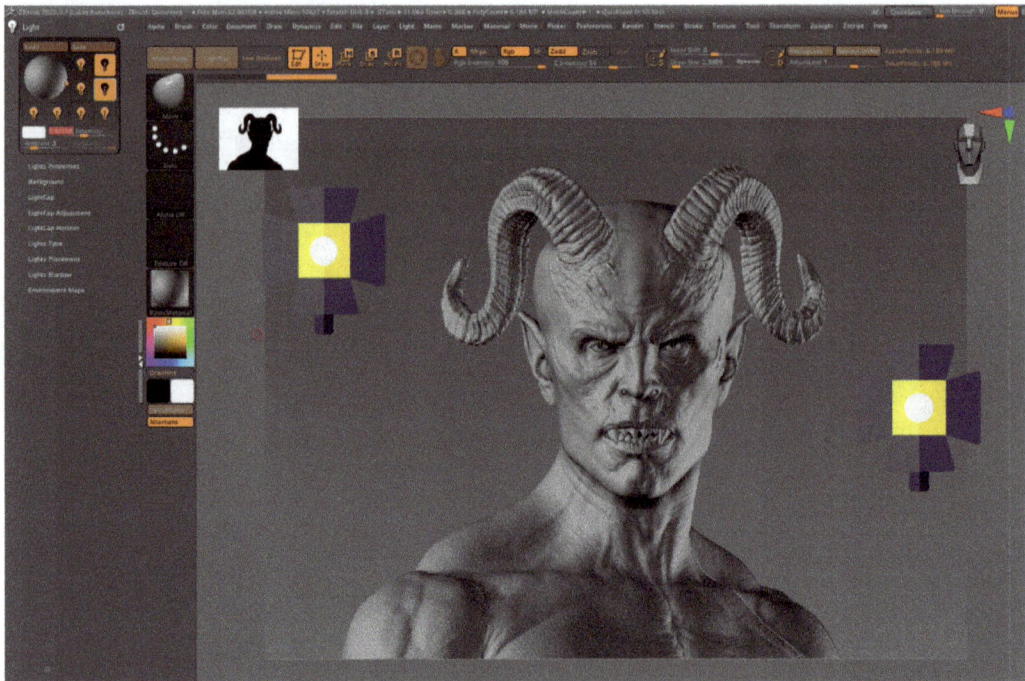

Figure 7.5 – Multiple light sources illuminating the model

Once you click on the currently selected light source icon again, the icon will turn orange and the light is turned on. There are eight lights available to use. You can continue to set up lights using the remaining light-bulbs.

Light intensity, ambience, and color

Below the sphere and lightbulb icons, you can find the following light properties:

- The color picker, which lets you determine the light's color when you click on it.

- The **Intensity** slider, which determines how bright or dim the light is.

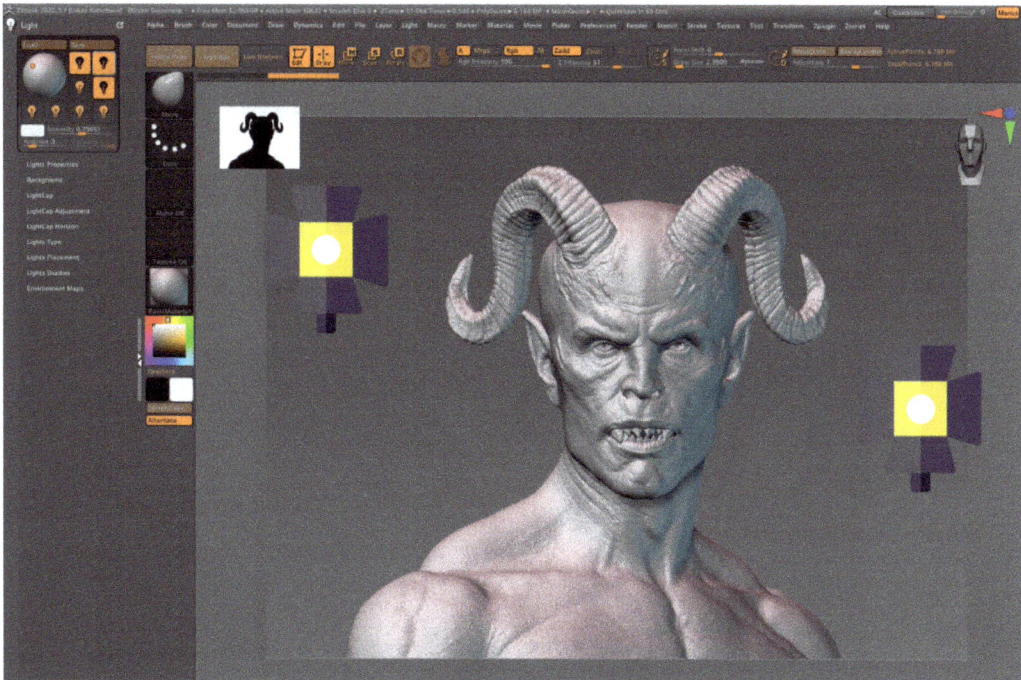

Figure 7.6 – Multiple lights with different colors and intensities

- Below that is the **Ambient** value. Similar to the **Ambient** modifier for materials, covered in *Chapter 6*, this attribute will increase global brightness. While intensity increases the light strength and casts shadows accordingly, ambient light will illuminate your model from every angle, increasing brightness uniformly. This makes the **Ambient** option useful for decreasing shadows that may be too dark.

Now that you have learned the technical side of lighting, including how to use lights and their modifiers in ZBrush, in the next section you will explore the more theoretical side, learning about lighting techniques that you can implement in ZBrush to achieve the desired presentation of your sculpture.

Exploring cinematography and lighting terminology

When looking for inspiration for your character's lighting, cinematography can be a great resource. Filmmakers understand the power of light and implement a variety of lighting techniques in their movies to achieve certain effects.

When it comes to the names of the lights, CGI (and ZBrush) uses similar terminology to that of cinematography. Let's take a look at some of the classic lighting techniques.

> **Important note**
>
> Before you start working on your light setup, make sure that you have one of the standard materials applied to your mesh, and not a MatCap material, as these have light "baked in," and do not react to changing light conditions.

Key lighting

Key lighting is the main light. It creates the strongest effect in the scene, setting up the mood and overall feel:

Figure 7.7 – Key lighting

It makes sense to build your light setup starting with this light. This light could come from any angle, depending on what you're going for.

Fill lighting

The purpose of **fill lighting** is to illuminate the shadow side of your model.

Generally, if you have just the key light in the scene, there is some area of the model that is concealed by shadow. If the shadow side is so dark that you can't recognize shapes clearly anymore, it makes the model less visually appealing. Fill lights are a great way to solve that issue, which is why they are used to some extent in most lighting techniques.

Figure 7.8 – Fill lighting

Backlighting/rim light

Backlighting is a common technique used to separate a subject from the background and highlight its silhouette. This type of light can either be very strong or very subtle and works well with any combination of other lights.

Figure 7.9 – Backlighting/rim light

Three-point light

The immensely popular **three-point lighting** setup features the already mentioned key light, fill light, and rim light. The goal is to emphasize a three-dimensional look in the subject, which is why it uses lights from different directions to highlight the forms.

Figure 7.10 – Three-point light

The key light is the strongest light and will often illuminate the subject at a slight angle in order to create a shadow side. This shadow will be brightened by the fill light. Then the rim light will separate the subject from the background and highlight its silhouette.

Rembrandt lighting

The **Rembrandt lighting** technique, named after the Dutch painter Rembrandt van Rijn, is one of the most dramatic, moody lighting setups used in portraiture. A strong key light positioned slightly above and to one side of the subject casts a strong shadow on one side of the face while illuminating just a little area around the eye in a triangular shape.

Figure 7.11 – Rembrandt lighting

Side lighting

Side lighting is a strong light coming from the side. The aim is usually to create a high-contrast look, with only subtle fill lighting for the shadow side. This type of lighting can be used to convey a sense of drama and enhance the mood in your portrait.

Figure 7.12 – Side lighting

High-key lighting

High-key lighting is a low-contrast lighting technique that uses a high exposure to illuminate the subject. Using a diffuse light source, the subject is evenly lit and has little shadow areas. This kind of light can be good for showcasing your model and making it easy to see.

Figure 7.13 – High-key lighting

Low-key lighting

Low-key lighting is the opposite of high-key lighting – it creates high contrast and strong shadows, making it ideal for the thriller or horror genre. Low-key lighting uses a strong key light and very limited fill light to keep a lot of the subject concealed in shadows.

Figure 7.14 – Low-key lighting

Uplighting

Uplighting is a classic lighting technique used in horror films. With the key light coming from below the subject, this setup creates unsettling shadows on the face and makes for a spooky look.

Figure 7.15 – Uplighting

This wraps up this section about lighting. Now that you are able to set up multiple lights with custom properties, and you have some inspiration from famous lighting techniques, you can create your lighting setup with purpose.

In the next section, you'll learn how to render your model and export different render passes that will allow you to composite them into a final image that showcases your sculpture in the best possible way.

Rendering in ZBrush

ZBrush has a feature called **Best Preview Render** (**BPR**). This function generates high-quality renders with shadows, **anti-aliasing** (**AA**), **subsurface scattering** (**SSS**), and **ambient occlusion** (**AO**) in a short amount of time. In this section, you'll learn how to create these kinds of renders and export them for further editing in Photoshop or another image editing software.

> Important note
>
> While rendering times in ZBrush are fairly quick, ZBrush is not known for the high quality and realism of its renders. If high fidelity and photorealism are the goal, external render engines are recommended. **Maya's Arnold Renderer**, **V-Ray**, and **Cycles** in **Blender** are all powerful render engines that allow you to get more out of your model. Here is the difference in realism and quality between a ZBrush BPR render, and an Arnold (Maya) render:
>
>
>
> Figure 7.16 – ZBrush BPR render (left) versus Arnold render (right)
>
> That being said, rendering in ZBrush can still be useful to evaluate the forms of the model in a new way, through the shading and shadow effects the BPR renders bring.

Preparing for rendering

Before you begin, dock the **Render** palette to the left side of the canvas so you can access its menus easily during this section. You can dock it by left-clicking and dragging the circular palette icon in the top-left corner onto any side of the canvas.

Then you should set up the resolution of your images by completing the following steps:

1. Go to the **Document** palette and adjust the **Width** and **Height** options to your liking.

Figure 7.17 – Resizing the document

2. Click **Resize** and confirm the prompt. This will transition you from **Edit** mode to **Draw** mode.

3. Then press *Ctrl + N* to update the canvas.

4. Now drag the model onto the canvas and return to **Edit** mode by pressing *T*.

If the new document size is larger than your canvas, you can use the **Zoom Document** button on the right side of the canvas to zoom out and position your model inside the document.

Creating BPR renders

To create a BPR render, simply press *Shift + R*, or click on the **BPR** icon on the top of the icon list, on the right side of the canvas:

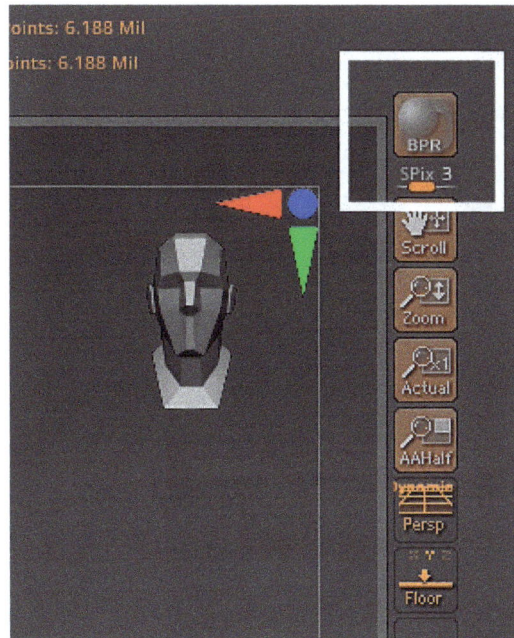

Figure 7.18 – BPR render and SubP level

Below the **BPR** button, you'll find the **SPix** slider – this controls the AA quality, which smooths out the visible pixels on the edges of the model. A value of 0 means no AA and a value of 7 gives the highest possible quality of AA. Rendering time is not heavily affected by high values so you may choose to use a high sample if you prefer the smooth appearance:

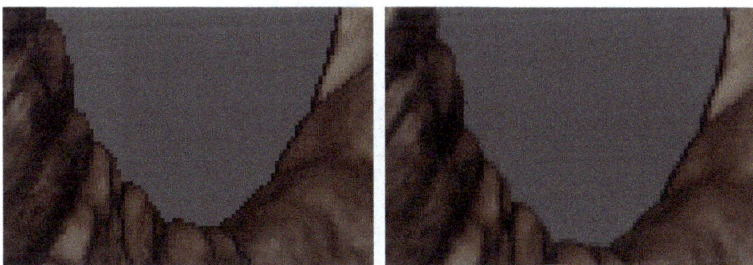

Figure 7.19 – SubP level 0 with no AA (left) and SubP 7 with AA (right)

Now click the **BPR** button to create a render pass. You can then export the renders from the BPR **RenderPass** menu in the **Render** palette. Depending on your render properties, you can see multiple available renders:

Figure 7.20 – BPR render passes

Let's take a look at what these are, and what they are good for:

- **Composite**: This render contains the shaded model, including shadows, as well as the canvas background. Shaded means that the model is rendered based on the lighting in the scene, adding shadow information that gives the model more depth and realism, which it does not have until BPR renders are created.

Figure 7.21 – Composite render

Here is how the model looked unshaded, and after rendering, shaded. Most noticeable is the added strong shadow coming from the light setup in your scene:

Figure 7.22 – Unshaded/before rendering (left) and shaded/after BPR rendering (right)

- **Shaded**: This contains the shaded model, including shadows, with a black background.

Figure 7.23 – Shaded render

- **Depth**: This render pass can be useful as a mask in Photoshop for creating effects such as depth of field.

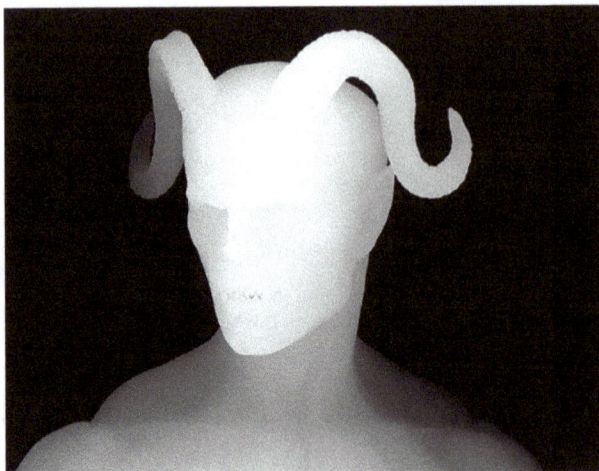

Figure 7.24 – Depth render

This mask stores the depth information, which lets you apply a blur on elements that are very close or far away. This can be useful to blur the background, like in the following figure:

Figure 7.25 – Using a Depth mask to create a blur effect for the background (right)

- **Shadow**: This pass extracts the shadows, which can be another useful pass to be used in Photoshop to manipulate the shadow parts of your model.

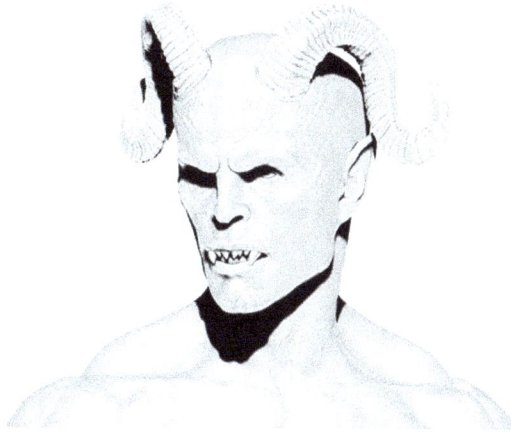

Figure 7.26 – Shadow render

You could add a brightened copy of your **Composite** pass and use the **Shadow** render to isolate and brighten up the shadow parts:

Figure 7.27 – Before (left) and after (right) brightening up the shadows, using the Shadow render

- **AO**: This pass contains self-shadowing information. This is useful for adding realism to large environments and can also be used to edit characters.

Figure 7.28 – AO render

You can see on this tractor model how the **AO** pass adds shadows on the floor and between close objects. This adds extra detail, and no lights are needed in the 3D software, which saves on render time:

Figure 7.29 – Adding AO (right) to add extra shadow detail to the composition

In Photoshop, the AO render can be added as a layer on **Multiply** blend mode, which will darken the image based on the map, creating shadows accordingly.

- **Mask**: This render pass contains a mask of your model that can be used to extract the model from the background, allowing you to add your own background image.

Figure 7.30 – Mask render

Here is an example of a model extracted from the background with some smoke added:

Figure 7.31 – Adding a background by isolating the model with a mask

- **SSS**: This render pass creates an SSS map of your mesh. SSS simulates how light is absorbed and scattered in translucent materials such as skin or fruits, creating the characteristic red "glow" of ears when lit with a strong light source from behind. You can use this render pass to create this effect of translucent materials in Photoshop.

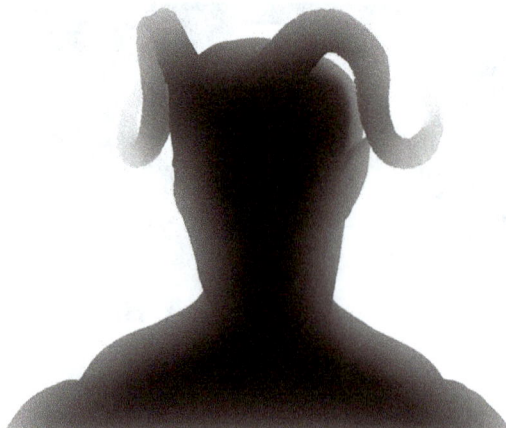

Figure 7.32 – SSS render

Here is the effect of SSS on ears, as described before:

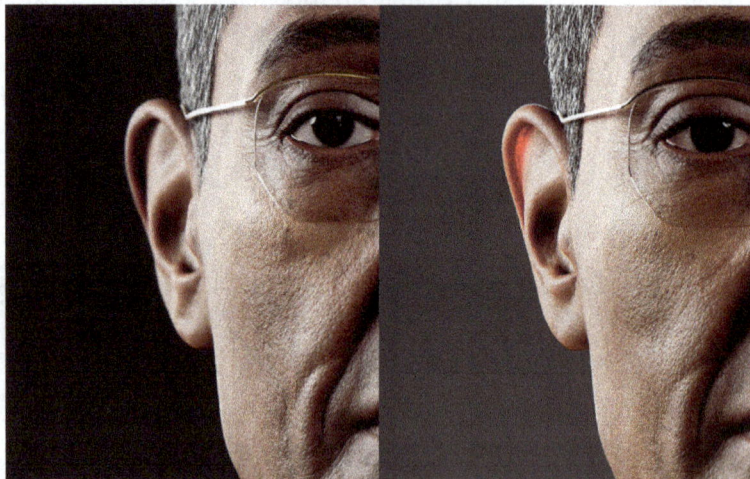

Figure 7.33 – A model without (left) and with SSS (right)

- **Floor**: This render pass captures the floor as a grayscale gradient texture.

Figure 7.34 – Floor render

Clicking on the thumbnails of these render passes will allow you to save them.

Note that some of these passes are not available by default and you have to enable them in the **Render Properties** menu, as you can see here:

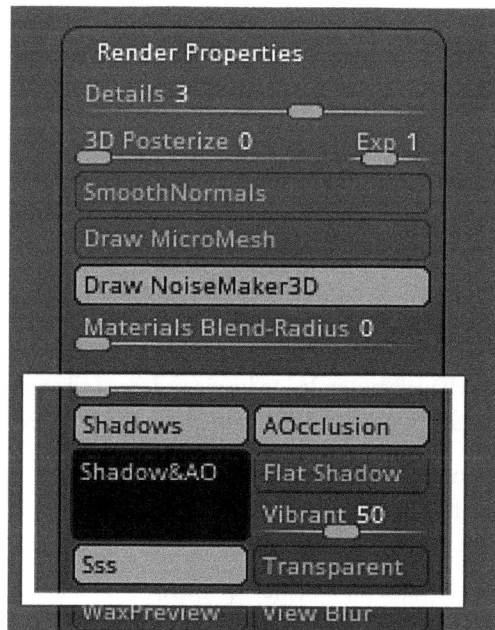

Figure 7.35 – Render Properties enabling render passes

If you want to dive deeper into compositing with ZBrush, there is a great, free resource by ZBrush artist Pablo Munoz Gomez – *The Cheat Sheet for Compositing*. You can download it for free here: `https://www.3dconceptartist.com/resources`.

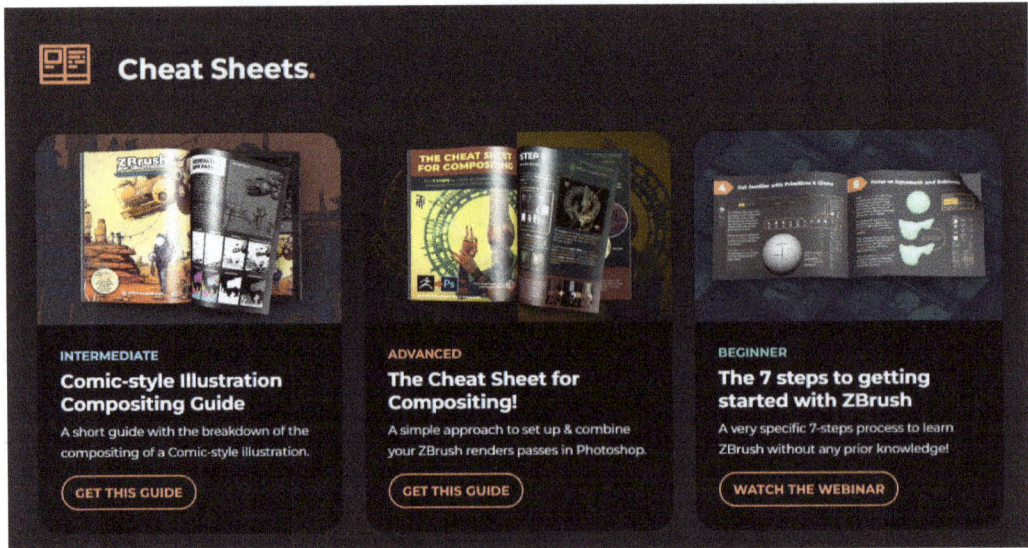

Figure 7.36 – Free learning resources for ZBrush artists on 3dconceptartist.com

Now that you know the most important render settings, you can export everything you need to composite a final image of your sculpture.

Images are a great format for showcasing your work on your portfolio and social media, but there is another great format in particular for sculptures: video. In the next section, you will learn how to create and export a 360-degree (turntable) video of your sculpture.

Creating turntable videos

Since a sculpture or 3D model is usually intended to be viewed from all angles, turntables make sense as a way to present your model. ZBrush makes it easy to create these videos!

To create a turntable video, open the **Movie** palette (shown in *Figure 7.27*) and select **Doc** instead of **Window**. This makes sure that it is only the canvas with your model being recorded instead of the whole window (including all the menus and icons).

Below the **Doc** option, you can choose the movie size. If you want your video to be the size of the document, select **Large**. **Medium** will result in a video that is 50% of your document size and **Small** only 25%.

Figure 7.37 – Basic turntable settings

The next menu contains the main turntable settings: **Modifiers** (shown in *Figure 7.28*). The default settings here are mostly fine, though one attribute you can experiment with is **SpinFrames** – this determines the number of frames in one spin and therefore determines how quickly or slowly the model will appear to spin, with higher values resulting in a slower spin speed. You will find a value of around 600 gives a reasonable speed that allows enough time to observe the model while it spins.

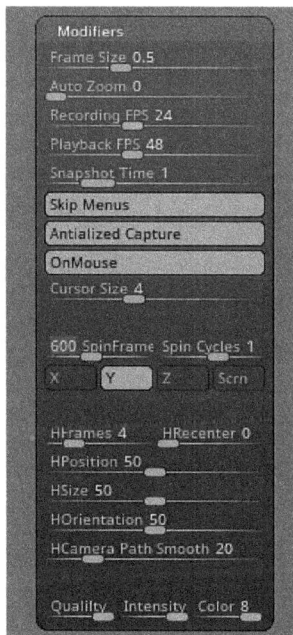

Figure 7.38 – Movie modifiers

To remove the ZBrush logo that is added by default from your video, open the **Overlay Image** menu and set **Opacity** to 0.

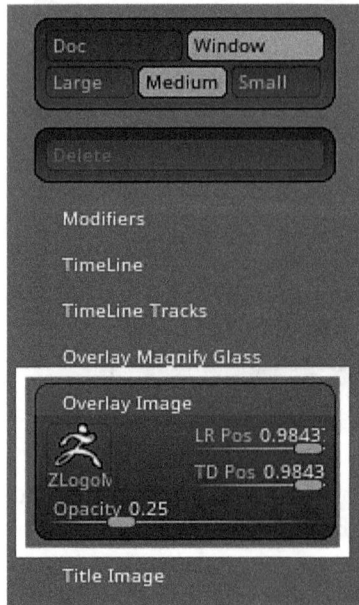

Figure 7.39 – Overlay Image option

Below this is the **Title Image** menu (shown in *Figure 7.40*). Open it and set **FadeIn Time** and **FadeOut Time** to 0 – this will remove a ZBrush logo fade-in sequence that would be added to the video otherwise.

Figure 7.40 – Title Image option

Once you've set up everything, go ahead and select **Turntable**. To export the video, select **Export** and make sure to enable **H** (high quality) in order to export it in the highest quality.

Let's recap what you learned in this section. You are now able to export both images and videos of your models. You know how to export a simple screengrab of your canvas, as well as rendered images and various render passes that you can use for compositing.

Summary

In this chapter, you learned about various lighting techniques and how to recreate them in ZBrush using multiple lights with different properties. You also now know how to create high-quality BPR renders that you can export for further editing with image editing software such as Photoshop. Finally, you learned how to create turntable videos, a great way to add video resources to your projects.

This is the last chapter about concepting in ZBrush and wraps up the creation of the demon bust. Congratulations if you followed along from *Chapter 2* or if you created your own concept sculpt! You covered a lot of ground in these chapters, which included a lot of different tools and techniques. Regardless of your level of experience in ZBrush, I hope that you were able to add a tool or two to your tool belt that might come in handy in a future project!

The next three chapters will take you through the process of creating a print-ready gladiator sculpture from start to finish. The first of these will cover human anatomy sculpting, after which you will learn about a variety of different tools and techniques to model armor and accessories.

Part 2:
Creating Characters
from Scratch:
A Comprehensive Guide

Part 2 features another practical example that you can follow along with if you wish. This time, the subject is a full gladiator character, and you will learn how to create him from scratch, including sculpting realistic anatomy, creating a complex costume, and finally, preparing the model for 3D printing.

This part includes the following chapters:

- *Chapter 8, Sculpting Human Anatomy*
- *Chapter 9, Creating Costumes, Armor, and Accessories with Classic Modeling Techniques*
- *Chapter 10, Preparing and Exporting Our Model for 3D Printing*

8
Sculpting Human Anatomy

In the next three chapters, you will learn how to create a full character that is ready to be 3D printed. This book will show the creation of a gladiator. It is recommended that you follow along, but you can also pick another subject and apply the lessons to your own model.

In this chapter, we'll start by creating realistic anatomy, a key skill that many professional ZBrush artists need to master to be attractive candidates for jobs in fields such as VFX or collectibles.

You'll begin by gathering references and resources. This is a crucial step that will save you time later on and will help you sculpt an anatomically correct and believable body.

Next, you will block in the body using anatomy knowledge and plenty of reference images. After that, you will go through the individual muscle groups that affect the look of the body the most, showing examples and providing sculpting tips.

Finally, you will pose the model and add asymmetry to make it more lifelike. Then, you can add veins and skin texture to finish up.

This chapter will cover the following topics:

- Preparing anatomy references
- Blocking out the anatomy
- Refining the anatomy
- Finalizing the sculpture's anatomy

Technical requirements

For the best experience, it is recommended that you have a strong PC that meets the minimum requirements described in the *Technical requirements* section of *Chapter 1*. However, you can work on this chapter with just a mouse, a functional PC setup, and a ZBrush license.

Preparing anatomy references

In this section, we'll gather references and resources to create our gladiator model, which will make your life easier later on; the preparation time will pay off in the final result. *Figure 8.1* gives you an idea of what we are aiming for:

Figure 8.1 – Preview of the gladiator creation process

> **Important note**
>
> Remember that human anatomy cannot be learned at the same pace as you would learn a new software. Although you might be able to understand the information easily, translating this into a sculpture requires sculpting skills and precision with your brush strokes, which develop over time, so be patient with yourself if results don't show immediately.

Gathering references

Let's take a look at the different types of references you should be collecting.

Proportions and measurement charts

Proportions and measurements are essential for our task because they will help us create a realistic-looking human body. A character with legs that are too short, or a head that is too big, will look odd no matter how nicely sculpted it may be.

Artists commonly use 7.5-head or 8-head figure measurements to establish the proportions of the human body:

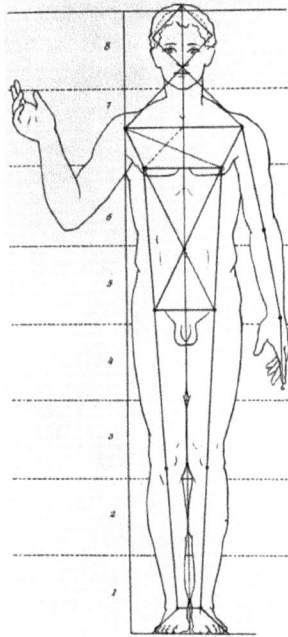

Figure 8.2 – 8-head figure measurements

There are many images of 7.5- and 8-head measurements that you can find via Google Images. Later, you can import these pictures into ZBrush so that they match your model.

Bodybuilder and athlete pictures

This type of reference can be useful when you're blocking out the body, but it will be especially helpful when you're refining the sculpture and adding details to the muscles.

Make sure you collect plenty of reference material, covering different angles and muscle groups. Your reference board will depend on how you want your character to look. Bodybuilders make for great reference because you can see the muscle so clearly, which helps you understand anatomy more easily.

Even if you plan to sculpt a slim or unathletic body type, this type of reference will still be useful because after you have sculpted a "flexed" and more muscular and defined body, you can simply tone down the muscles and make them only appear very subtly. This way, your character looks "normal" and in a more relaxed state, but at the same time, the body has some detail and forms through the subtle muscle shapes that are left, resulting in a more complex and realistic sculpture.

Of course, additional references that are closest to the desired body type will be important to add to your reference collection.

You can view and assemble your references with a free piece of software called PureRef, which you can find here: `https://www.pureref.com/download.php`. With Pureref, you can group your reference images based on muscle groups or other categories of your choice. Here's what this might look like:

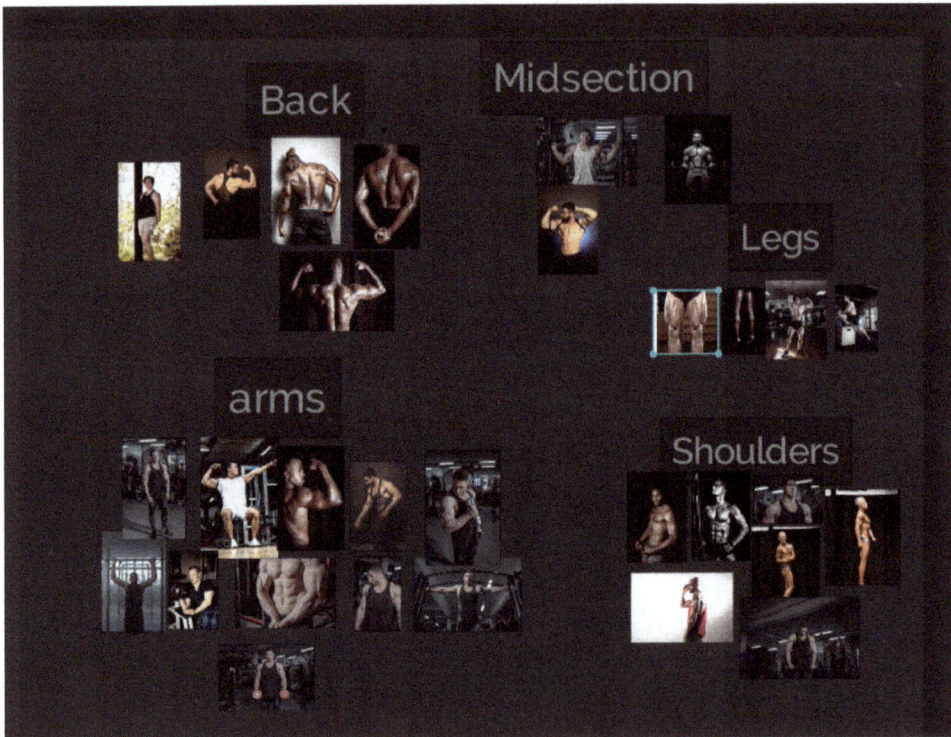

Figure 8.3 – Using Pureref to collect reference pictures

Anatomy illustrations, including muscles and skeleton

When studying anatomy, it's important to collect illustrations that show the origin and insertion points of muscles. These are the points at which muscles attach – the origin point is the part of the muscle that does not move during muscle contraction, and the insertion point is the part that is moved by contraction. Having an understanding of how the underlying muscles attach will help you understand why the human body looks the way it does. You can get some good results by just looking at photos, but unless you understand the underlying anatomy, the results will be rather random.

You could create a new Pureref file to collect those reference images, or you could add them to the previous Pureref file, which already contains real-life photo references:

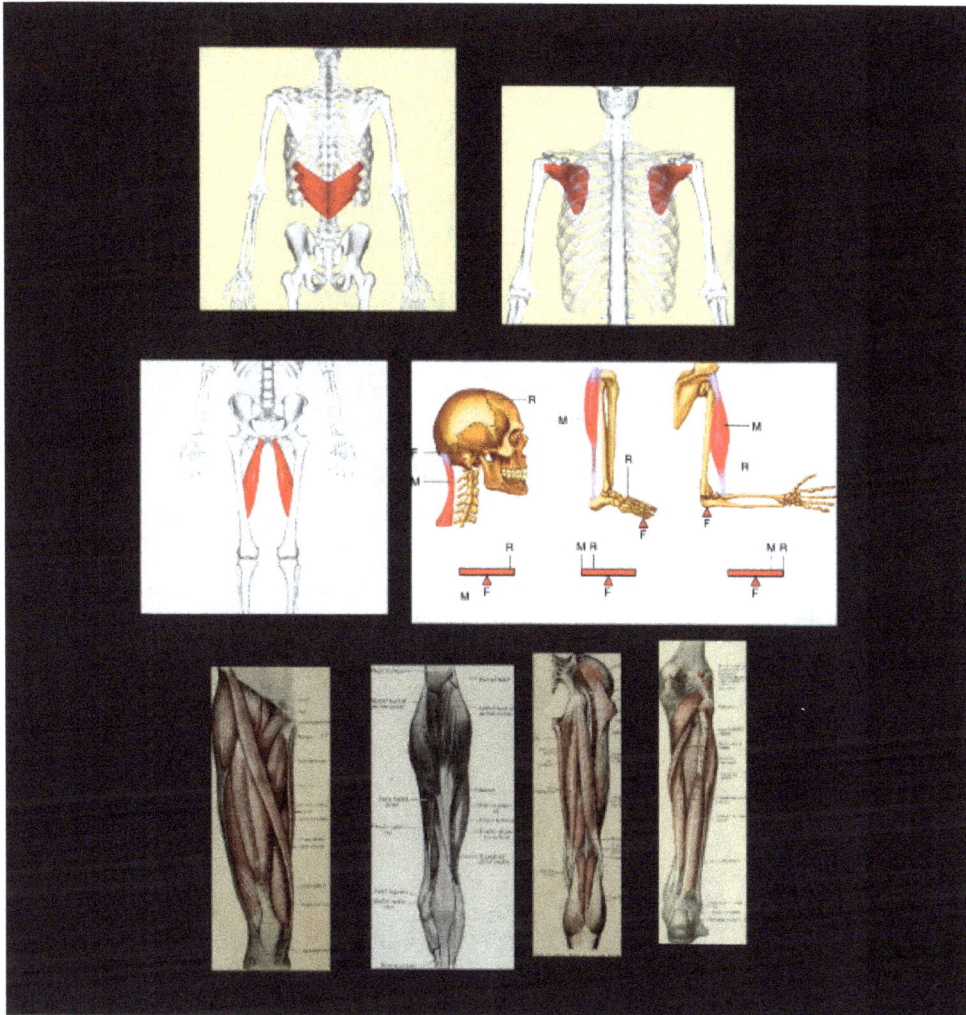

Figure 8.4 – Anatomy illustrations

With your reference collection ready, you can start to load some additional references in the form of 3D models.

Loading resources

Images are a great resource and are probably the most commonly used reference, but there is another type of reference that is extremely valuable: 3D models. Being able to view a model from any angle and zoom in on specific areas can be a great way to visualize and understand it:

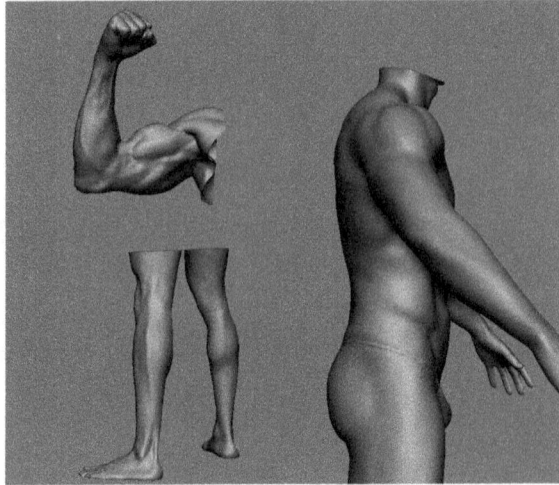

Figure 8.5 – Using 3D scans to have an anatomy resource that can be viewed from any angle

Scans are especially great because you can isolate parts of the model to see the cross-section of certain features. This can help you sculpt a more accurate shape, and it is a type of information that you can not get in pictures in the same way:

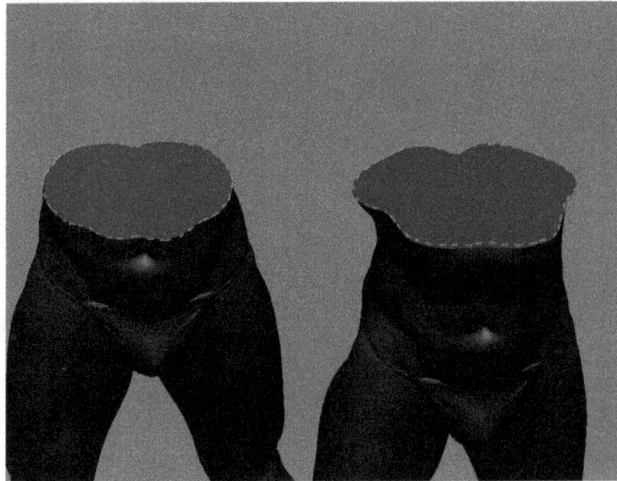

Figure 8.6 – Isolating parts of a scan model to observe the cross-section
and shape of different areas of the body

Loading an écorché

For reference, ZBrush has a female écorché model in its content library, LightBox. An **écorché** is an anatomical drawing or figure that depicts a human or animal with its skin removed. The muscles can be seen with ease, making it a great learning resource for artists and sculptors:

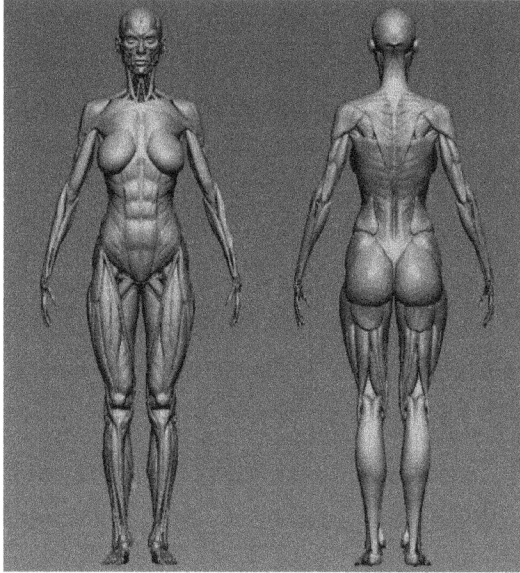

Figure 8.7 – The écorché model from LightBox

You can find the model in LightBox by pressing , and then navigating to the tool menu. There, you can select `Ryan_Kingslien_Anatomy_Model.ZTL`.

The model contains all the relevant muscles as individual subtools, which you can isolate, to check how they attach to the skeleton. This will help you understand the anatomy better, and why the body looks the way it does:

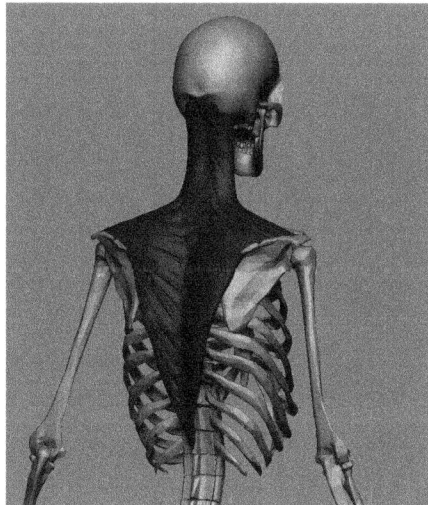

Figure 8.8 – Isolating muscles on ZBrush's écorché model to check how the muscles attach to the skeleton

Although the gladiator in this book is male, the écorché is still useful because the muscle origin and insertion points are similar in men and women. If you prefer to have a more accurate reference in terms of proportions and muscle size, feel free to purchase a male écorché model from a 3D marketplace such as CGtrader.com.

Loading a base mesh

Next, you will need a base mesh. A **base mesh** is a generic human character that has good overall proportions and good topology. You can create a character from scratch, but we will import one – this will save you time and allow you to focus on the most important skills, which are anatomy and sculpting-related.

If you already have a base mesh you prefer to use, you can import that one into ZBrush by going to **Tool | Import** and selecting your model. Otherwise, there is a suitable model in LightBox called **Nickz_HumanMaleAverage**, which is available in the tool menu of LightBox as the écorché:

Figure 8.9 – ZBrush's Nickz base mesh

Now that you have collected plenty of reference images and loaded the écorché and base mesh, you are ready to begin establishing proportions and blocking out the character.

Blocking out the anatomy

In this section, you will block out the character. To do this, we will start by working with the skeleton and adjusting its proportions, after which we will work with the bash mesh, which includes working with bony landmarks. So, let's get started.

> **Important note**
>
> There are advantages to blocking out the human body from scratch instead of using a base mesh. To show a more efficient and goal-oriented way of doing this, this book shows the workflow with a human base mesh.
>
> If you want to study anatomy in more depth on a more fundamental level, there are many great resources available, but an especially popular and proven one is the website of artist Stan Prokopenko: `https://www.proko.com/`.

Starting with the skeleton

You should now have two subtools in your ZBrush session: **Nickz_HumanMaleAverage** and **Ryan_Kingslien_Anatomy_Model**. You can keep the anatomy model file open while you work on other files so that you can switch to it whenever you want to check the anatomy, while the base mesh file can be your main file to work in:

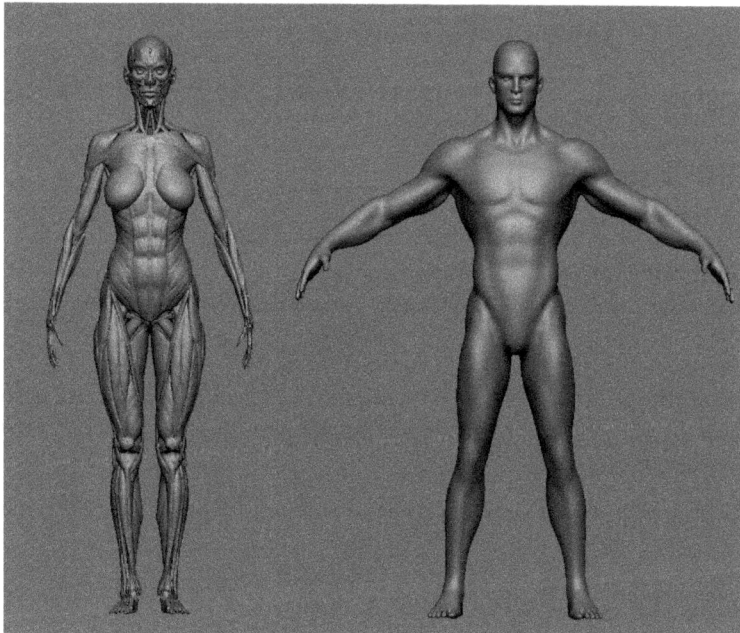

Figure 8.10 – Beginner anatomy resources loaded in ZBrush

Next, select the skeleton in the anatomy model, and then switch to your base mesh tool. From there, navigate to **Tool | Subtool | Append**, and pick the skeleton subtool to add it to the tool. After that, you can scale up the skeleton and position it in the center of the scene, where the base mesh is:

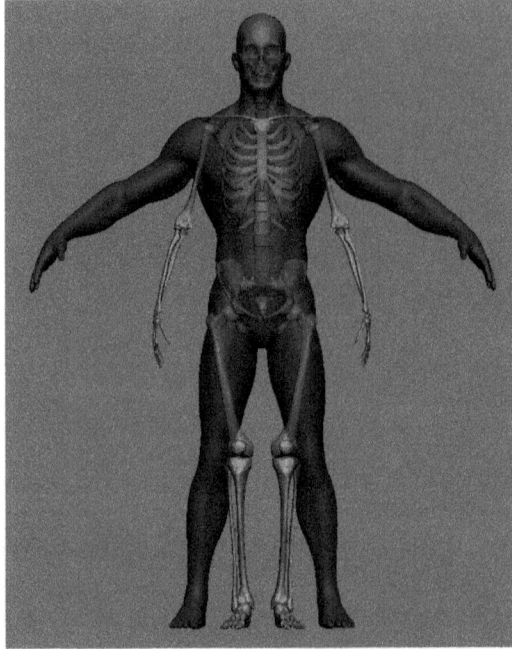

Figure 8.11 – Adding the skeleton to the Nickz base mesh ZTool

Adjusting the skeleton

Since we have a female skeleton, we need to make some tweaks so that it will work with our male base mesh. Let's start by making sure all the measurements are correct. To do this, you'll need to import the 8-head figure measurement image into ZBrush. There are several ways to do this, but here is a convenient way:

1. First, make sure the picture has square dimensions. You can crop or add space in software such as Photoshop to ensure that there will be no stretching or compression in your image when it is imported into ZBrush.

2. Then, in ZBrush, go to **Tool | Subtool | Append** and pick **Plane3D** from the basic shapes:

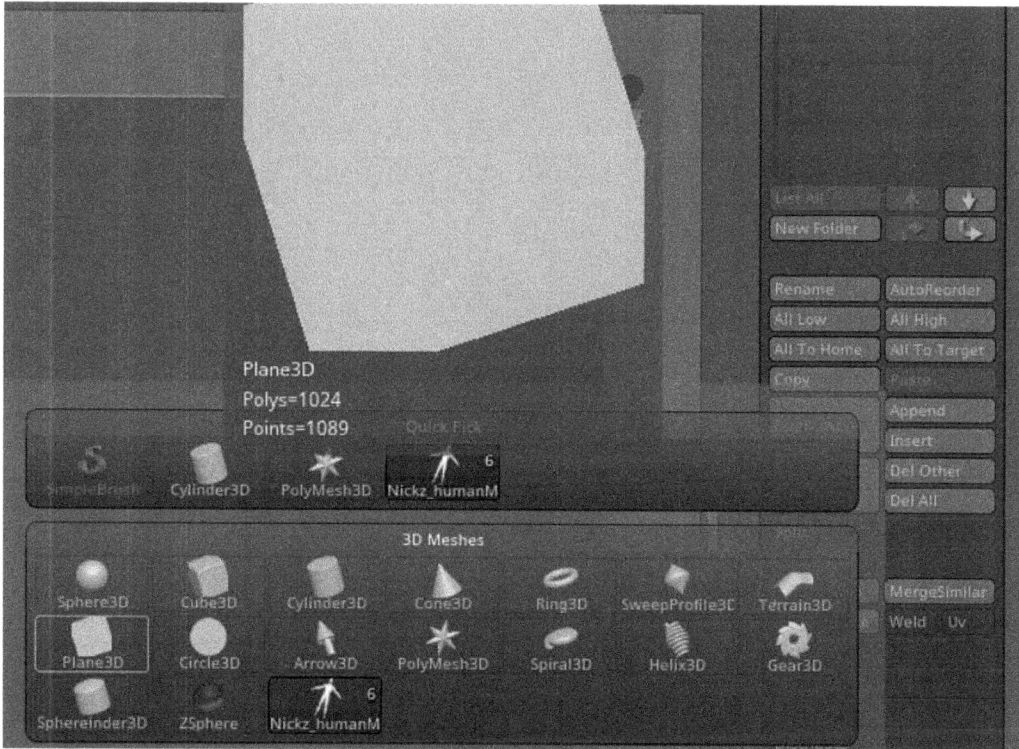

Figure 8.12 – Appending a Plane3D model

3. With the plane selected, go to **Tool** | **Texture Map**, click on the empty thumbnail in the top-left corner, select **Import**, and load your 8-head picture.

4. Rotate the plane 180 degrees to orient it correctly.

5. Scale the plane and position it so that it matches your base mesh.

6. Place the plane behind the skeleton and turn off all the other subtools so that you are left with the plane and the skeleton model:

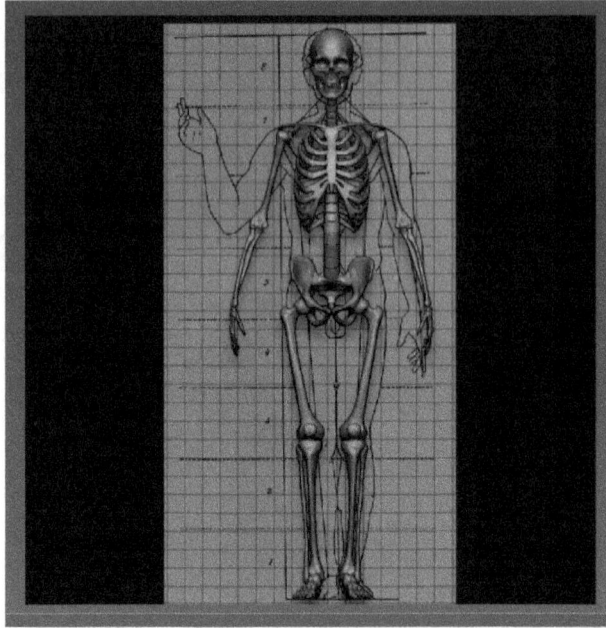

Figure 8.13 – Skeleton model and reference on a plane model

7. Now, you can start adjusting the skeleton, giving it male proportions. Here are some of the adjustments you need to make to give the skeleton male characteristics:

    - Broaden the shoulders

    - Make the ribcage longer

    - Make the legs longer, and position the pelvis higher

    - Make the arms longer

    - Make the ilium less flared:

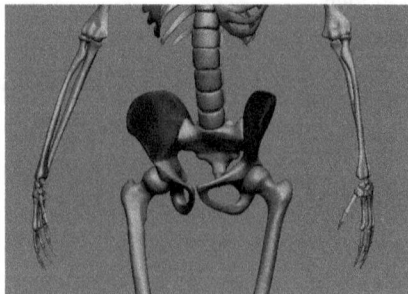

Figure 8.14 – The ilium, marked in red

A useful function for easy adjustment of the bones is to assign a PolyGroup to each bone so that you can isolate parts easily. Go to **Tool | Polygroups** and select **Auto Groups** to assign a PolyGroup to each separate part of the mesh. Now, you can hold *Ctrl + Shift* and left-click on parts to isolate bones and use the **Move** brush or Gizmo to adjust them.

8.   Since male bones are thicker, go to **Tool | Deformation** and apply a slight **Inflate**. The effect of this deformation can be seen in the following figure:

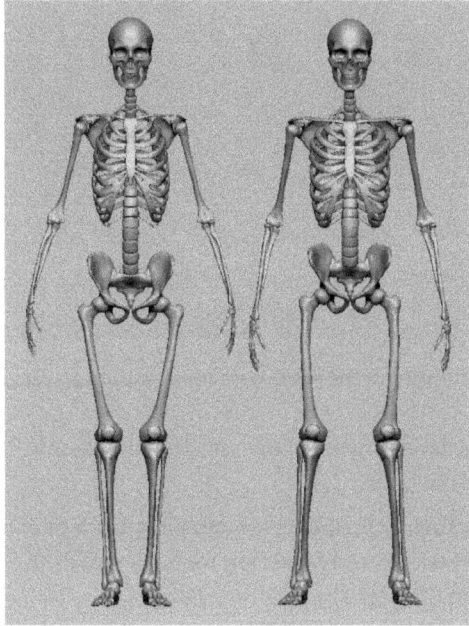

Figure 8.15 – Before (left) and after (right) changing the skeleton from female to male

These are the main adjustments you should make. The skeleton does not have to be perfect because you can adjust the proportions of your sculpture at any point anyway. With the skeleton in place, you can switch to the base mesh and match it to your skeleton.

Correcting the base mesh

Now that you have a fairly accurate male skeleton, you can match the base mesh to the skeleton so that you have an anatomically accurate mesh to work with.

Before adjusting the shape of the base mesh, you can lower the arms so that they match the skeleton more; the body of the Nickz ZTool has a layer saved that lets you lower the arms:

1.   Go to **Tool | Layers** and turn on the visibility of the first layer, called **ArmsDown**, by clicking on the eye icon. Now, the arms are straight at the sides, but we want them slightly raised. Adjust the **Layer** strength, which is currently at **1**, to **.8**:

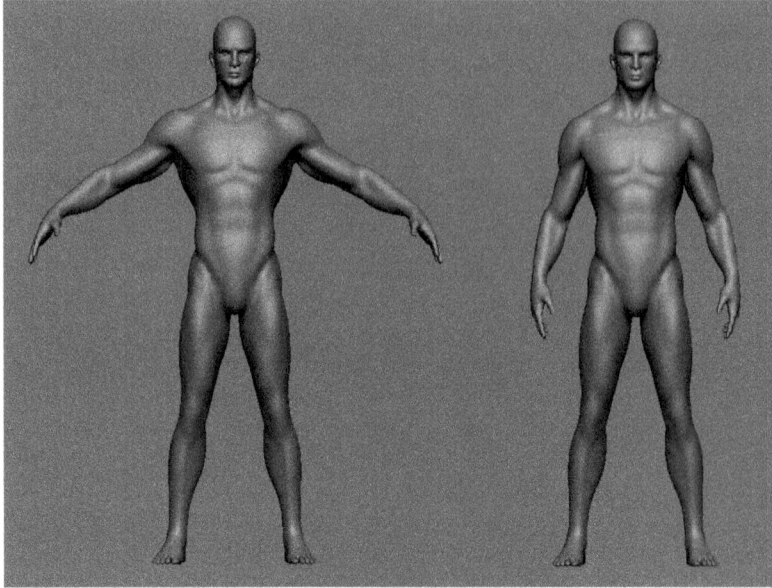

Figure 8.16 – Lowering the arms of the Nickz base mesh with the layers saved in the model

2. Since we do not need the layers anymore, we can click on **Bake All**. This will apply the changes to the model and delete the layers.

3. Place the skeleton's legs slightly further apart and place the legs of the base mesh a bit closer together by using the **Masking** and **Selection** tools, as well as the Gizmo (this process was covered in more detail in *Chapter 3*).

4. Move the feet of the skeleton back to match the feet of the base mesh so that your model has a more natural stance:

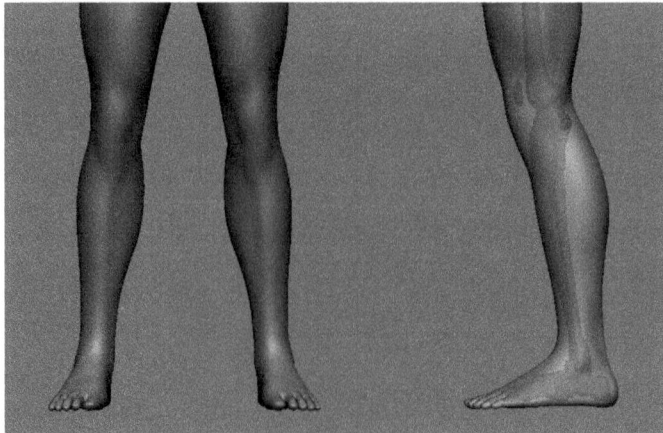

Figure 8.17 – Adjusting the model's stance

5.   Navigate to the **Color** palette, pick a red color, and click on **FillObject** to apply the color to your skeleton:

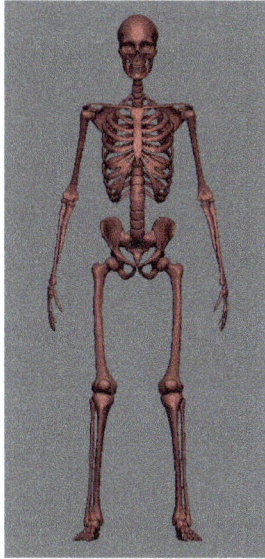

Figure 8.18 – Filling the skeleton with a red color

6.   Select your base mesh model and enter transparency mode by clicking on the **Transp** icon on the right-hand side of the canvas. Now, you should be able to see the red skeleton through your model:

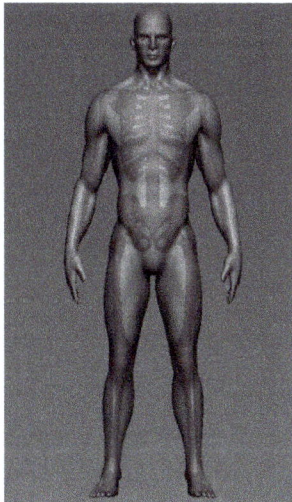

Figure 8.19 – Base mesh and fitting skeleton in transparency mode in ZBrush

While working on this, you can also look at references of male skeletons to make adjustments to the skeleton. It is up to you how much focus you want to put on accurate anatomy.

You will notice that some points of the skeleton naturally intersect with the base mesh. This is useful because these points, which are called bony landmarks, help you achieve more accurate anatomy.

Bony landmarks

Bony landmarks are areas on the skeleton where there is little space between bone and skin. These areas are easily identifiable, even in individuals with large muscle or fat mass.

Since the bony landmarks are the origin and insertion points for many muscles, they provide a convenient starting point for sculpting the anatomy and adding muscles in between:

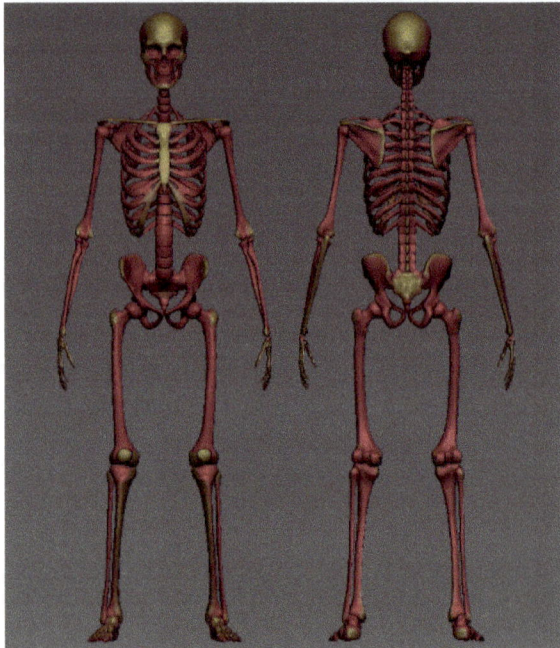

Figure 8.20 – Bony landmarks of the skeleton

Here are some of the important bony landmarks you can focus on while adjusting your base mesh:

- *Clavicle*: The clavicle is one of the most noticeable bones. Many muscles attach here but do not cover the top part of the bone, which makes the shape of the bone show quite clearly. Pay attention to the unique shape of the clavicle when sculpting, especially to the area on the shoulders where it meets the scapula:

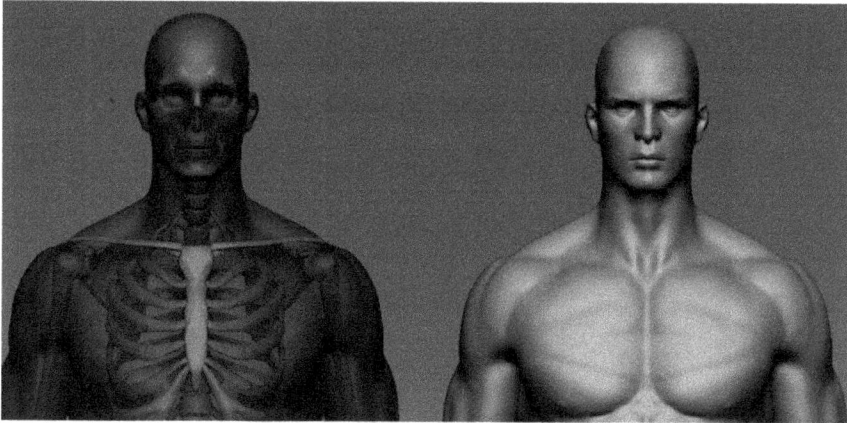

Figure 8.21 – The clavicle in X-ray view

- *Scapula and ulna*: Depending on the position and orientation of the scapula, its spine can be pronounced, while in a relaxed pose, it is more subtle but still recognizable. As many back muscles attach to it and create mass around it, the scapula is an excellent landmark for sculpting the surrounding anatomy. When analyzing reference images, you will find that by identifying the orientation and position of the scapulae, you can understand the shape of the muscles, which will help you sculpt them as realistically as possible:

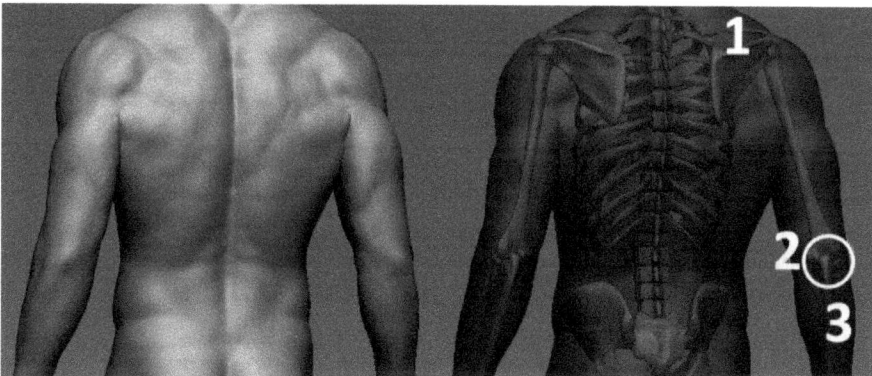

Figure 8.22 – The bony landmarks of scapulae (1), olecranon (2), and ulna (3)

The upper end of the ulna contains the olecranon, which is the bumpy part of the elbow. You may sculpt the elbow and then continue the sculpting along the shape of the ulna bone toward the wrist.

- *Fibula*: The upper part of this bone is the attachment point for many of the calf muscles, as well as the biceps femoris of the upper leg. The protruding upper-end part of the fibula bone is visible as a bump on the outside of the leg slightly below the knee. Being aware of this will help you use this as a guide in constructing the leg anatomy around this area:

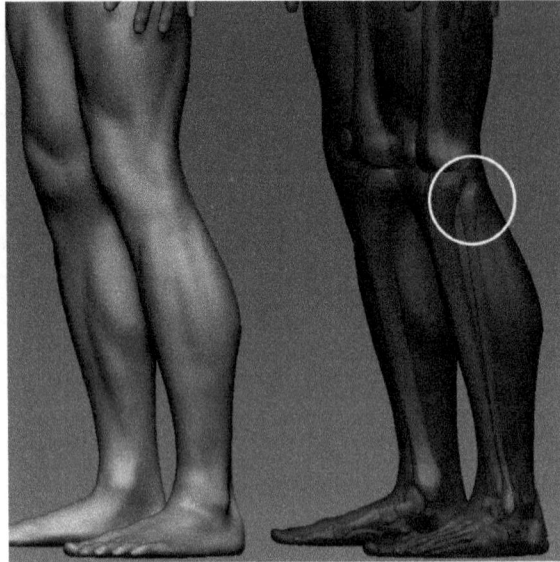

Figure 8.23 – Bony landmark of the fibula

With this information and preparation being done, you now have a great starting point to begin sculpting and refining the sculpture's anatomy.

Refining the anatomy

In this section, you will explore the main muscles when you're sculpting a believable human body. To begin, you will learn about some of the most useful sculpting brushes for creating muscles. Then, you will go through the individual muscle groups of the upper and lower body; references and sculpting tips will be included.

Useful brushes for anatomy sculpting

There are many approaches and tools you can use for creating anatomy in ZBrush, but as with most organic sculpting tasks, a few of the basic brushes will go a long way. Let's look at some of the brushes that are perfect for beginners and advanced artists alike.

ClayBuildup

This brush is excellent for building up forms and creating unique shapes. Like with any sculpting work, this will be the primary brush of many artists when it comes to sculpting muscles.

By default, **ClayBuildup** comes equipped with a square alpha, but you can switch it to **Alpha6** to get a smooth fall-off, which makes for a better sculpting experience when creating organic shapes, as you do when sculpting a human body.

This brush is great for adding chunks of digital clay and quickly building up forms. Even though the result can look rough initially, in combination with other brushes, it is the perfect tool for creating an organic model:

Figure 8.24 – A rough sculpture when using the ClayBuildup brush

DamStandard

DamStandard is perfect for dividing forms. Especially in the beginning stages, this brush is great for quickly breaking up forms and establishing the overall appearance of the body. Here, it can be used to sculpt the transition between whole muscle groups, or smaller bunches of muscle fibers.

It is important to note that the results of the **DamStandard** brush are quite harsh, so it is recommended to use a lower brush intensity and to break up the sharp lines into more irregular and smooth transitioning shapes. This will help you achieve a more natural and believable look:

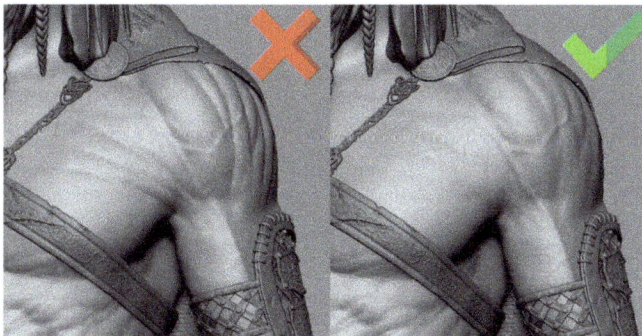

Figure 8.25 – Applying too much intensity and excessively defined lines with the DamStandard brush (left) versus proper intensity and break up (right)

StandardBrush

This is another classic sculpting brush that is perfect for organic sculpting, and great in this case as well. While the **ClayBuildup** brush lets you create new forms, **StandardBrush** is perfect for enhancing existing shapes and giving them more volume and contrast. This brush's strength is that it keeps the detail and forms intact because it doesn't fill crevices as much as other brushes; because of this, you can use it without worrying about "destroying" forms.

It is best used on a very low Z intensity of around 10–14 so that it does not create new forms as easily but, rather, just enhances what is already sculpted.

You could use this brush to give the biceps a bigger peak or to emphasize certain muscle bellies on the legs or back. It is also great for sculpting veins, which will be covered later in this chapter.

The following figure shows how **StandardBrush** can be used to give flat-looking muscles more "pop" by increasing their volume, which results in a brighter highlight:

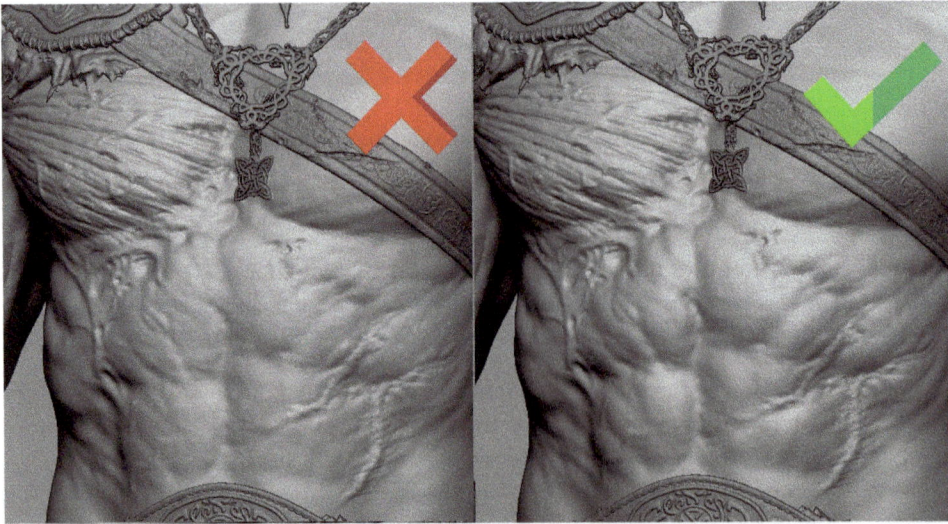

Figure 8.26 – Giving flat-looking muscles (left) more "pop" with StandardBrush (right)

Move and Move Topological

These two brushes will be useful when you're changing proportions and making big changes to silhouette and muscle volumes. When working on the pose of your model, **Move** brushes are also often used. To learn more about the difference between these move brushes, check out *Chapter 4*.

You only need these four brushes in your repertoire so that you can start sculpting confidently.

Sculpting the upper body

We will start with simple anatomy on the front of the upper body, where muscle groups are easier to sculpt than on some other regions of the body. Sculpting these parts now will give you some early success, which will motivate you to finish the remaining parts later on.

> **Important note**
>
> It's better not to work on one area for too long. This is more efficient because spending too much time on a small area will make you lose sight of the bigger picture. Once you come back to an area, you can look at it again with fresher eyes and notice flaws more easily.

An effective approach to sculpting the human body is to define the muscles very clearly and slightly exaggeratedly, building a sort of blueprint from which you can soften the forms again and refine them until they match your concept. Here is what I have done:

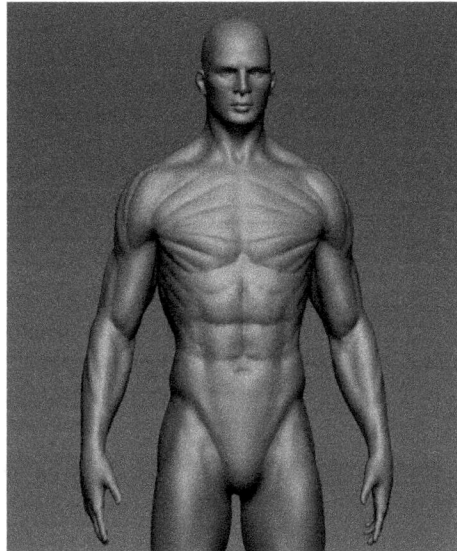

Figure 8.27 – Carving out a muscles blueprint

> **Important note**
>
> Make sure you use Symmetry Mode so that you can work faster and fix issues with proportions without having to address sides individually. You can enter Symmetry Mode by navigating to **Transform | Activate Symmetry** or by pressing *X* on your keyboard.

Once you have sketched the muscles, you can study each muscle a bit more in-depth and sculpt them with precision.

Chest

When sculpting chest muscles, you should pay attention to the different groups of muscle fibers that form separate shapes and overlap in their insertion point on the upper arm bone. This creates the characteristic look of chest muscles in that area:

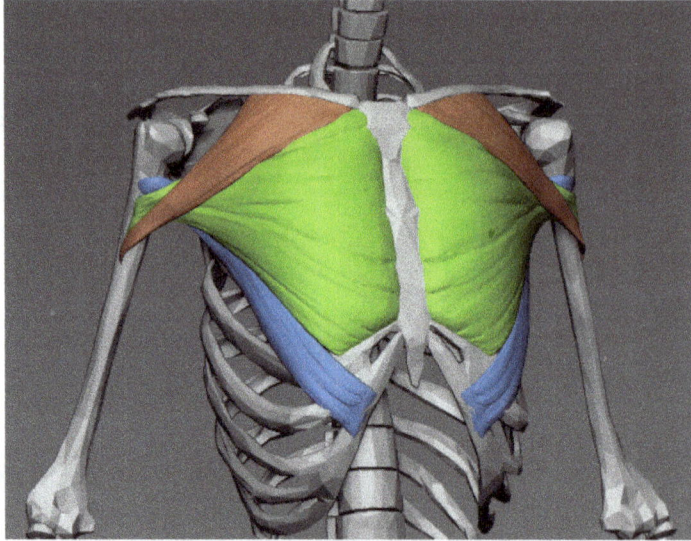

Figure 8.28 – Layering the upper, middle, and lower chest muscles

A subtle but important point to keep in mind is that the pectoralis minor, a small chest muscle that lies underneath the larger pectoralis major, pushes on the chest muscles above it, creating more mass in that area of the chest. Make sure you translate that in your sculpture, giving this area more volume than the inside of the chest:

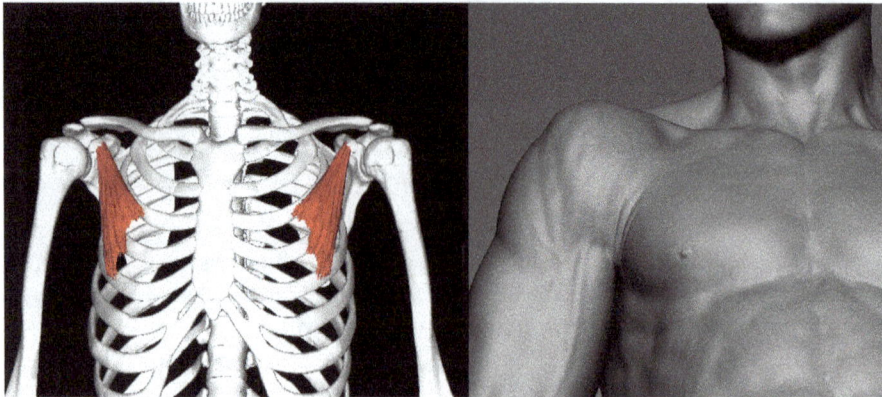

Figure 8.29 – Pectoralis minor

Deltoid

This muscle has three parts: the anterior, lateral, and posterior portions. Many of the muscle fibers interlock and overlap with each other. Capturing the appearance of each part in your sculpture will improve realism:

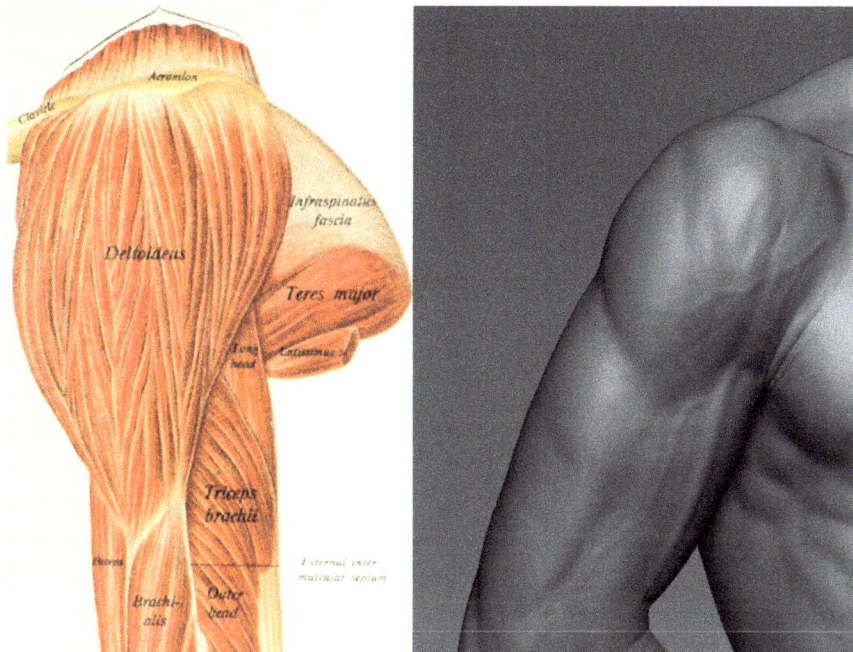

Figure 8.30 – Shoulder muscles

Midsection

There are three important muscles in the midsection: the rectus abdominis, the external obliques, and the serratus anterior. Make sure you pay attention to their exact location and orientation:

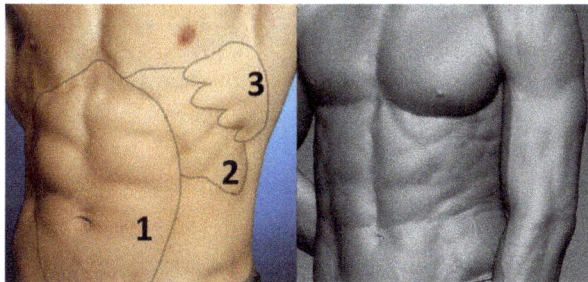

Figure 8.31 – Rectus Abdominis (1), external obliques (2), and serratus anterior (3)

A common mistake of beginners is to sculpt the midsection with clear, distinct lines between each muscle group. Instead, try to create smooth transitions where it makes sense so that there is a balance between smooth surfaces and high-contrast areas. As always, reference is key in finding areas where it is realistic to implement this:

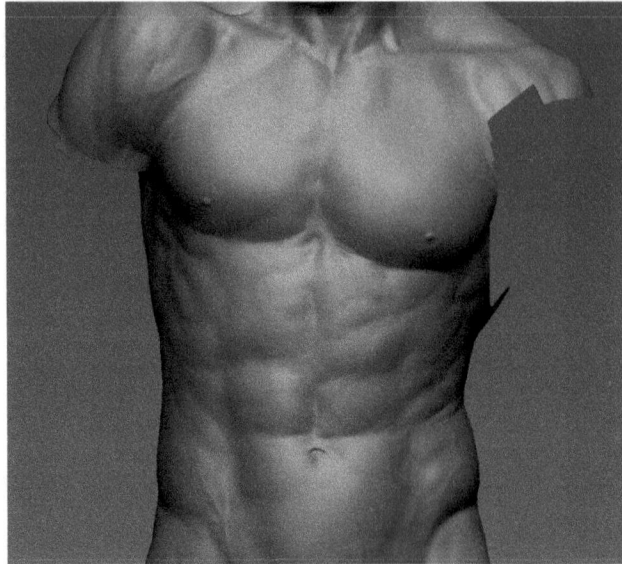

Figure 8.32 – Asymmetry and variety of contrast in the sculpt

Back

Back muscles are complex – the appearance of the back will vary based on not only the amount of muscle mass but also on the pose. The position and orientation of the shoulder blades – where various back muscles attach – will influence the look of the back, so make sure to consider that when sculpting.

To get the best results, you should find reference images that show the exact pose you are going to sculpt. Of course, you can add other pictures, but make sure you understand the anatomy and function of the muscles before copying their shapes; otherwise, your results will be inconsistent.

For beginners, it makes sense to start with a muscular but neutral and relaxed back pose, with no extreme flexing or stretching of muscles. Let's look at some of the back muscles that affect the appearance of the back the most.

Iliocostalis and longissimus

The iliocostalis and longissimus muscles run like two parallel tubes on the lower back, with the iliocostalis lying above the longissimus. These muscles stand for power and strength, making them essential muscles to emphasize when sculpting superheroes or athletic characters.

Make sure you sculpt a lot of asymmetries and break up the shapes into multiple smaller shapes of varying sizes. Muscles rarely show as one clean, continuous shape in real life, so your sculpture becomes more realistic when you incorporate some of these small differences and varied muscle fibers:

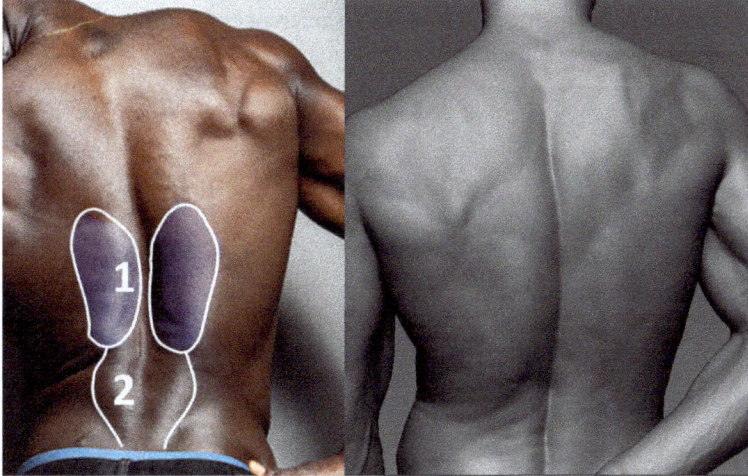

Figure 8.33 – Iliocostalis (1) and longissimus (2)

Latissimus dorsi

The latissimus dorsi is the largest muscle of the upper body. When it is well developed, it creates a classic V shape that is great for creating a strong silhouette in your sculpture. This muscle does not require much sculpting of detail, but make sure to add enough volume and width to this area, even on the lower part, because that will ensure that your character's muscular back is visible from the front as well:

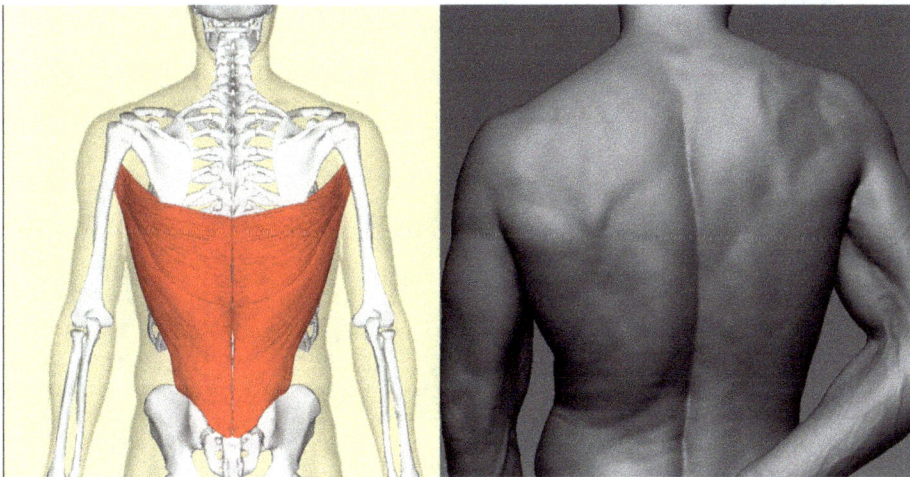

Figure 8.34 – Latissimus dorsi

Infraspinatus and teres major

These are some of the smaller muscles of the back, but they can have a bulky appearance. In a relaxed pose, the teres major appears bulkier and the infraspinatus flatter; when contracted, however, the infraspinatus becomes more pronounced as well. Using references for a specific pose will help you determine how to sculpt these two muscles:

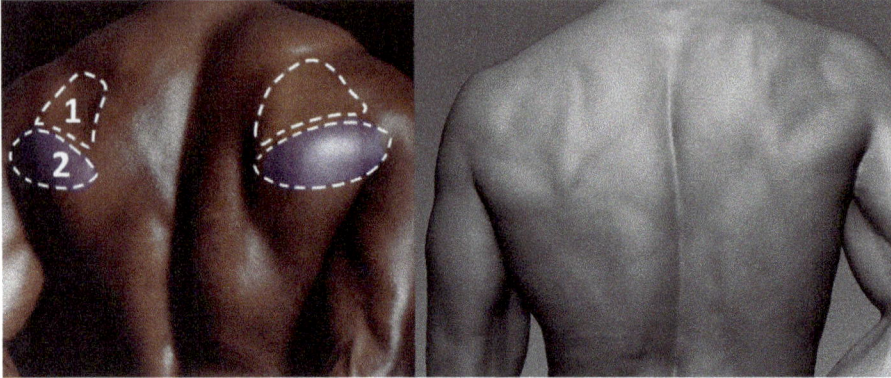

Figure 8.35 – Infraspinatus (1) and teres major (2)

Trapezius

The trapezius is visible from the front and back. Depending on how muscular you want your character to be, you can sculpt this muscle very pronounced. Similar to the muscles in the center of the back, it also needs a lot of asymmetry:

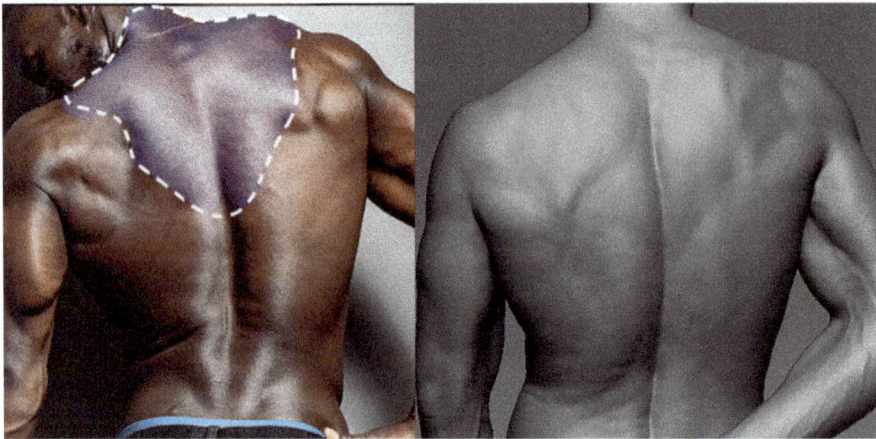

Figure 8.36 – Trapezius

Break up the muscle into smaller shapes using the **DamStandard** and **ClayBuildup** brushes to create a natural-looking sculpt.

Upper arms

Sculpting the upper arms is relatively simple. As with any muscle group, the better your understanding of anatomy, the better your sculpture will look. Small details can have a big effect, so pay attention to things such as the origin and insertion points of muscles.

Biceps

The biceps are a muscle that consists of two parts, which are also called "heads" – the inner and outer heads. The outer head is shorter, which is especially noticeable in muscular individuals. At times, the separation of the two heads of the biceps can be visible, although this is mostly the case for flexed biceps and for people with low body fat:

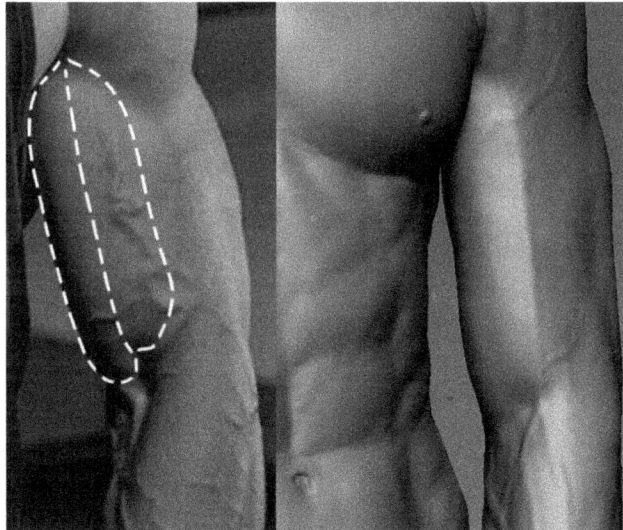

Figure 8.37 – The two-headed biceps muscle

Triceps

The triceps consist of three heads: the lateral and long heads of the triceps are straightforward and rather easy to sculpt; it is only the medial head, the smallest of the three heads, that is more subtle. Reference pictures will be crucial to sculpting it accurately.

It is important to note that a relaxed triceps does not show the separation of the three heads, and it blends more into one shape. It is still important to be aware of the shape of the individual triceps heads to sculpt them accurately:

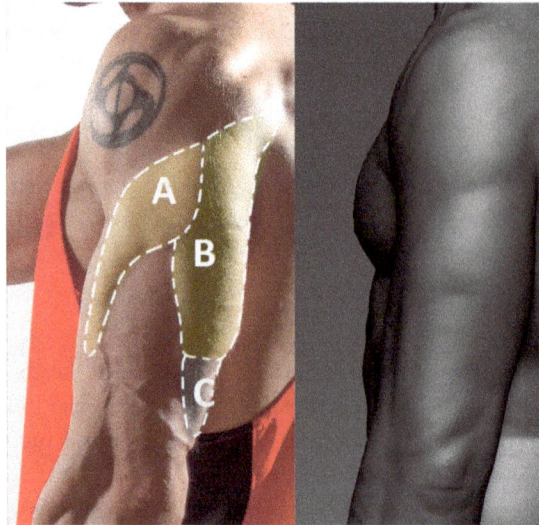

Figure 8.38 – Lateral (A), long (B), and medial (C) heads of the triceps

Brachialis

The brachialis muscle lies between the biceps and triceps. It tapers toward the shoulder and runs diagonally along the biceps. Make sure you capture that tapering look when you're sculpting this muscle instead of depicting it as two parallel shapes:

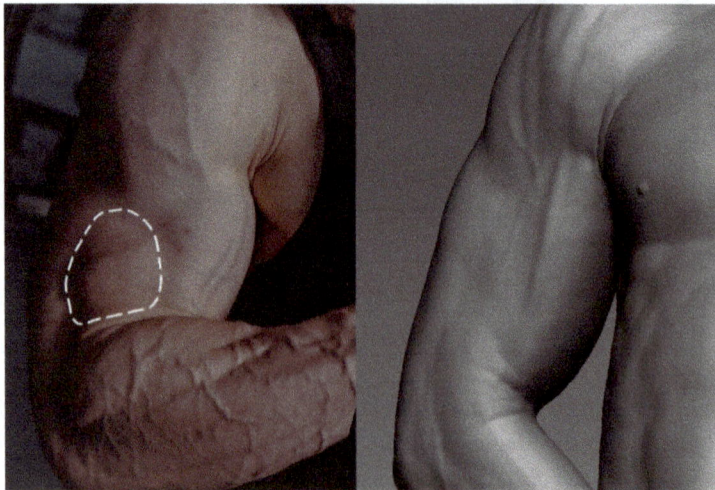

Figure 8.39 – Brachialis

Lower arms

Forearms are a more challenging muscle group to sculpt properly as they consist of many muscles, with some of them being small and subtle. However, the main challenge is that their appearance changes based on the rotation of the wrist, with different movements and actions activating the muscles to a different degree. To keep it simple, we'll focus on two groups of forearm muscles – the forearm extensors and the forearm flexors – and see how they determine the look of the forearms.

Forearm extensor muscles

The forearm extensor muscles originate from the outer portion of the humerus (upper arm bone) and create movement in the forearms, hands, and fingers. Here are some of the more noticeable of these muscles that contribute most to its shape and appearance:

- **Brachioradialis and extensor carpi radialis longus**: These two forearm muscles are the most noticeable because of the muscle mass they add to the top of the forearms. They blend and appear as one shape in most cases:

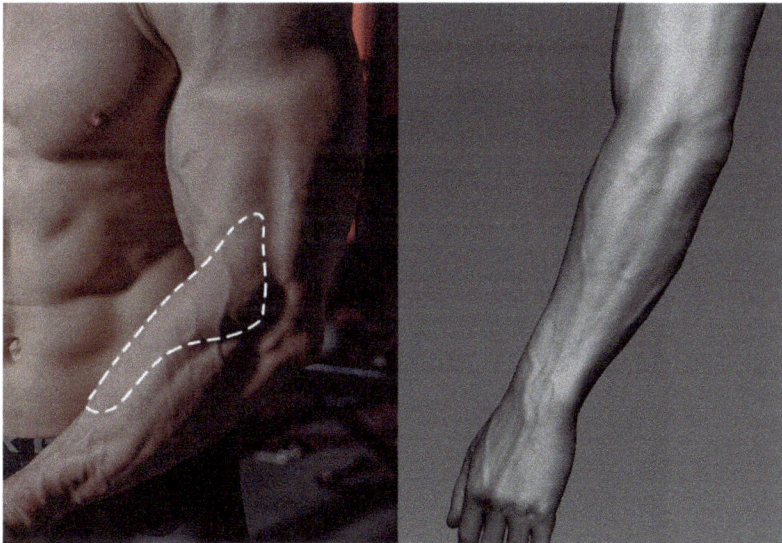

Figure 8.40 – Brachioradialis and extensor carpi radialis longus

- **Extensor carpi radialis brevis**: Depending on the rotation of the wrist and muscle activation, this muscle can be pronounced and stand out. In a relaxed pose, it blends in with the carpi radialis longus. The right reference images will help you get the subtleties of this muscle right:

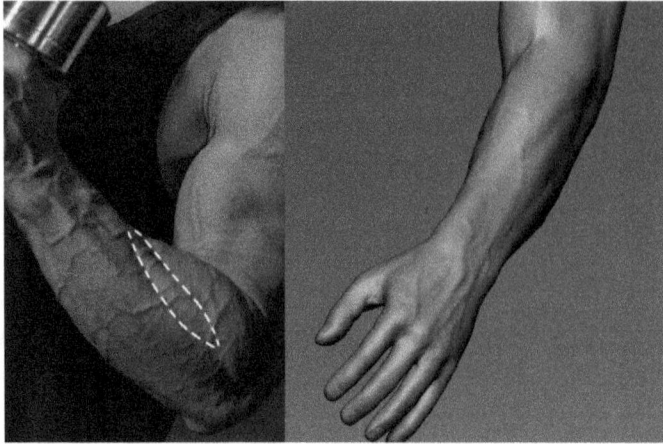

Figure 8.41 – Extensor carpi radialis brevis

- **Extensor digitorum**: This is one of the bigger forearm muscles and adds to the arm's silhouette with its sweeping profile. It becomes especially pronounced when the hand is gripping something forcefully, which activates the muscle:

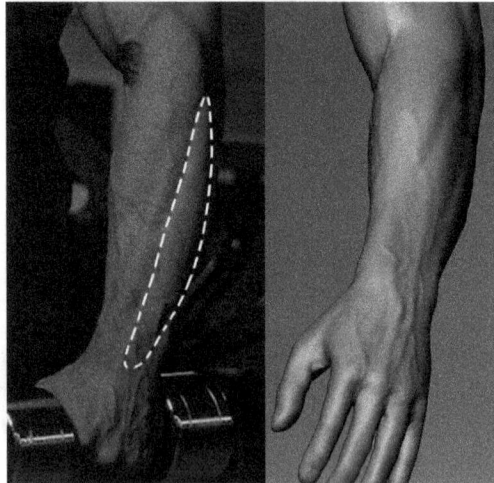

Figure 8.42 – Extensor digitorum

Forearm flexors

In a relaxed state, the forearm flexors appear visually as one shape. Here, you should try to capture the silhouette as it bulges in the upper third of the forearm and tapers toward the wrist, while the forearm extensor muscles create a sweeping shape across the full length of the forearm. To sculpt realistic flexors, make sure you add joints showing some muscle separation and add veins – **StandardBrush** is great for that!

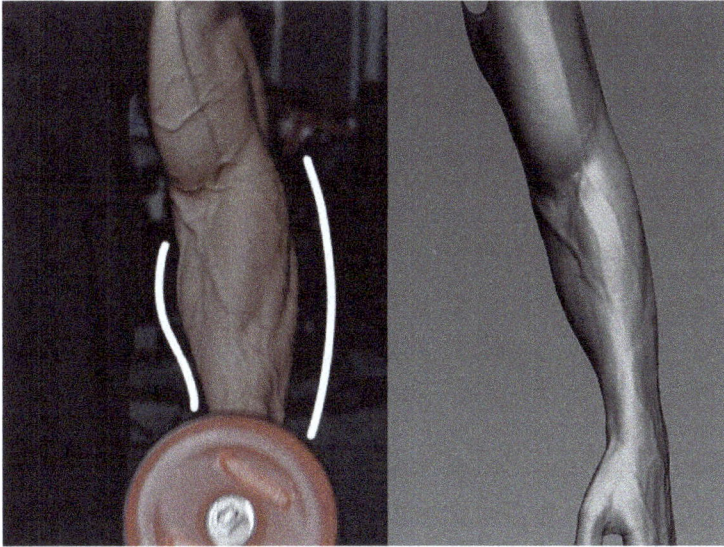

Figure 8.43 – Curves/silhouette of the forearm

Sculpting the lower body

Sculpting the lower body can be more challenging than the upper body because its muscles tend to be more subtle and blend visually. Flexed, as well as generally more muscular legs, show the anatomy better, which makes bodybuilding reference images very useful:

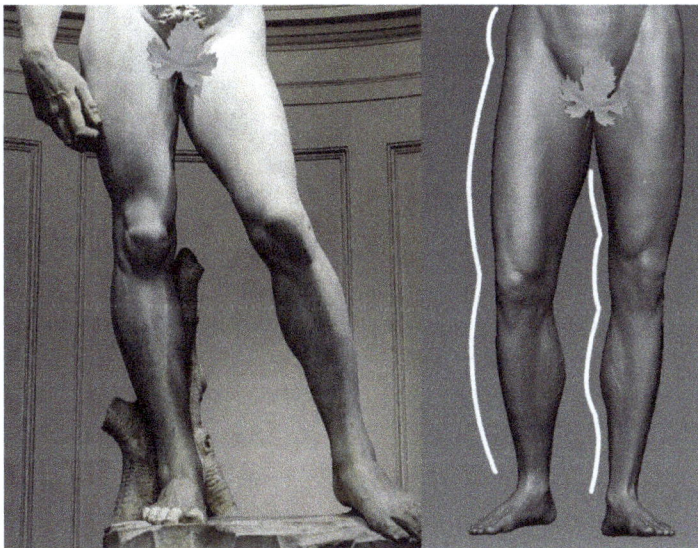

Figure 8.44 – The legs' silhouette

Paying attention to the curves and shape of a leg will help you create realistic and visually appealing legs. The mix of long and short curves on the inside and outside of the legs creates the classic leg silhouette.

> **Important note**
>
> Remember to take a look at the écorché model to study the origin and insertion points of individual muscles so that you understand how they attach and layer with other muscles.

Upper legs

The front of the upper legs can be simplified as two masses: the large quadriceps muscle, located in the middle and outside of the leg, and the adductors, which make up the inside of the upper thigh. Both large masses are separated by the sartorius muscle:

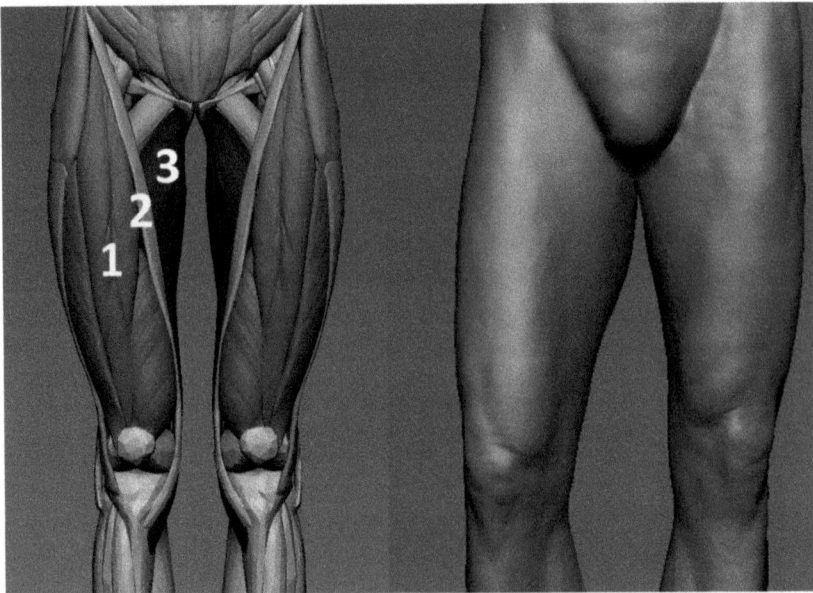

Figure 8.45 – Quadriceps (1), sartorius (2), and adductors (3)

The quadriceps muscle itself consists of four parts: rectus femoris, vastus lateralis, vastus medialis, and vastus intermedius (which is not visible since it is lying underneath the rectus femoris):

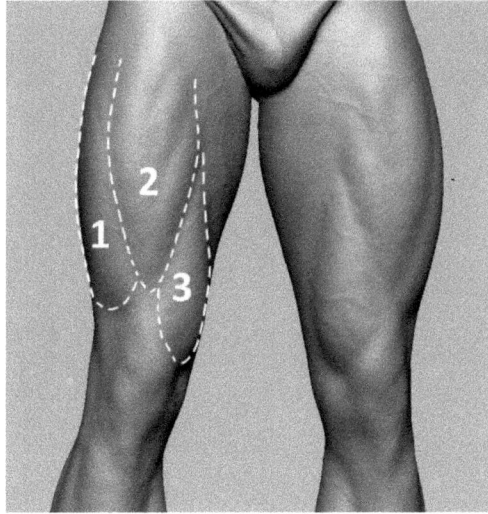

Figure 8.46 – The three visible heads of the quadriceps: vastus
lateralis (1), rectus femoris (2), and vastus medialis (3)

The vastus medialis is the teardrop-shaped muscle that sits right above the knee toward the inside of the leg. Because it is one of the more pronounced masses of the leg, even in a relaxed state, it makes sense to pay extra attention to that area in your sculpture.

It is only when the leg's extensors are flexed that the rectus femoris and the vastus lateralis become more defined, in which case you need good reference and attention to detail to properly capture these muscles.

Hamstrings

This muscle group makes up the back part of the upper leg and consists of the following three muscles:

- Semimembranosus
- Semitendinosus
- Biceps femoris

These muscles are less pronounced and tend to blend, making them hard to sculpt. Looking at an écorché, you can see how the muscles wrap around both sides of the knee, creating mass on the inside and outside of the thighs that can be observed in reference images:

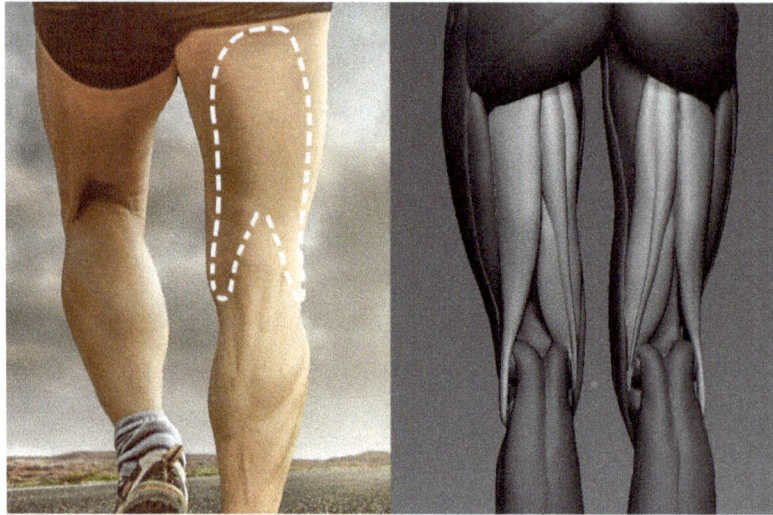

Figure 8.47 – Hamstring muscles

Additionally, you can observe how they affect the silhouette from the side, where they can create a curved shape on the back of the leg, especially in muscular individuals:

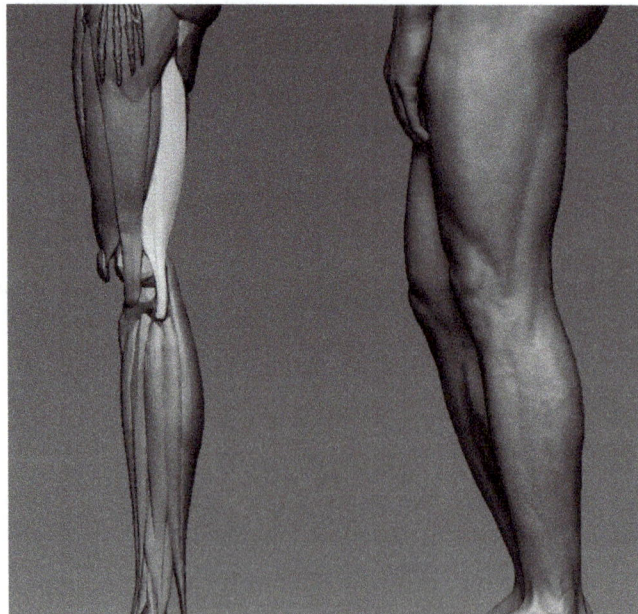

Figure 8.48 – Hamstring profile

Lower legs

Although the calves are made up of smaller muscles, they are a bit easier to sculpt since they have a very defined and characteristic appearance. The silhouette is similar to the upper leg as it consists of a mixture of short curves on the inside of the leg and long curves on the outside of the leg:

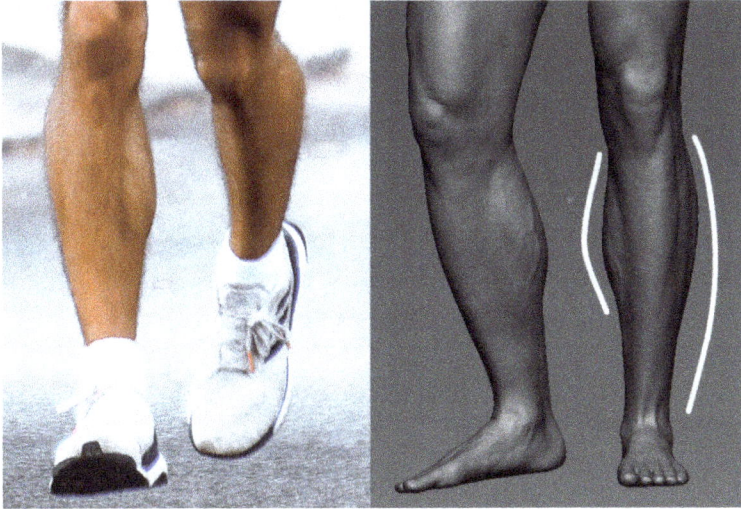

Figure 8.49 – Calves silhouette

Gastrocnemius and soleus

These two muscles are the most prominent and contribute to the silhouette of the calves the most. Especially when flexed, they have a very distinct look, and reference images will help to provide the right guide for sculpting them accurately:

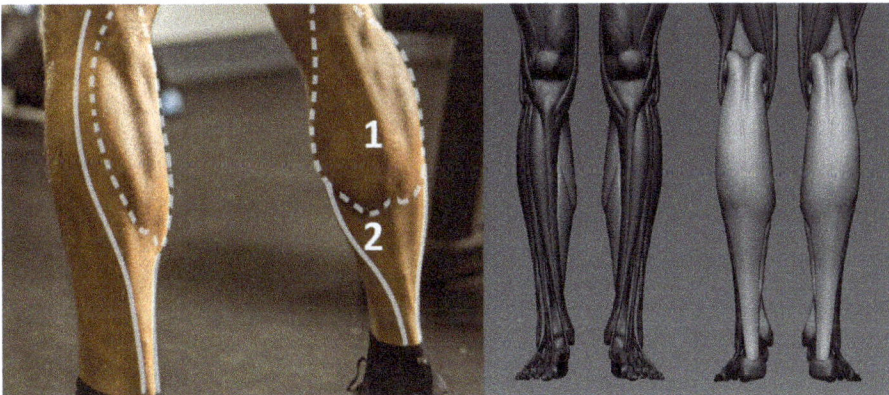

Figure 8.50 – Gastrocnemius and soleus

Tibialis anterior

Running along the outside of the shin, this muscle is one of the more pronounced muscles in the lower leg, besides the larger gastrocnemius and soleus muscles. You can use **StandardBrush** to create some volume on the upper shin while keeping the transition to the surrounding muscles smooth:

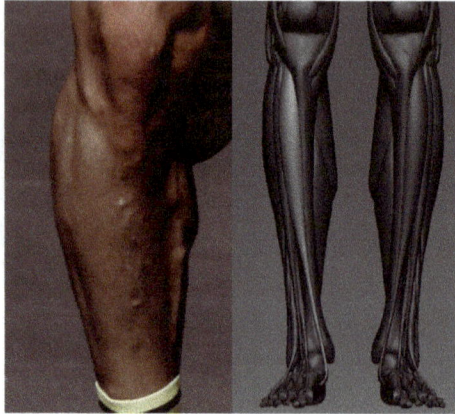

Figure 8.51 – Tibialis anterior

Peroneus longus

This muscle is visible on the outside of the calves, especially when the lower leg muscles are engaged. You may decide to emphasize this muscle if your character is especially muscular. Again, **StandardBrush** is a great choice to add form on the outside of the shin, to indicate this muscle:

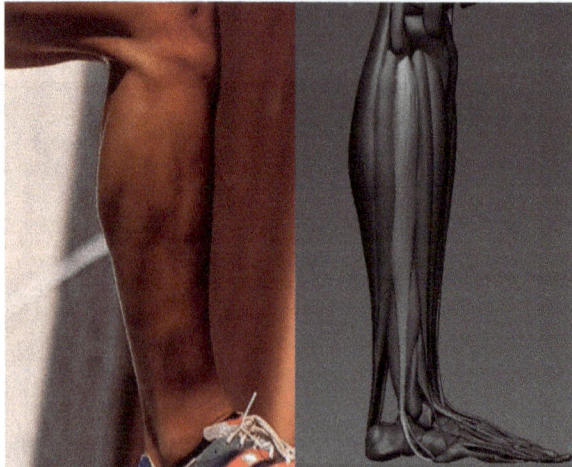

Figure 8.52 – Peroneus longus muscle

Next, let's move on to the extremities of the body and take a look at how to sculpt the hands and feet.

Sculpting hands and feet

Two more areas contain several small muscles: the hands and feet. The bones in these areas play a large role in how the hands and feet, and especially their digits, look. We can simplify these complex areas by looking at their shape a bit closer than at the underlying muscles.

Hands

Hands are one of the most challenging subjects for artist, but they are equally important because of their ability to add a lot of expression to the character. When sculpting hands, anatomy knowledge is essential but it also helps you look at the characteristic shapes that make up the hand.

Note the contrast between the bony surfaces of fingers and knuckles on the back of the hand and the fleshy rounded shapes on the palm. This translates to a more square shape of fingers on top and a rounded shape on the bottom. A common mistake of beginner artists is to make fingers completely round instead, which leads to an undesirable "sausage" look.

On the palm, you can see how the hand is divided into three larger masses. Of course, a lot of refinement and detailing is needed to achieve more realism. Later in this chapter, we'll add veins, which is especially important for improving the realism of the hands further.

Here, you can see the hand broken down into simple shapes, showing the mix of flat and rounded surfaces:

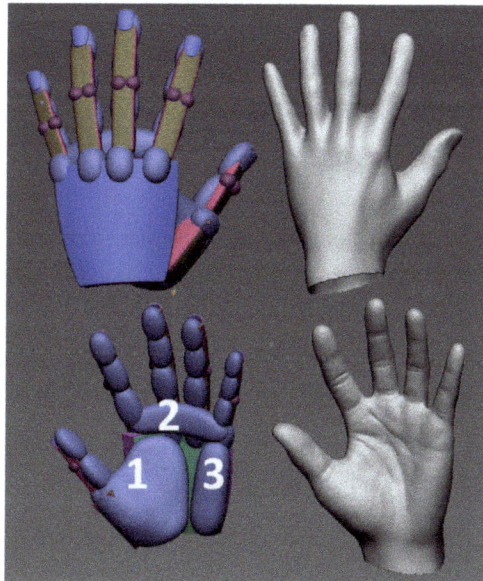

Figure 8.53 – Hands simplified, with the palm consisting of three larger shapes

Feet

Feet are similar to hands in that the skeleton determines their shape more than muscles. The bones of the feet, as well as those of the leg that make up the ankles, can help you capture their appearance more easily. Break them down into simple shapes, as shown in the following figure:

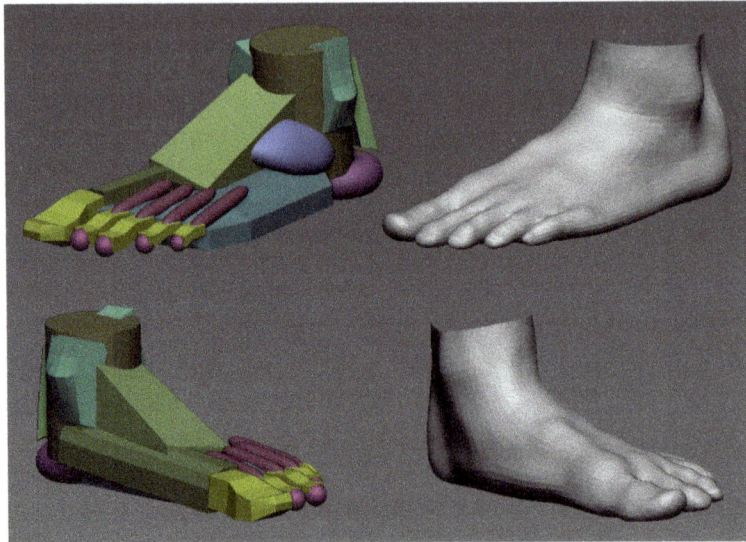

Figure 8.54 – Feet simplified and sculpted

At this point, we have covered the most essential muscles of the upper and lower body. Now, we can move on to another important anatomy element: fat.

Adding fat deposits

After looking at the skeleton and muscles, one important element that's missing from your anatomy knowledge is the subcutaneous fat deposits. This fat is stored under the skin and is responsible for the shape and appearance of the body.

Even though this area can be somewhat neglected for an athletic physique like that of a gladiator, it is worth looking at the fat distribution and gender-specific properties of fat so you know how to sculpt an overweight person as well.

First, let's look at what the increase in body fat looks like:

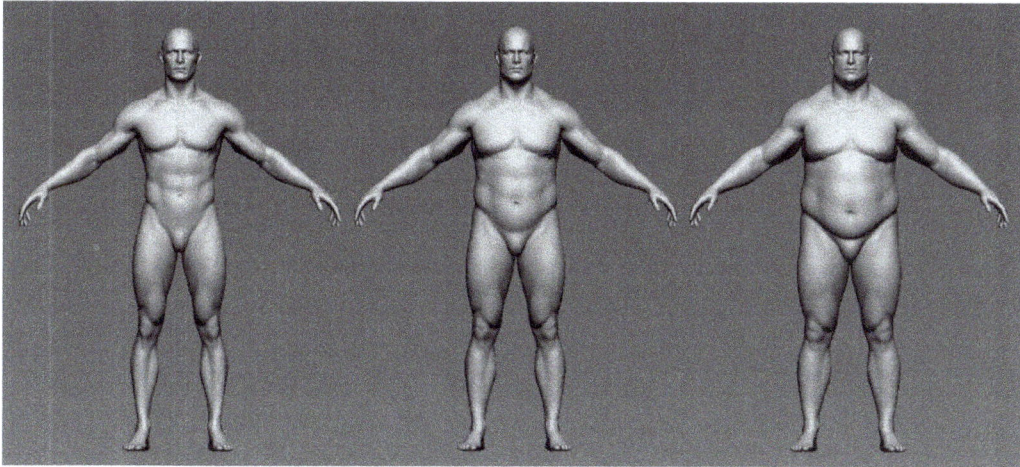

Figure 8.55 – Increasing body fat

As you can see, some areas show a bigger change in volume than others. In the following figure, you can see some of the areas that are not too affected by increasing body fat:

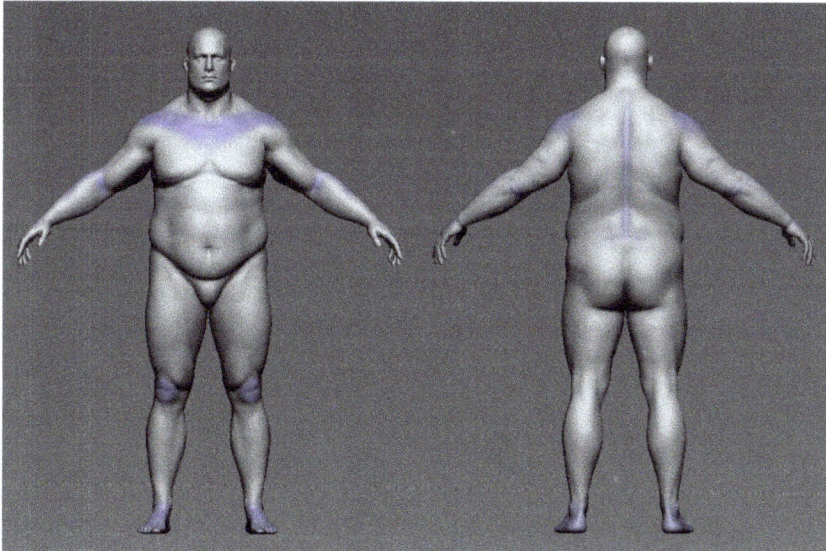

Figure 8.56 – Areas that show relatively little increase in body fat volume

These areas are close to the bone and have little subcutaneous fat deposits. On the other hand, there are some areas of the body with significant fat deposits. The distribution and size of these are different in the male and female body, which is something to keep in mind when you're sculpting the body of the relevant gender. This is visualized in the following figure:

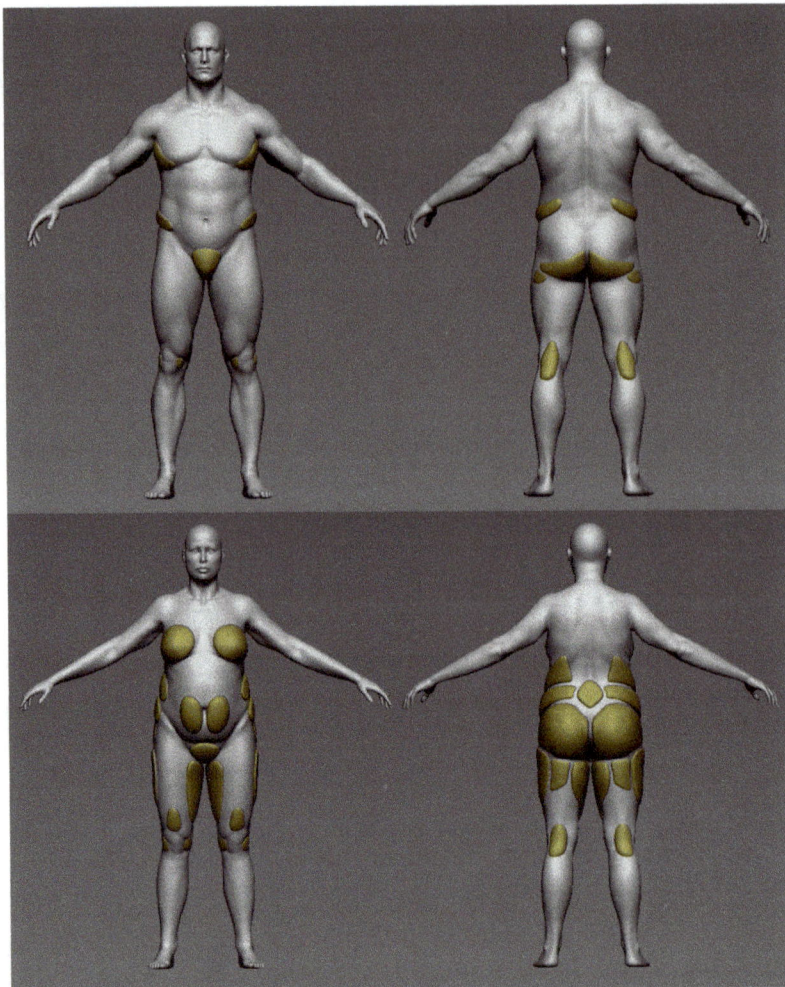

Figure 8.57 – The subcutaneous fat deposits of a man (upper) and a woman (lower)

When you are sculpting a regular or overweight person, paying attention to the characteristic distribution and shape of fat deposits can make the difference between a believable sculpture and a body that simply looks inflated, so make sure you include this type of information in your anatomy studies.

At this point, after working on all of the body parts we have discussed, you should have a symmetrical model standing in a neutral pose, something like this:

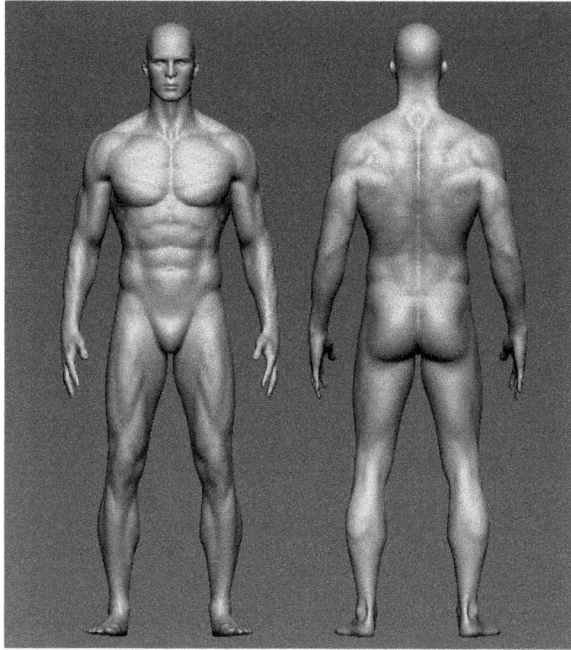

Figure 8.58 – The symmetrical model in a neutral pose

> **Important note**
>
> When sculpting complex and challenging forms, it is always helpful to break them down into separate areas that share a direction. You will find that if you sculpt those areas and can match the direction they are facing, you will get close to capturing the reference.

In this section, you sculpted all the major muscles of the upper and lower body, as well as the hands and feet, and last but not least you learned about the subcutaneous fat deposits.

Right now, your model should still be in its original neutral pose, and its muscles symmetrical. You will change both in the next section.

Finalizing the sculpture's anatomy

When you are satisfied with the basic anatomy of your model, it is time to refine it, shifting your attention from anatomical correctness and accuracy to visual appeal. In this section, you will introduce asymmetry and change the pose from a neutral stance to one that is more dynamic and interesting. Finally, you will add veins and apply a skin texture using the **NoiseMaker** plugin.

Posing your model

At this point, you can go ahead and pose your model. Here's a look at the difference between a neutral pose and a more dynamic one; you can see how introducing a shift in shoulders and hips can improve your character's presentation:

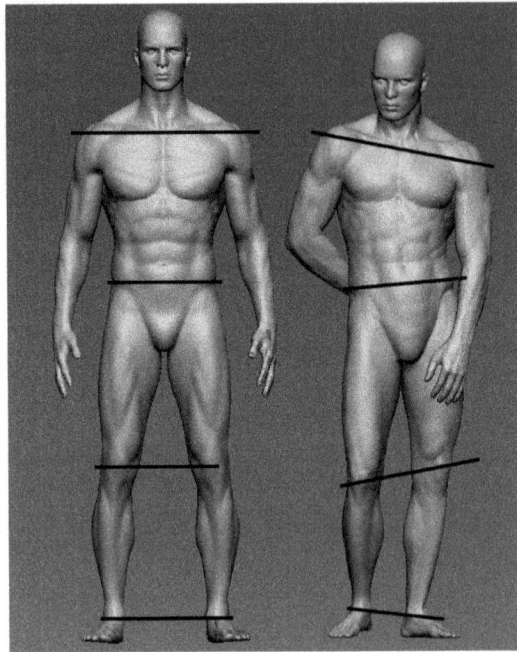

Figure 8.59 – Neutral versus dynamic pose

While there are a couple of ways to pose your model in ZBrush, a good way of posing a simple model is to use masking and the Gizmo, as follows:

1. Switch to the lowest subdivision level.

2. Using **MaskPen** or any other masking tool you prefer, mask off a limb you want to rotate (to learn more about masking, please go back to *Chapter 2*).

3. *Ctrl* and left-click on an empty space on the canvas to invert the mask.

4. *Ctrl* and left-click on the mask to blur it.

5. Hit the *W*, *E*, or *R* key to activate the Gizmo tool.

6. Move the Gizmo on the border where the mask ends and position it to the point you want to use as the pivot for the rotation.

7. Rotate the Gizmo, moving the limb in the desired direction.

Continue this process until your model is posed in the way that you want:

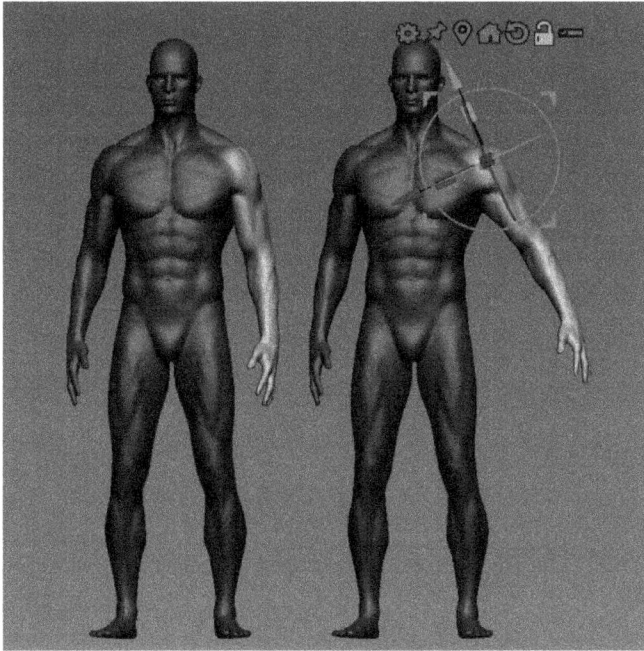

Figure 8.60 – Using masking and the Gizmo to pose the model

You could use the **Move** brush for quicker and simpler deformations instead, but this technique is also more prone to unwanted distortion. Saying that, regardless of which method you choose, some cleanup and fixing will be required.

Posing will require some patience and good reference images, but it is an important step to properly present your character. With a strong pose in place, you can continue to refine the muscles, reducing symmetry even further and making them look natural and believable.

Adding asymmetry

Posing the model is an important step in reducing symmetry, but it can be taken further by adjusting certain muscle groups so that they are not exactly mirrored shapes. This is especially important for the midsection and chest as they lie on the center of the body, which makes it easier to spot flawless symmetry than it is on muscles that lie separated on both sides of the body, such as the legs and arms. The abdominal muscles are quite asymmetrical by nature, so make sure your sculpture reflects that:

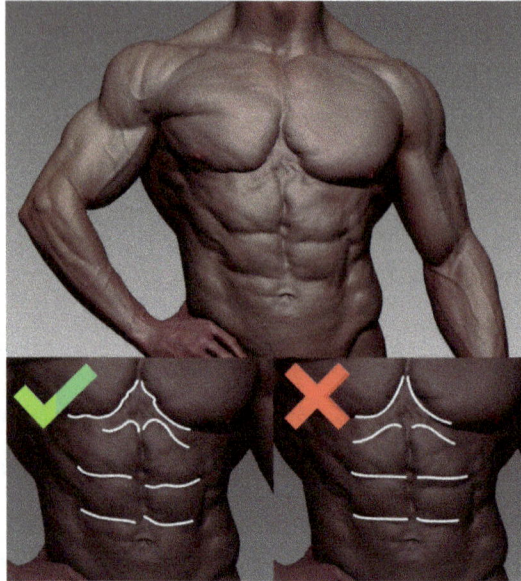

Figure 8.61 – The abdominal and chest muscles should be asymmetrical

Of course, the limbs should not look the same either, but luckily, it does not need a lot of work to introduce asymmetry.

Just some small changes to the shape of the muscles will go a long way. Since there is some flexibility in how you sculpt the muscles, without making them anatomically inaccurate, you can just go ahead and tweak small aspects.

You may add a bit more volume to some muscles with **StandardBrush** or use the **Move** brush to "pull" a muscle to make it a bit longer or shorter than its counterpart. Another option is adding finer muscle separations using the **DamStandard** brush. The veins, which will be added later in this chapter, will add further asymmetry.

When you are happy with the asymmetry, you can start to blend some of the muscles and refine the transition to make the body even more realistic.

Working on the transitions of shapes

The human body shows a mix of high and low-contrast areas, consisting of pronounced muscles standing out, and others that visually blend smoothly. Taking your sculpture to the next level will require extra attention to establishing appropriate transitions between shapes:

Figure 8.62 – A mix of high and low contrast

Beginners often create too many harsh transitions between muscles, forgetting to dial back the intensity of their shapes after the initial anatomy blockout. Reference images will be essential here to see how skin, fat, and muscles give certain appearances to different areas of the body.

Let's take a look at the following figure:

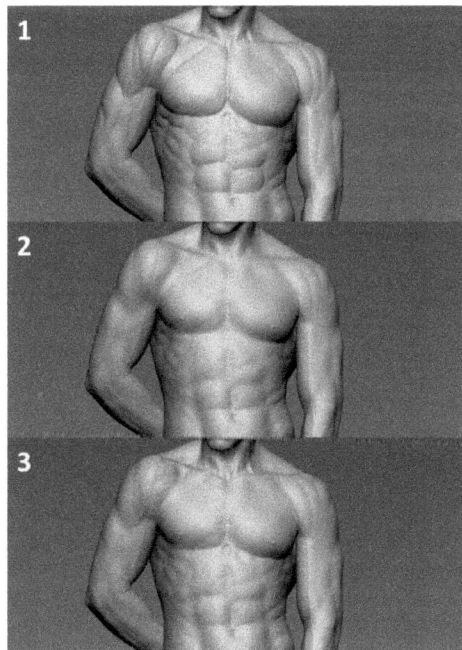

Figure 8.63 – Dialing in the transition of shapes/muscles

At the top, the muscles are separated too clearly and harshly.

When softening the transitions between muscles, try to avoid using the **Smooth** brush since it tends to soften and distort the shape of the muscles. This leads to a blurry, muddy look, like what's shown in the middle image.

Finding a balance between areas that blend and areas that are separated clearly will improve the quality of your sculpture, as illustrated in the bottom image. Try to look for reference pictures to inform your sculpting, instead of doing it randomly.

Once you have addressed the transitions and are happy with how your sculpture looks overall, you can apply the finishing touches by adding veins and skin surface texture to your model.

Adding veins

Although veins are lower on the priority list of your sculpture, they can add that extra bit of realism and complexity to the model. Especially when you're sculpting athletic and muscular characters with a low body fat percentage, you should consider adding veins.

Reference images are important for learning where veins occur, what they look like, and how they are shaped. Some parts of the body are more likely to have visible veins – such as the forearms, hands, and feet – but they can appear almost anywhere, depending on the body type and age.

While it makes sense to add an appropriate and realistic number of veins, you can certainly consider adding them from a design and aesthetic perspective. This means adding them where they would look good. This could be across muscles to introduce a different direction of lines, as shown in the following figure, where horizontal veins cross the vertical muscle fibers of the shoulder. Alternatively, they can create a focal point when added to a low-contrast area, such as the hips:

Figure 8.64 – Veins to add visual (guiding) lines (1) and focal points (2)

Adding skin detail

The final and optional step for sculpting the body is to give it a skin texture. You should look for a tileable skin displacement texture, which you can find on 3D asset sites such as the Artstation marketplace (`https://www.artstation.com/marketplace/game-dev`).

To apply the texture to the whole model, you can use the **NoiseMaker** tool. Before using it, make sure your model has UVs. You can check this by navigating to **Tool | UV Map**. If **Morph UV** is grayed out, your model has no UVs. In that case, you can simply follow the UV creation process, as described in *Chapter 7*.

Additionally, you need to make sure your model has enough resolution to hold the kind of skin detail you want. This means switching to the highest subdivision level, and potentially adding more subdivision levels. If you've been following along and using ZBrush's Nickz model, you will need at least **6 subD** levels, which results in a mesh with 13.5 million polys.

Once you've ensured your model has UVs and enough resolution, go to **Tool | Surface** and click **Noise**:

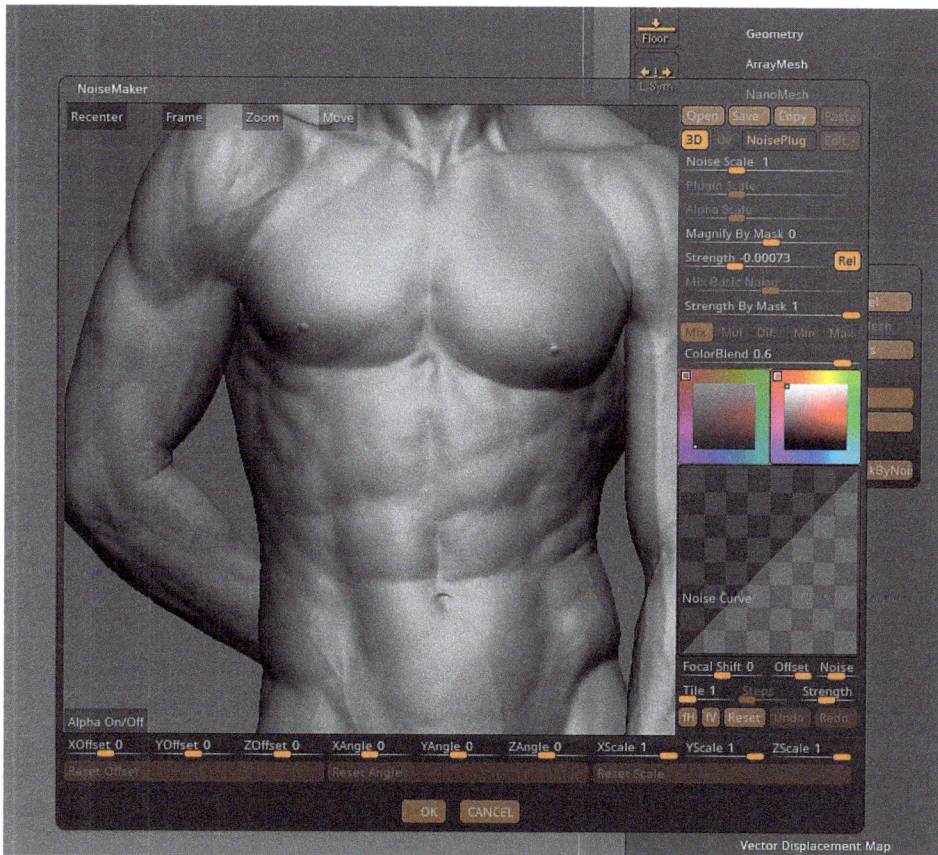

Figure 8.65 – NoiseMaker

The **NoiseMaker** window will open. Here, you will see a preview of the model with surface noise applied to it, as well as several options to adjust the effect.

The first thing you want to do is switch from **3D** to **UV** mode. You can see the buttons on the second line on the right-hand side of the window. This will make the plugin apply the texture based on the UVs, instead of 3D space, which would create stretched results when using a tileable texture.

Then, to apply the skin texture, do the following:

1. Click on **Alpha On/Off** to load your texture map.
2. Adjust the **Strength** value until you see an effect on the preview.
3. Adjust the **Alpha Scale** value to determine how many times the texture will be repeated. Try to find a value that has a frequency of skin detail that looks natural and good to you.
4. You can adjust the **Noise Curve** value and see what effect that has on the appearance of the skin detail.
5. Once you are happy with the preview, click **OK** to apply the skin detail.

You may find that you need to go through some experimentation with **Alpha Scale** and **Strength** until you achieve the desired result:

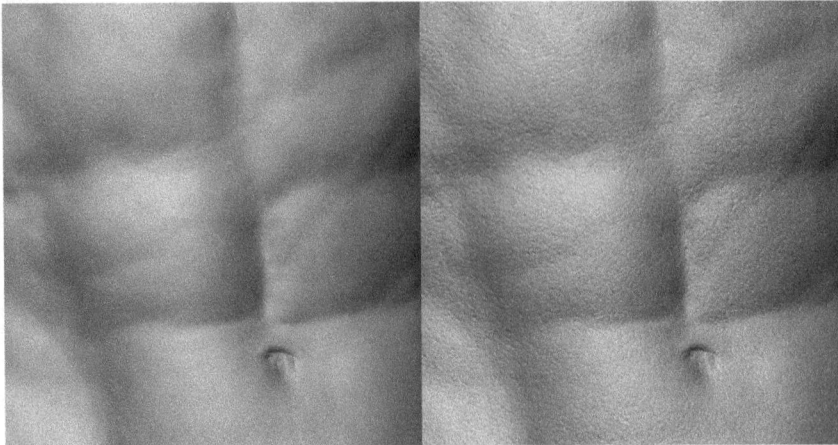

Figure 8.66 – Before (left) and after (right) adding skin detail

With the skin details added, you have finished sculpting the body. You may continue working on each of the steps, from finding additional reference images to refining individual muscle groups, until you are satisfied with the result.

Summary

Well done on finishing this chapter – it was probably the most information-dense and challenging one so far!

This chapter began with collecting various anatomy reference materials, including athlete pictures, the 8-head figure chart, and an écorché model to help create a more realistic sculpture. You learned about the importance of the skeleton, which you used alongside ZBrushs Nickz base mesh to establish an anatomically accurate starting point.

After that, you explored and sculpted many of the large, superficial muscles of the human body, and learned how they affect the appearance and shape of different body parts in the process. Next, you posed the body in a new way, making it more dynamic and engaging. This new pose, along with additional sculpting, added much-needed asymmetry to the model, increasing its realism and design quality.

Creating a well-sculpted human is one of the most challenging tasks for 3D modelers and ZBrush artists, so you can be pleased that you've finished the hardest part of this project!

In the next chapter, you will explore different modeling and detailing techniques that will help you create the gladiator costume. After that, all that's left to do is to prepare the model for 3D printing by merging, cutting, and keying it.

9

Creating Costumes, Armor, and Accessories with Classic Modeling Techniques

ZBrush offers many tools for creating new geometry, each suited for different objects. The more tools you know, the more efficient you will become as a modeler and sculptor. In this chapter, you will learn simple and effective workflows that allow you to create almost any object. More specifically, you will create a costume and accessories for the gladiator that you started creating in the previous chapter.

The first section focuses on primitive shapes and the use of **ZModeler**, ZBrush's polymodeling tool, perfect for creating precise, clean meshes. It is especially useful for creating the foundation of clothing, armor, and accessories.

The following section explores the creation process of more complex pieces, such as the gladiator's shoulder armor and helmet, for which a combination of kitbashing and DynaMesh will be used.

Finally, the last section will guide you through the detailing process. Here, you will be adding fine surface details to your costume, as well as adding damage and imperfections.

This chapter will cover the following topics:

- Creating simple costume pieces using the ZModeler brush and primitive shapes
- Creating complex costume pieces using DynaMesh, IMM brushes, and kitbashing
- Adding detail to the costume pieces

Technical requirements

For the best experience, it is recommended that you have a strong PC that meets the minimum requirements described in the first chapter's *Technical requirements* section. However, you can work on this chapter with just a mouse, a functional PC setup, and a ZBrush license.

If you completed *Chapter 8* and sculpted the body of a gladiator, the lessons in this chapter will be easier to follow. However, as always, the workflows and tools listed here can be applied to any subject.

Creating simple costume pieces using the ZModeler brush and primitive shapes

The **ZModeler** brush is one of Zbrush's best tools for creating and modifying simple shapes and creating clean meshes. Here, you can see the many parts of the costume that were created with the help of **ZModeler**:

Figure 9.1 – Parts created with the ZModeler brush

In this section, you'll learn about the creation process behind each of the pieces, with a focus on how the **ZModeler** brush is used.

Skirt and hip cloth

Clothes and fabrics are an important element in most character designs. Software such as Marvelous Designer can simulate fabric, which is great for creating realistic cloth wrinkles and folds, but ZBrush offers more control and precision, making it a useful tool for many clothing pieces.

Skirt

When you are thinking about how to make a certain object, it can help to break it down into simple shapes. In this case, the skirt is shaped roughly like a widened cylinder, so you may choose a simple cylinder shape as your starting point:

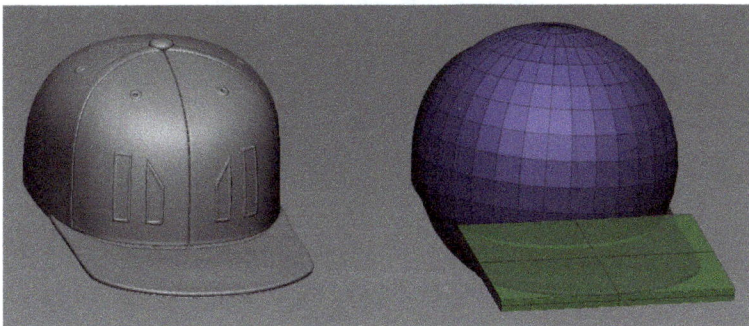

Figure 9.2 – Breaking down an object into simple shapes

Let's get started:

1. To add a cylinder to your active ZBrush tool, under **Tool | Subtool**, select **Append** and pick **Cylinder3D**:

Figure 9.3 – Appending a simple cylinder shape

2. Use the **SelectRect** brush to select the middle part of the cylinder, hiding the top and bottom parts. It should look like this:

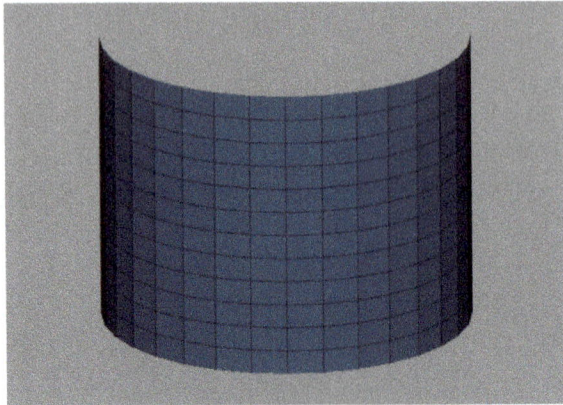

Figure 9.4 – Cylinder for the skirt

3. Go to **Tool | Geometry | Modify Topology** and select **Del Hidden** to delete the hidden part. At this point, you can use your **Move** brush or your preferred tool to shape the cylinder to match the shape of the gladiator's hips and legs more closely:

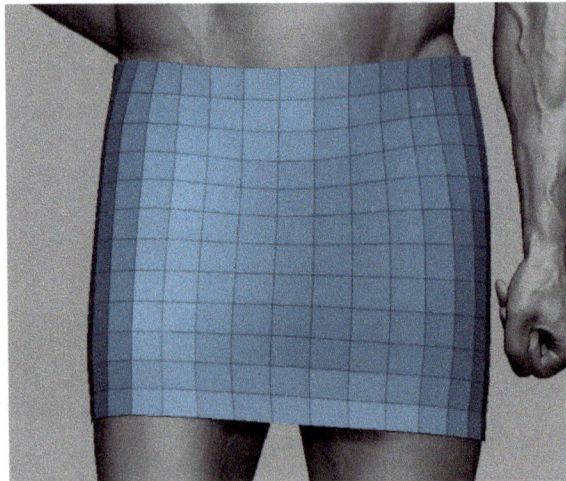

Figure 9.5 – Matching the cylinder to the body

4. Next, you need to equip the **ZModeler** brush. Hover over a polygon with your mouse, then press the spacebar. A menu will appear in which you can select the desired operation. We want to give the skirt some thickness, so we chose **Extrude**; in the second submenu titled **TARGET**, select **Polygroup All**:

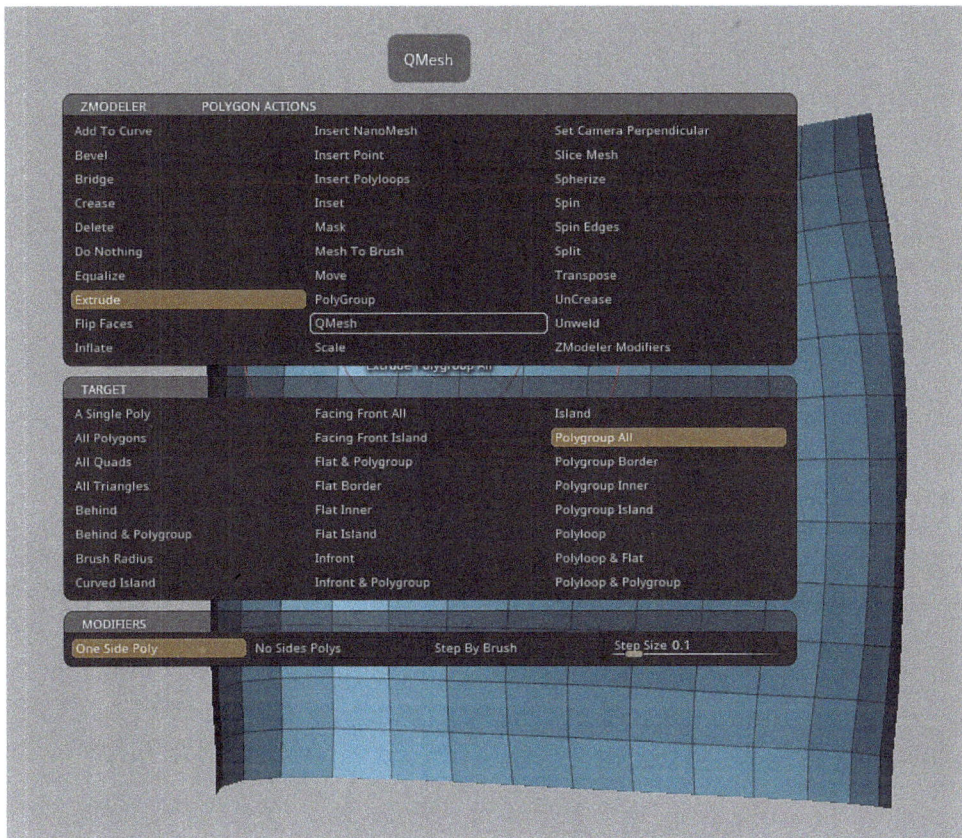

Figure 9.6 – ZModeler window

5. To ensure that the mesh has only one PolyGroup, assign a new PolyGroup by pressing *Ctrl + W*.

6. Left-click and *drag* to perform an **Extrude** operation and give the skirt a sufficient thickness so that there are no issues when 3D printing – 1.5 mm is often a good thickness:

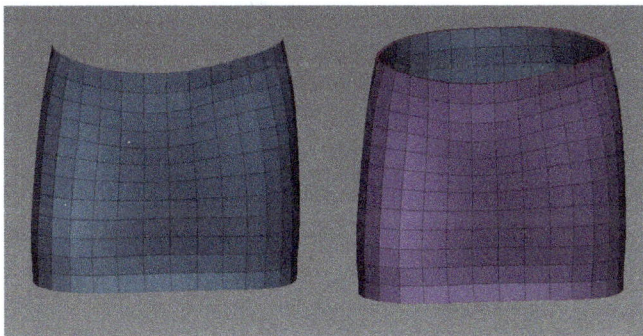

Figure 9.7 – Before (left) and after (right) performing an Extrude operation

7. Next, you'll want to add a seam along the top of the skirt. With the **ZModeler** brush active, hover over an edge. Then, in the upper window, select **Insert | Single EdgeLoop**.

8. Now, left-click on the edge where you want the seam to be and insert an edgeloop:

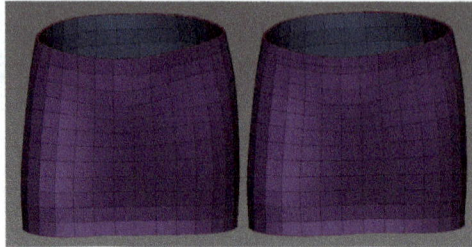

Figure 9.8 – Inserting an edgeloop

9. Hover over one of the polys on the highest polyloop, press the spacebar to select **Extrude** and **Polygroup All**, and apply the extrusion to the upper polyloop:

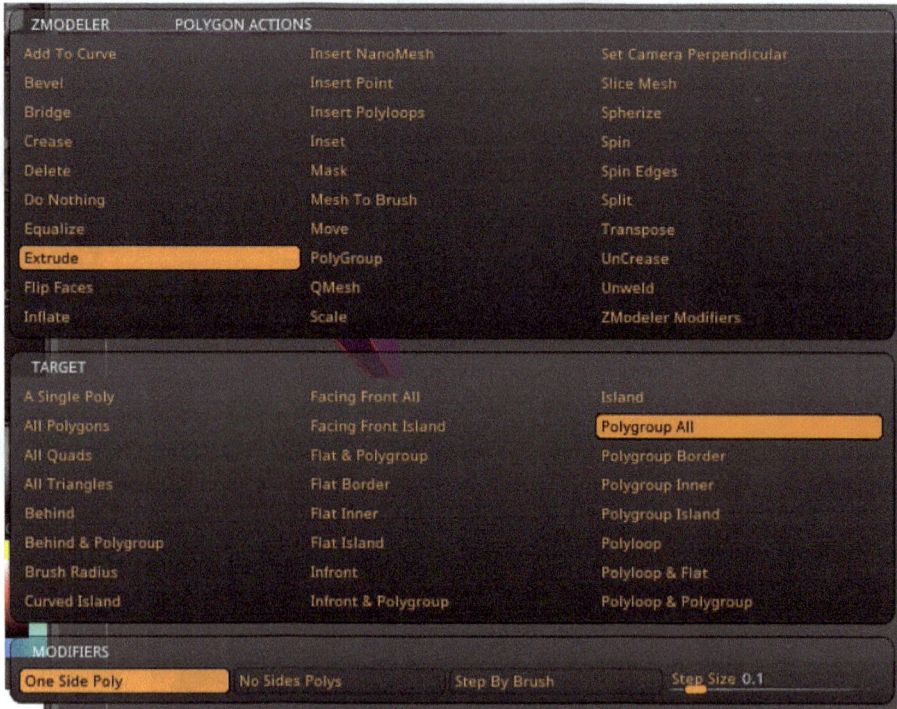

Figure 9.9 – Extrude polyloop operation

10. Finally, you can add some additional edgeloops on the seam and some just below it. This will ensure that the shape of the seam stays intact when you add subdivision levels to the mesh. Here is the result after polymodeling with the **ZModeler** brush:

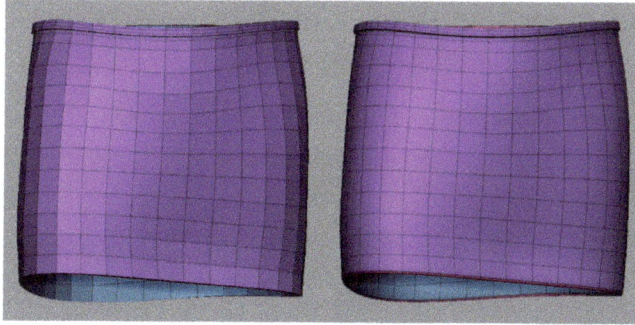

Figure 9.10 – Skirt model topology with Dynamic Subdiv mode
(right) and without Dynamic Subdiv mode (left)

> **Important note**
>
> You can press *D* to enable **Dynamic Subdiv** mode, which will make the model look like it was subdivided twice, without actually being subdivided or having an increased polycount. This is useful for having a preview of what your mesh looks like subdivided. This can also be useful for presentation and visualization purposes since the model appears smoother and does not show visible edges as much. By pressing *Shift + D*, you can disable **Dynamic Subdiv** mode again.

Hip cloth

Now, let's take a look at the piece of fabric that will be layered above the skirt. The fabric's wrinkling makes it a bit more tedious to model, but it is important to create an actual overlap instead of sculpting just superficially. This will ensure that it has proper depth and shadows, resulting in a more realistic accessory:

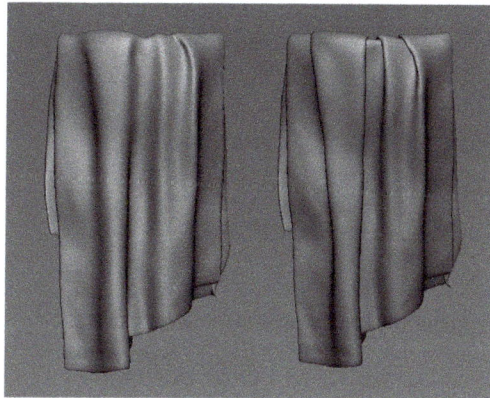

Figure 9.11 – The difference in realism and quality: sculpting superficial
folds (left) versus real overlap and depth (right)

The overlapping and wrinkled effect can be created with simple masking and transforming with the Gizmo, but it might require a little patience. Let's get started:

1.   Go to **Tool** | **Append** | **Plane3D**:

Figure 9.12 – Appending a basic plane model

2.   Mask the area at one of the lateral endpoints where you want to start creating bunched-up fabric.

3.   Hold *Ctrl* and left-click on the masked area to blur the mask:

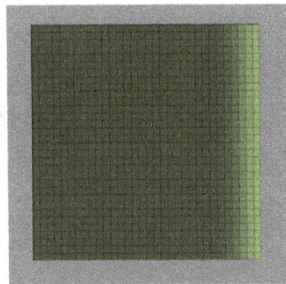

Figure 9.13 – Using (un)masked areas with a blur effect to deform the plane

4. Press *W* to access the Gizmo and position the Gizmo on the border between the masked and unmasked areas. Rotate the unmasked part to start building the overlapping structure:

Figure 9.14 – Folding the plane

5. Continue masking parts of the model and rotating the mesh until you have larger folds.

6. If you need more resolution, you can add edgeloops with the **ZModeler** brush using an **Insert Single EdgeLoop** operation.

7. Once the vertical wrinkling is established, you can use masking to fold over the plane along the horizontal axis so that it can be integrated into the skirt, like this:

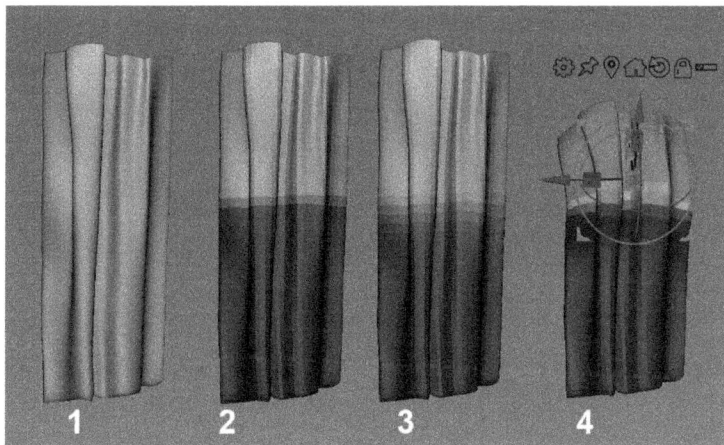

Figure 9.15 – Start point (1), applying a mask (2), blurring the mask
(3), using the Gizmo rotation to "fold over" the cloth

This is the start and end point of the process:

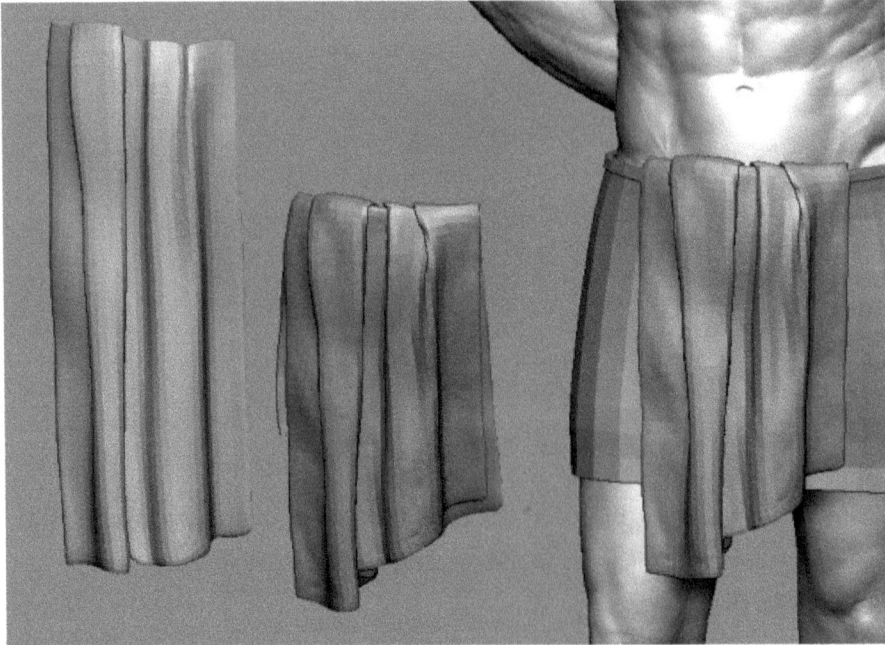

Figure 9.16 – Before folding (left), after folding (middle), and integrated into the skirt (right)

8.   After this, you can extrude the plane to give it thickness using an **Extrude** operation. Make sure to set the appropriate **ZModeler** target – if you have one PolyGroup assigned to the whole mesh, the target should be **Polygroup All**.

This concludes the polymodeling part for the skirt, but later in this chapter, you will continue to refine it and add more subdivision levels and detail. Next, you will create some simple straps.

Straps

ZBrush has a brush that's perfect for quickly adding straps to your models: the **CurveStrapSnap** brush. You can draw on any mesh without subdivision levels and create a strap along your brush stroke. To create leather straps on the model, do the following:

1.   With the **body** subtool selected, go to **Tool | Duplicate**. Alternatively, you may choose any subtool or append a new one since you will delete everything but the newly created straps anyway.

2.   Open the **Geometry** submenu and hit **Del Higher** and **Del Lower** to delete the subdivision levels, if your subtool has any.

3. Equip the **CurveStrapSnap** brush and draw a strap on the body, creating something like this:

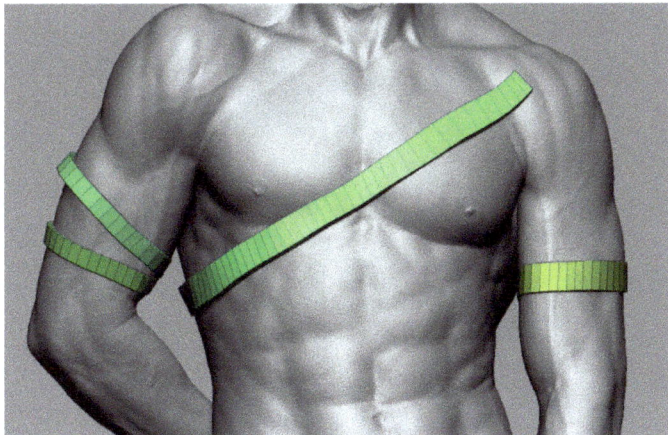

Figure 9.17 – Creating straps with the CurveStrapSnap brush

4. Use your **Selection** tool to hide everything but the straps. Then, go to **Tool | Geometry | Modify Topology** and select **Delete Hidden**, since you want to have just the straps in the subtool.

5. Now, add some edgeloops with the **ZModeler** brush so that the model will keep its shape when subdivided:

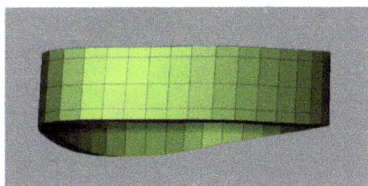

Figure 9.18 – Support edgeloops on a strap

You can add more detail to the straps if you like. In the last section of the chapter, you will learn how to refine straps and other leather elements by applying surface texture and sculpting detail. However, next, you will create armor plates on the arm.

Arm armor

For the arm armor plates, you have several options to start polymodeling. Here is one of them:

1. Go to **Tool | Subtool | Append | Cube3D**.

2. Then, go to **Tool | Polygroups** and select **GroupByNormals**. This will assign a PolyGroup to each side of the cube:

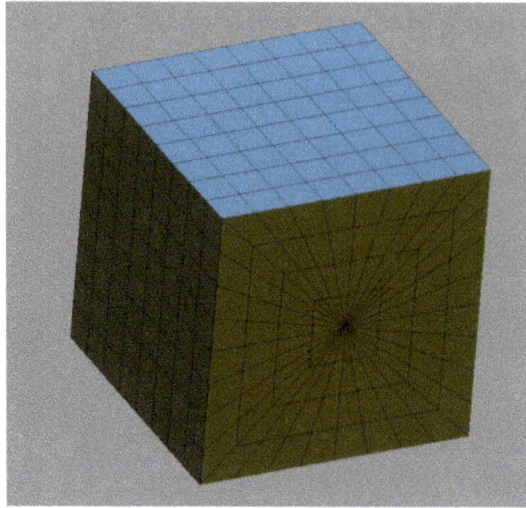

Figure 9.19 – Cube3D shape with PolyGroups for each side

3. Isolate one of the sides of the cube that has even quad topology. Then, go to **Tool | Geometry | Modify Topology | Del Hidden** to delete the hidden part:

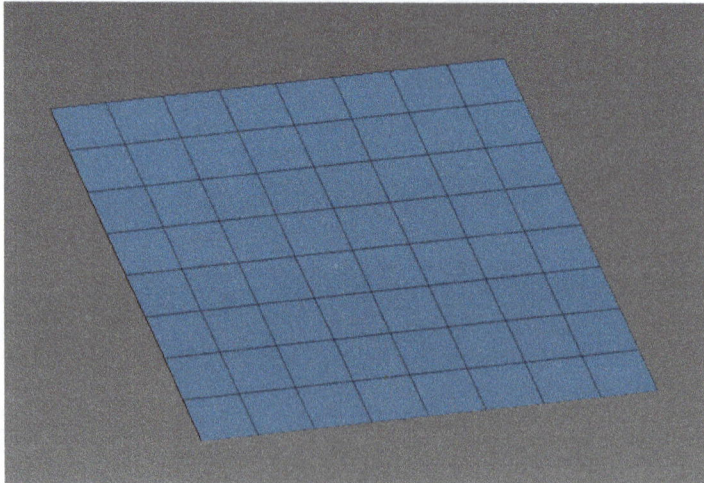

Figure 9.20 – Isolating a PolyGroup of the cube

4. At this point, you can adjust the design by tapering some of the edges, using the **Move** brush as follows:

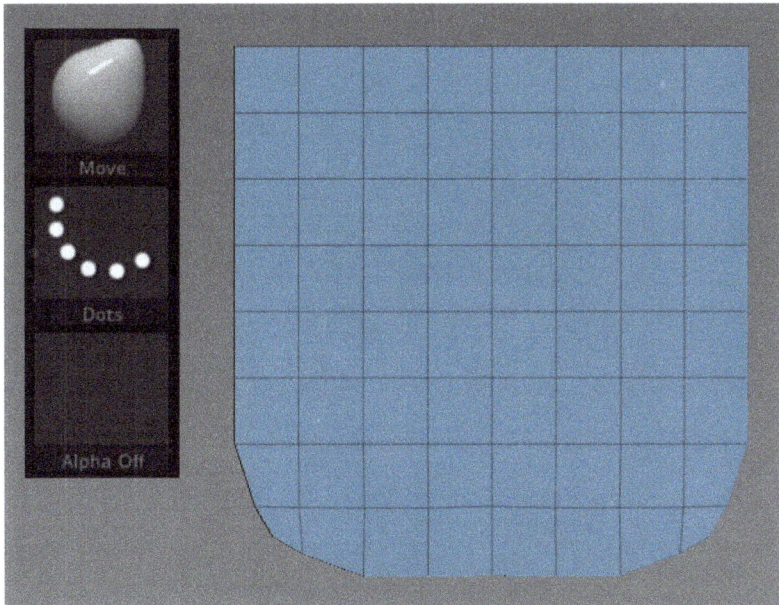

Figure 9.21 – Using the Move brush to taper the lower edges of the plane

5. Press *W* to access the Gizmo and click on the gear icon to access the **Gizmo Deformation** menu. From here, select **Bend Arc**:

Figure 9.22 – Bend Arc modifier in the Gizmo Deformation menu

6. Left-click and drag the green arrows that appear in order to bend the model in that direction. Give the armor plate a slight curvature so that it fits the arm a bit better:

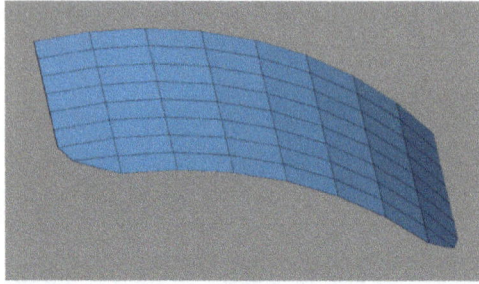

Figure 9.23 – The plane after adding a curvature

7. Use the **ZModeler** brush with an **Extrude + Polygroup all** operation to give the model some thickness:

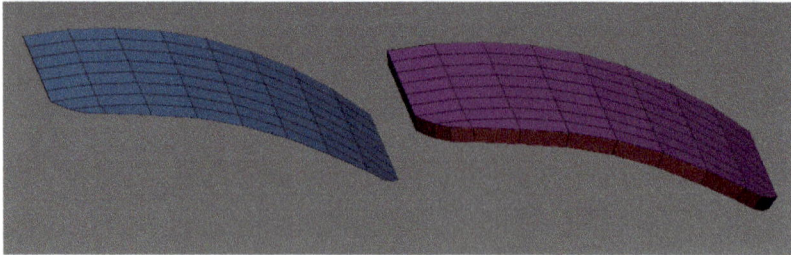

Figure 9.24 – Extruding the plane to give it thickness

8. Go to **Tool | Polygroups** and select **GroupByNormals**.

9. Use the **ZModeler** brush with **Inset**, **Polygroup all**, and **Standard** enabled, and apply it on the large outside plane of the model. This will create a polyloop toward the edge:

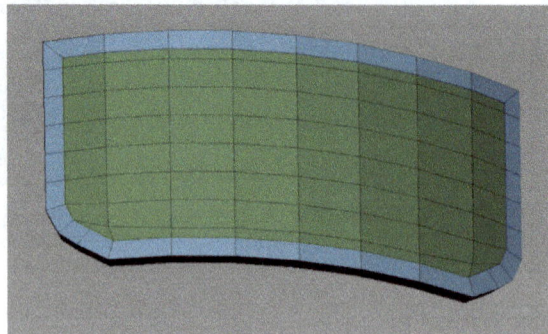

Figure 9.25 – Using an Inset operation to create a polyloop at the edge

10. Use the **ZModeler** brush with **Extrude** and **Polyloop** enabled and apply it to the outer polyloop of the plane to create a raised edge on the armor.

11. Finally, add supporting edgeloops with **ZModeler**'s **Insert EdgeLoop** option:

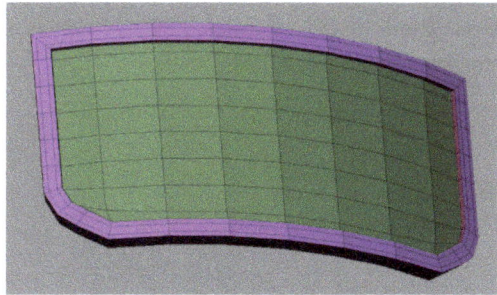

Figure 9.26 – Arm armor plate with raised borders

Metal ring

Next, you will add a simple metal ring that will contrast the leather parts nicely:

1. Go to **Tool | Subtool | Append | Ring3D**:

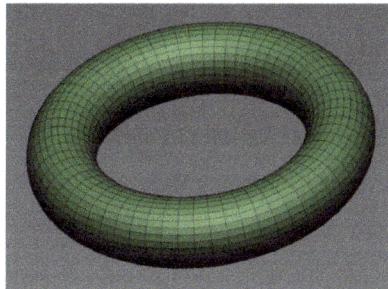

Figure 9.27 – Adding a Ring3D mesh

2. Use the **ZModeler** brush with **Insert + Single EdgeLoop**, then *Alt* and left-click on edges to delete them. Delete several edges on the ring so that there are flat planes:

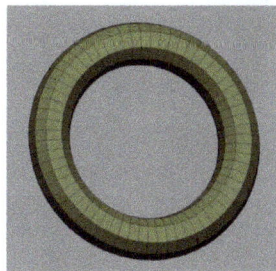

Figure 9.28 – Ring shape with deleted edgeloops

Shoes

In this section, we'll create some shoes, using a duplicate of the body as a starting point, and the **ZModeler** brush to do the remaining work:

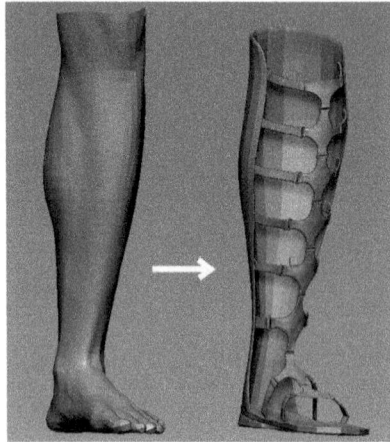

Figure 9.29 – Using the body as a starting point to create fitting shoes

Let's get started:

1. Duplicate the body by going to **Tool | Subtool | Duplicate**.

2. Go to the lowest subdivision level by pressing *Shift + D*.

3. Open the **Geometry** submenu and hit **Del Higher** and **Del Lower** to delete the subdivision levels.

4. Isolate an area of the calves and heel, then go to **Tool | Geometry | Modify Topology** and select **Del Hidden**:

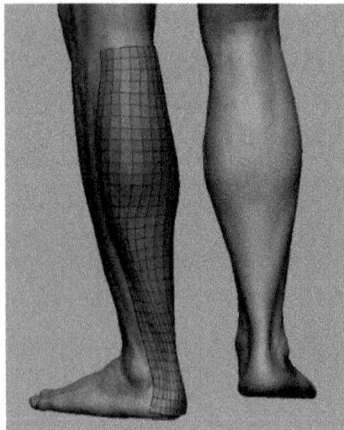

Figure 9.30 – Using the body to create clothing parts

5. As an optional step, if you need to improve the topology of this mesh, go to **Tool | Geometry** and use the **ZRemesher** tool to create a new topology (**ZRemesher** is covered in *Chapter 5* if you are unsure about how to use it):

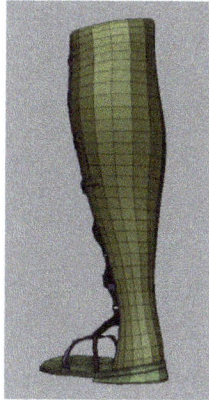

Figure 9.31 – Shoe after retopology

6. Use the **ZModeler** brush with an **Extrude + Polygroup all** operation to give the model thickness.

7. Proceed to create other pieces for the shoe, using basic shapes, masking, selection, and **ZRemesher**, as well as tools of your choice. The result may look similar to this:

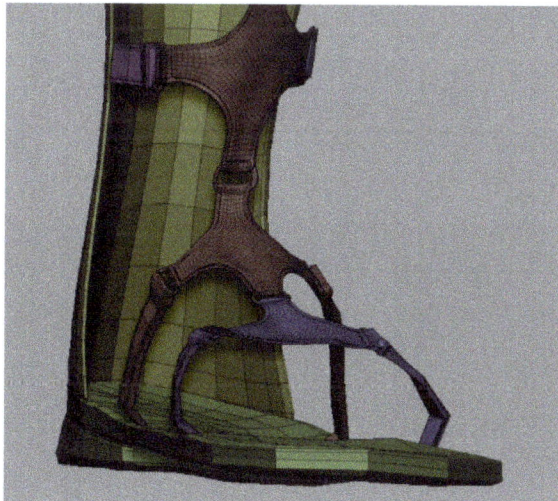

> **Important note**
>
> To learn more about ZBrush tools and techniques for creating a large variety of different objects, the best resource available is the *YouTube* channel of Michael Pavlovich. The large amount of straightforward and well-explained ZBrush tutorials makes it the perfect resource for anyone looking to get familiar with and get the most out of ZBrush. You can find the channel here: `https://www.youtube.com/@MichaelPavlovich/videos`.

With the shoe completed, next, you will learn how to model a simple spear with **ZModeler**.

Spear

In this part, you will create a typical Roman spear. First, you will create the lower part of the spear, which is a cylindrical shape that tapers toward the end. Here is how you create it:

1. Go to **Tool | Subtool | Append | Cylinder3D**.

2. Use the Gizmo to scale it in one direction, creating the lower half of the spear:

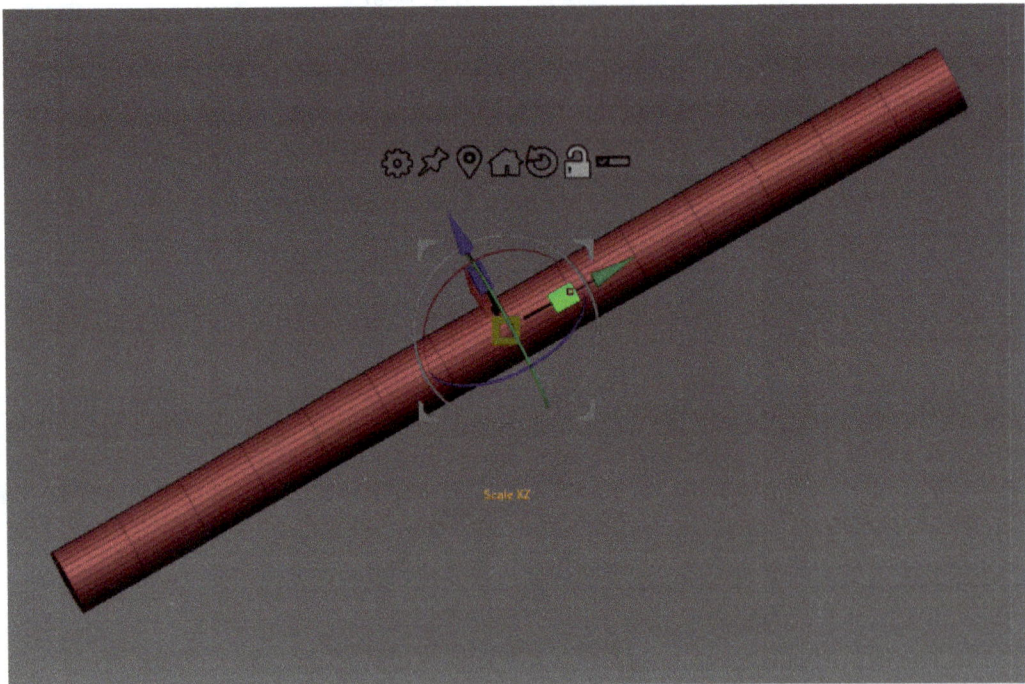

Figure 9.33 – Scaling a cylinder

3. Go to **Tool | Geometry** and use the **ZRemesher** tool to retopologize the spear so that the surface consists of more evenly sized quads again:

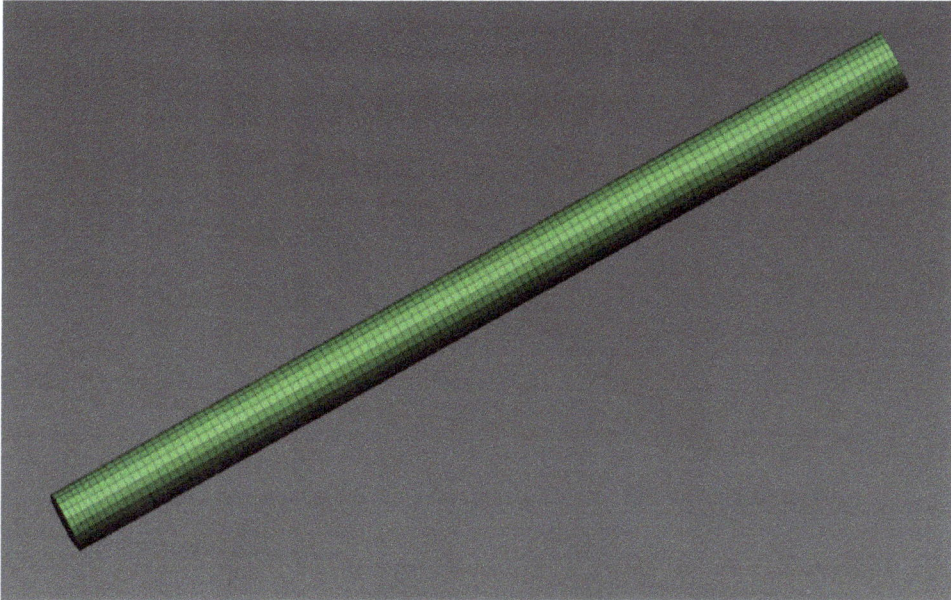

Figure 9.34 – Creating evenly sized quads through ZRemesher

4. Use your **MaskPen** brush and *Ctrl* + left-click + drag a mask on the end part of the spear.

5. Invert the mask by *Ctrl* + left-clicking on a space on the canvas.

6. Then, *Ctrl* + left-click on the mask a few times to blur it:

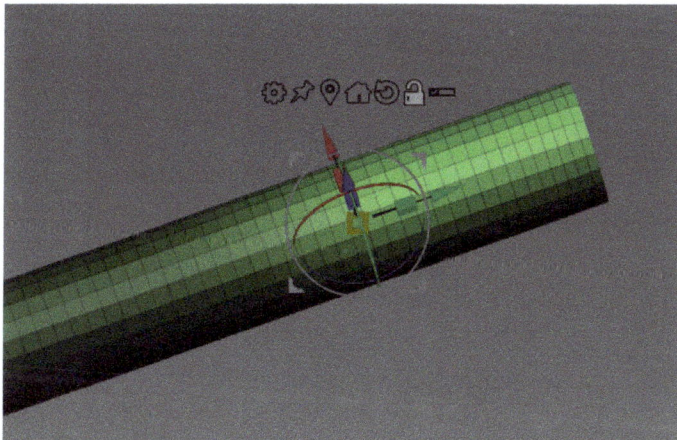

Figure 9.35 – Using masking to deform the end part of the spear

7. Next, hit *W* to access the Gizmo, then click on the gear icon to open the **Gizmo Deformation** menu. From here, select **Taper**.

8. Left-click and *drag* the arrows that appear on the end part of the cylinder to taper it. This is what the end result should look like:

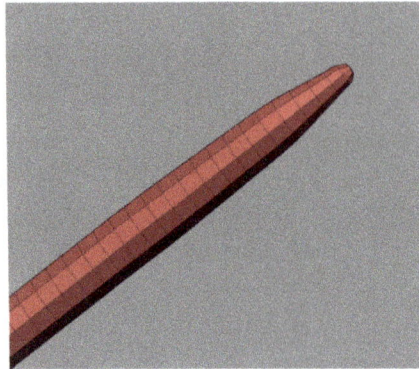

Figure 9.36 – The end part of the spear after using the Taper modifier

Next, you can create the tip of the spear:

1. Go to **Tool | Subtool | Append | Cylinder3D**.

2. Use the Gizmo to scale it in one direction, creating the upper half of the spear. Then, flatten the mesh slightly by scaling it with the Gizmo; that way, it is not rounded but rather flat.

3. Go to **Tool | Geometry** and use the **ZRemesher** tool to retopologize the spear so that the surface consists of more evenly sized quads again:

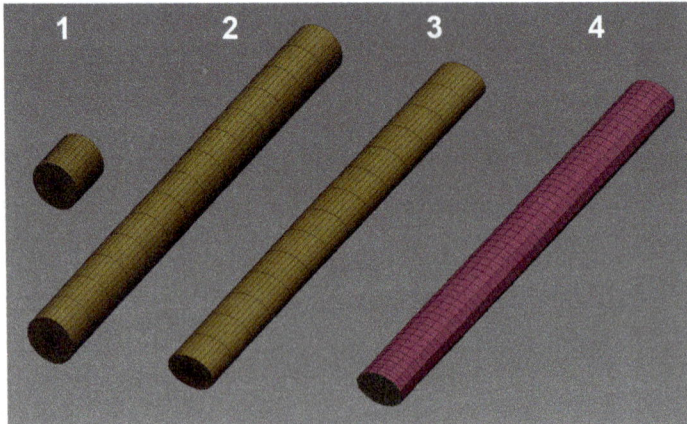

Figure 9.37 – Cylinder shape (1), scaling the cylinder (2), flattening
the cylinder (3), using ZRemesh on the cylinder (4)

4. Manually pull the vertices, using the **Move** brush, to create a spear tip shape:

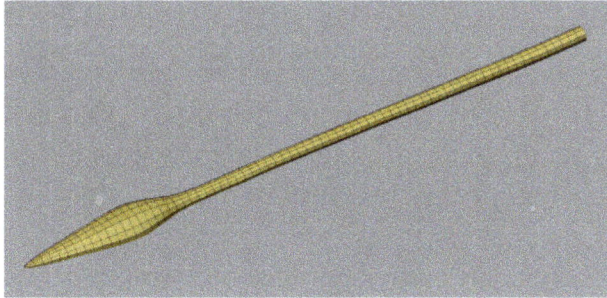

Figure 9.38 – Spear front part

Finally, let's build the middle part of the Roman spear, which is basically just a tapered cube, along with any detailing you might want to add:

1. Go to **Tool | Subtool | Append | Cube3D**.

2. Go to **Tool | Initialize** and set the dimension sliders to **4**, then press **QCube**. This will transform the **Cube3D** shape with its irregular topology into a cube with a perfect grid of quads on each side. This type of topology will work better when deforming and refining the object:

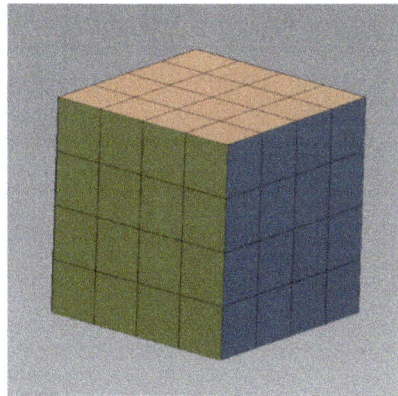

Figure 9.39 – Cube from the Initialize menu

3. Press *W* to access the Gizmo and click on the gear icon to open the **Gizmo Deformation** menu. From here, select **Taper**.

4. Left-click and *drag* one of the arrows that appears to taper the cube. Additionally, you might need to adjust the **Exponent** value through the white arrow:

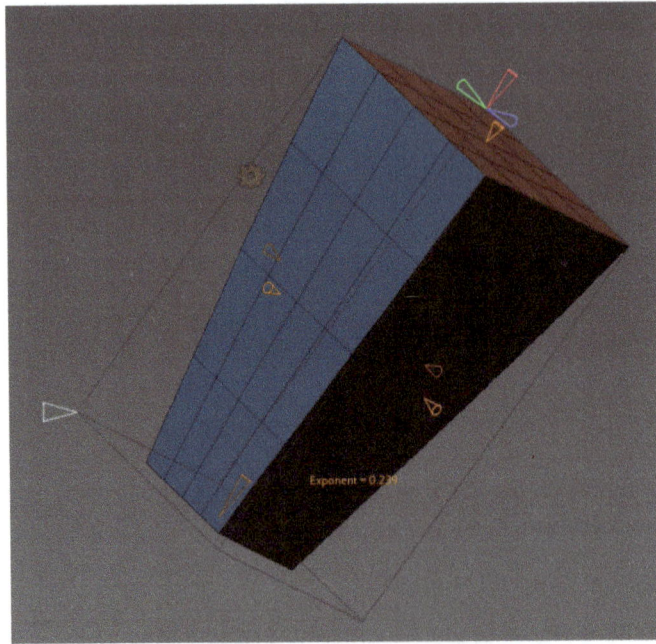

Figure 9.40 – Using the Taper function

5.   At this point, you may also decide to further refine this part of the spear, adding metal details or ornaments, based on your concept and reference:

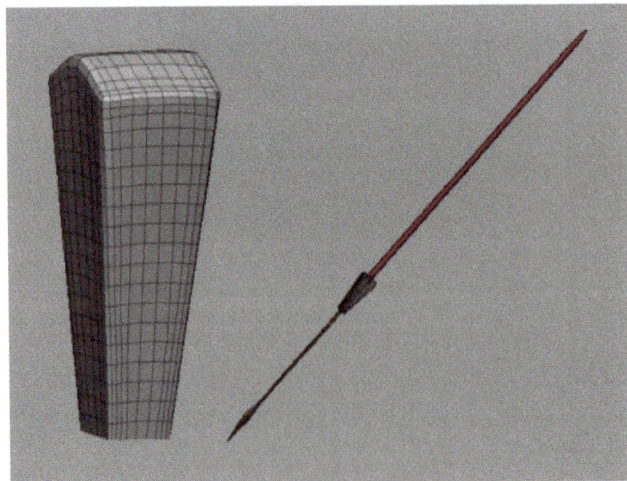

Figure 9.41 – Spear middle part and complete spear

Once you've created the spear, it's time to turn your attention to creating shoulder armor.

Shoulder armor

Now, you will create the base part of the shoulder armor, which will be refined later in this chapter into one of the most intricate parts of the gladiator. For now, let's get started:

1. Go to **Tool | Subtool | Append | Sphere3D**.

2. Press *X* to enable **Symmetry** mode and sculpt the sphere into a shoulder armor shape:

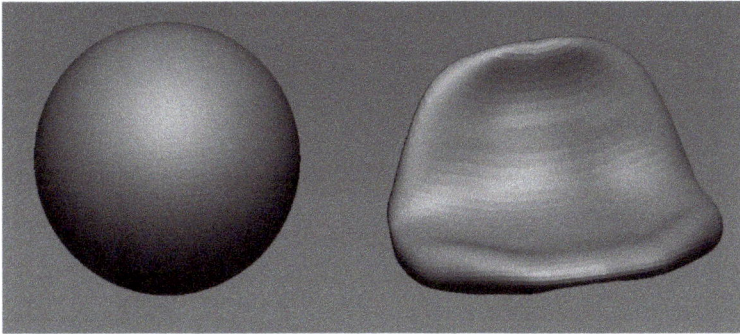

Figure 9.42 – Turning a Sphere3D mesh into the rough shape of shoulder armor

3. Isolate the main, outward-facing side of this mesh, and delete the rest by going to **Tool | Geometry | Modify Topology | Del Hidden**:

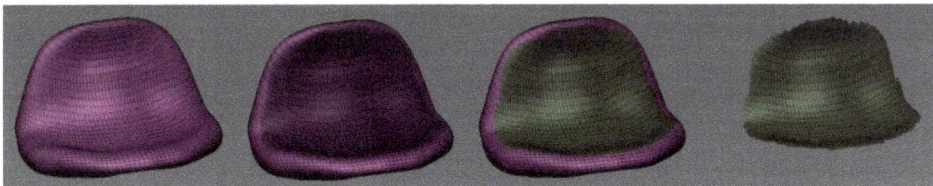

Figure 9.43 – Using masking to isolate a part of the mesh

4. Next, go to **Tool | Deformation** and apply a **Polish** deformation to smooth out the border of the mesh:

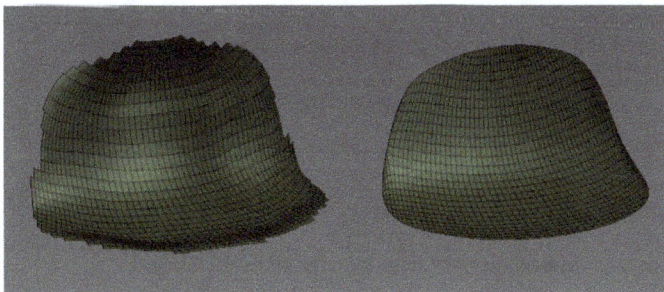

Figure 9.44 – Before (left) and after (right) applying a Polish deformation

5. Then, go to **Tool | Geometry** and use the **ZRemesher** tool to retopologize the mesh so that the surface consists of more evenly sized quads:

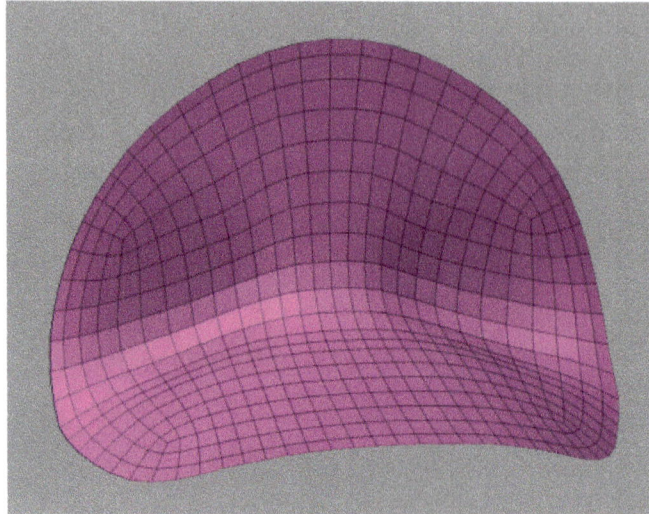

Figure 9.45 – Creating the main piece of the shoulder armor

6. Use the **ZModeler** brush, hover over an edge, and press the spacebar to pick an **Extrude + Mesh Border** operation. Extrude the outer edgeloop until you have the thickness you want for the raised edge of this piece:

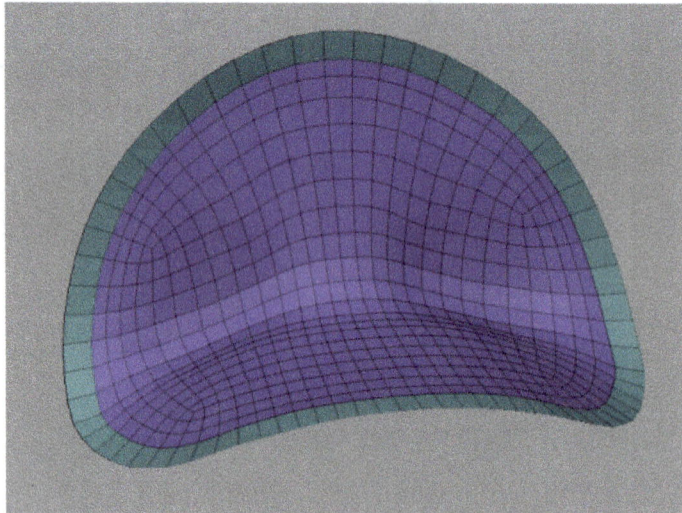

Figure 9.46 – Mesh with extruded mesh borders

7. Next, use the **ZModeler** brush to extrude the surface and give it thickness.

8. Then, use the **ZModeler** brush with **Extrude** and **Polyloop** enabled, and apply it to the outer edge of the plane you inset before to create a raised edge on the armor.

9. Add supporting edgeloops with **ZModeler**'s **Insert EdgeLoop** option:

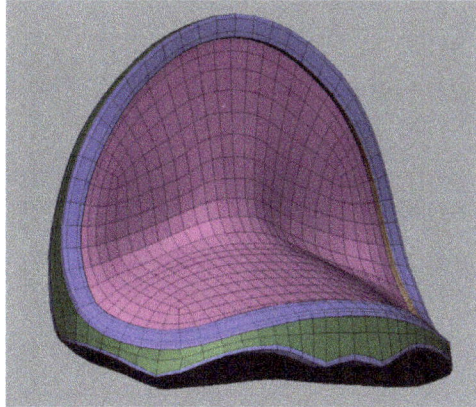

Figure 9.47 – Shoulder armor with raised borders

10. With your **ZModeler** brush, hover over a vertex and press the spacebar to open the **ZModeler** menu. Then, use **Split + Point** and apply that operation to areas where you would like to add bolts.

11. You can insert edgeloops and extrude faces to create bolts or similar parts:

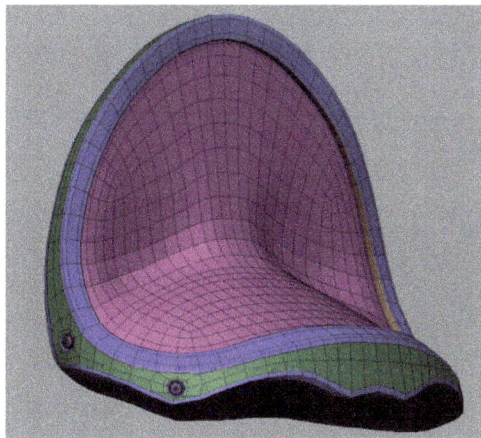

Figure 9.48 – Finished shoulder armor base model

Later in this chapter, you will use kitbashing on the shoulder armor to create a sophisticated, detailed model. Next, you will create the eye protection part of the helmet.

Helmet

The larger portion of your helmet will be built later in this chapter, but for creating the eye protection piece of the helmet, **ZModeler** is the best tool:

1. Go to **Tool | Subtool | Append | Cylinder3D**.

2. Use the **ZModeler** brush to add an edgeloop on the outside of one of the end parts. This will create a polyloop that can be isolated and used for the frame of the eye protection piece:

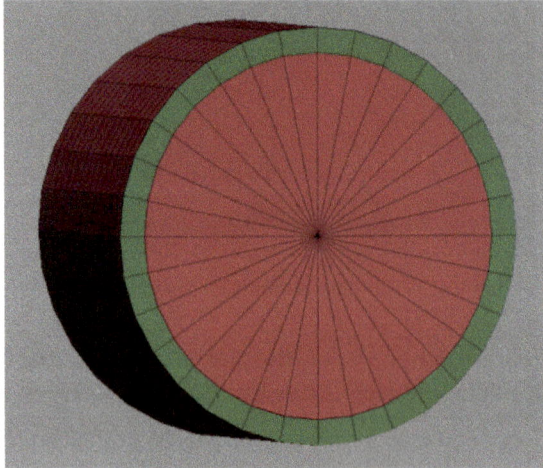

Figure 9.49 – Using the cylinder as a starting point

3. Isolate the outer polyloop and delete the rest by going to **Tool | Geometry | Modify Topology** and selecting **Del Hidden**.

4. Then, use the **ZModeler** brush to extrude the polyloop:

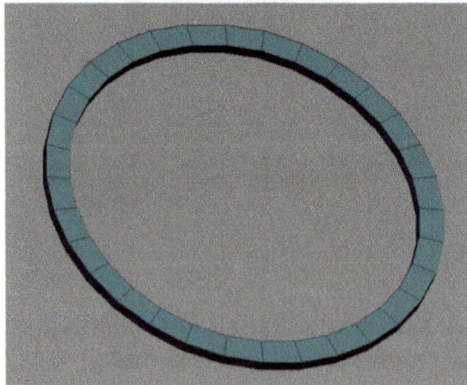

Figure 9.50 – Ring created from a basic cylinder shape

5. Go to the **Tool** palette and duplicate the ring for later use. There, you will be able to reuse it on the inside.

6. Next, you will want to create space to add bolts and smaller rings on the inside, to create the grid structure of this part.

   With the **ZModeler** brush, *Alt* and left-click on polys on the inside and outside, where you want to add this space. This creates a special, white PolyGroup for these polygons, marking them as an active PolyGroup.

   Now, the next **ZModeler** operation will be applied to all these white parts, similar to other PolyGroup operations:

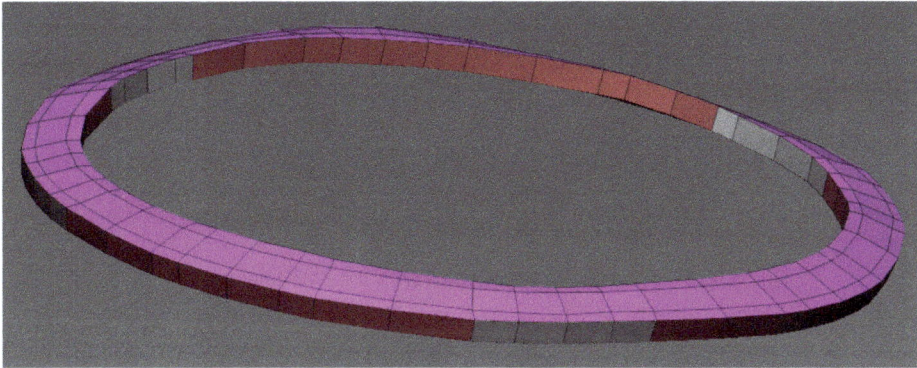

Figure 9.51 – Applying an active PolyGroup to refine the piece

7. Extrude all areas around the edge of the ring where you want to create extrusions, like the ones marked in the following screenshot:

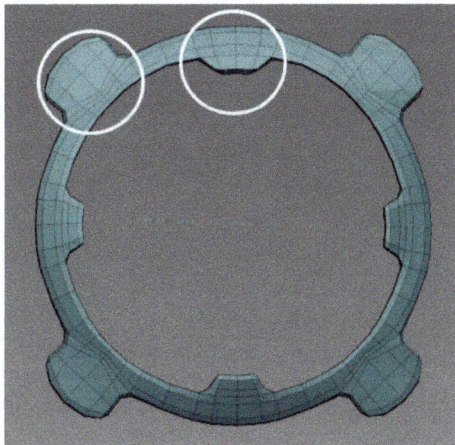

Figure 9.52 – Extruding parts of the ring

8. With your **ZModeler** brush, hover over a vertex and press the spacebar to open the **ZModeler** menu. Then, use **Split + Point** and apply that operation to areas where you would like to add bolts. You can insert edgeloops and extrude faces to create any kind of bolt or structure:

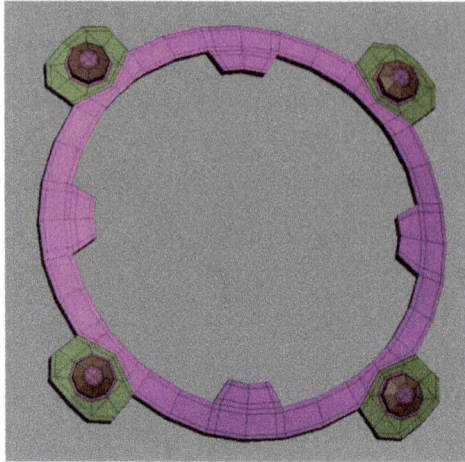

Figure 9.53 – Refining the helmet part with the ZModeler brush

9. Next, take the ring you duplicated, and make eight copies (you have nine rings in total). You can do this by pressing *W* to access the Gizmo, *Ctrl* + left-clicking + dragging the **Move** icon, and then releasing *Ctrl*.

10. Assemble the nine new rings in a square inside of the larger ring, like so:

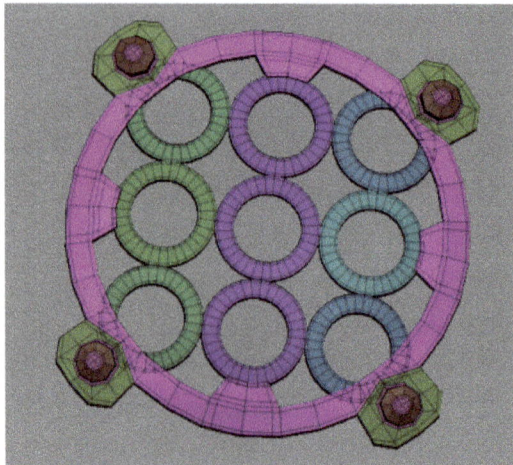

Figure 9.54 – Adding the inside rings of the eye protector

11. Press *W* to access the Gizmo, and click on the gear icon to open the **Gizmo Deformation** menu. From here, select **Deformer**.

12. You can now use the grid of points (lattice) to deform the mesh. You can left-click and drag on the sliders next to the grid to adjust the number of points. Then, you can move any of the points, distorting the mesh in its proximity:

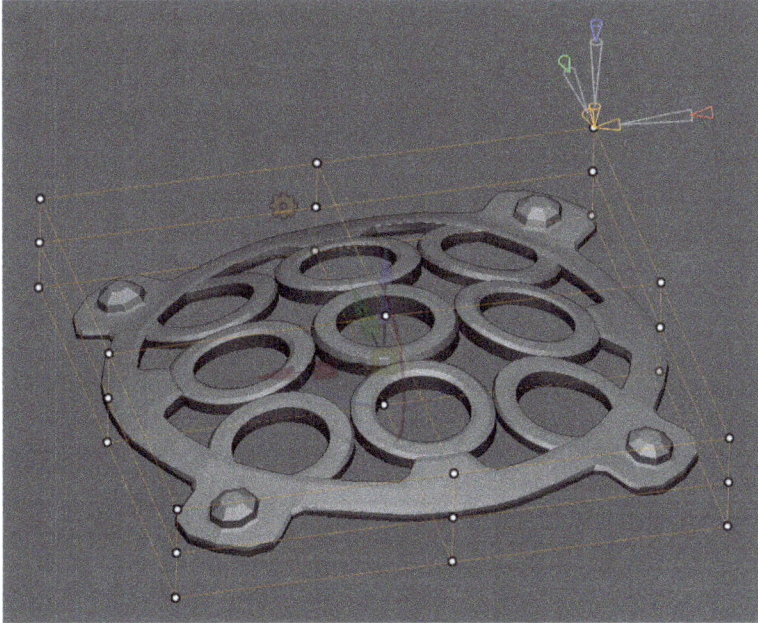

Figure 9.55– Using a Deformer operation to change the shape of a mesh

13. Mask the points on the outside, then switch to the Gizmo and move the points on the inside upward so that the mesh gets a rounder shape, as is typical in gladiator helmets:

Figure 9.56 – Armor piece with curvature

This concludes the section on armor blockout. You learned how **ZModeler** can help you create simple, precise models that will serve as the base model for further refinement. Next, you will learn how to create shoulder armor and helmets, which are more complex accessories. For these, you will use a workflow that includes DynaMesh, IMM brushes, and kitbashing.

Creating complex costume pieces using DynaMesh, IMM brushes, and kitbashing

DynaMesh is a great choice for creating new designs for accessories as it lets you create complex, detailed models quickly without having to commit to the topology of a mesh, as is the case with regular polymodeling. In this section, you will learn how to design complex armor parts for your gladiator model using DynaMesh, a custom **IMM Curve** brush, and **kitbashing**.

Kitbashing is the process of taking pieces from previously created models and combining them to create a new model, or simply adding them to a design to create additional detail and visual noise. This technique is great because of its efficiency and speed in creating complex objects without having to create everything from scratch. The bigger your 3D asset library, the more powerful this becomes. Here are some examples where kitbashing is used to add interesting detail to character designs:

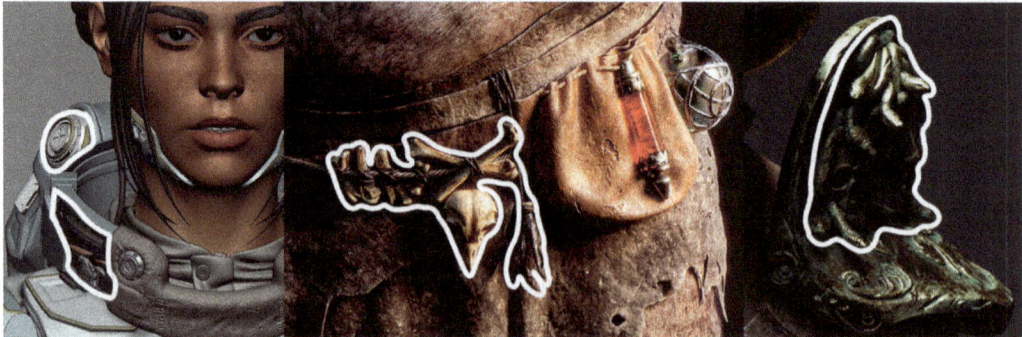

Figure 9.57 – Using kitbashing to create complex objects

Creating shoulder armor

The shoulder armor will be the most elaborate of all the armor pieces. It will include a Medusa head with several snakes, as well as skulls and ornate decorations:

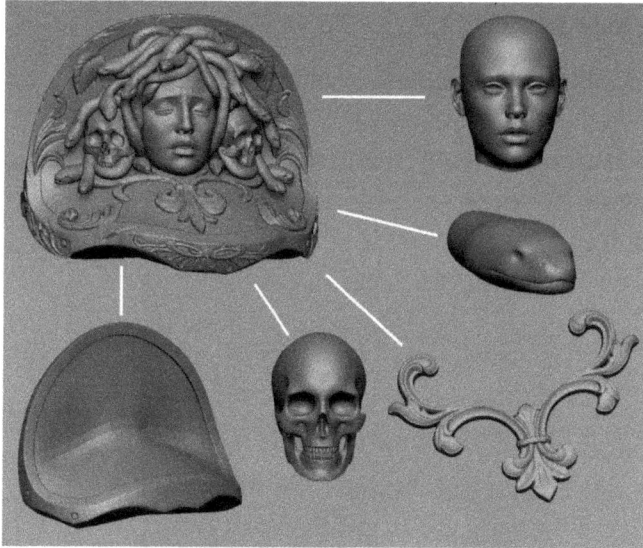

Figure 9.58 – Breakdown of the shoulder armor

The base of this piece was created in the first section of this chapter, using **ZModeler**, and now you can start kitbashing to create a more complex model.

Creating ornamental detail from alphas using the Make3D option

When you need to create any kind of ornament or pattern, ZBrush has a very efficient tool that can turn grayscale images into 3D models, called **Make3D**:

Figure 9.59 – Turning a (grayscale) image into a 3D model using Make3D

For this example, we will be using a baroque-style ornament texture, but you can use any ornamental pattern you like (there are many sites for this, but a good choice is always `artstation.com`). Once you have your textures gathered, do the following:

1. Open the **Alpha** palette and click **Import** in the upper-left corner to load your ornament image.

2. Next, open the **Make 3D** submenu:

    A. Increase the value of **MRes** (Mesh Resolution) to increase the polycount of the mesh that will be created.

    A. Make sure **DblS** (double-sided) is enabled so that the mesh has volume and can be used more conveniently later on.

    B. Press **Make 3D**:

Figure 9.60 – Alpha palette

3. The resulting mesh may need to be scaled, using the Gizmo, to get the thickness you want:

Figure 9.61 – The resulting mesh of a baroque Alpha turned 3D

Depending on your character design choices, you may create multiple pieces in this fashion. You will use them shortly, after adding some other elements.

You can also use this method to add detail for the arm armor part:

Figure 9.62 – Detail added to the arm armor

Next, you will load models from ZBrush's LightBox library, as well as from your own content, to add them to the shoulder piece in a kitbashing approach.

Kitbashing the shoulder armor

Once the shoulder armor base is complete and some ornament pieces are created, you can think about other elements to add to your design. ZBrush's content library, LightBox, has some good options, but of course, you might have your own 3D assets that you would like to use instead.

For this example, we'll load the demoHeadFemale.zpr file, which is intended to serve as a Medusa-like head. To load a .zpr file, open the LightBox library (or use the , keyboard shortcut) and navigate to the **Project** tab. Keep in mind that loading a .zpr file will open a new ZBrush session, discarding your active tools, so make sure to save your file before loading the head:

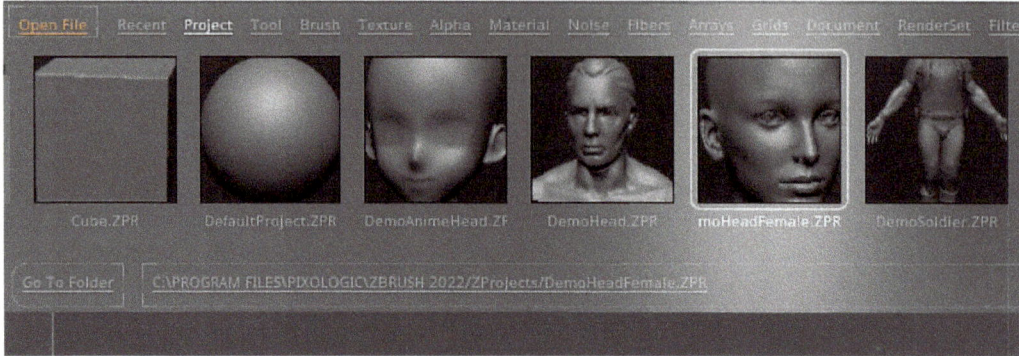

Figure 9.63 – Demo female head in LightBox

At this point, you can make any changes to the head, altering the proportions and facial expression. One important step is to scale the face in one dimension, making it flatter so that it will fit on the armor better. These are the rough steps:

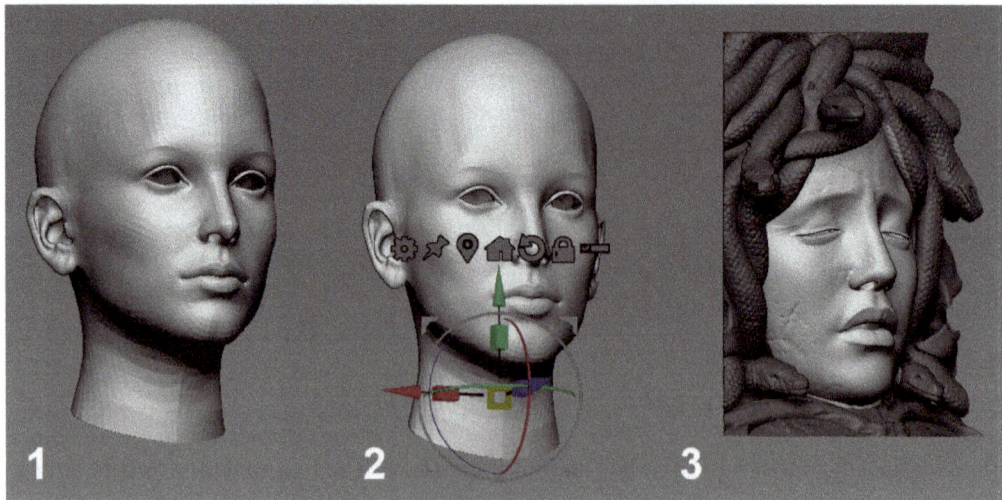

Figure 9.64 – ZBrush's female head model (1), flattening it with the
Gizmo (2), and adjusting the facial expression (3)

Additionally, you can add a skull using ZBrush's **Ecorche** model. You can find it in the **Tool** tab under `Ryan_Kingslien_Anatomy_Model.ZTL`:

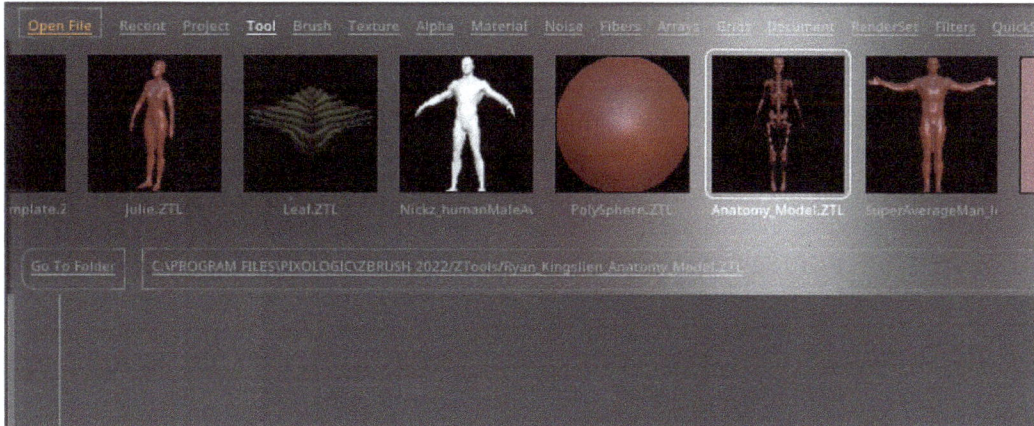

Figure 9.65 – ZBrush's Ecorche model containing a skeleton

Now, you can start assembling all the models you have prepared on the shoulder armor base model created in the previous section:

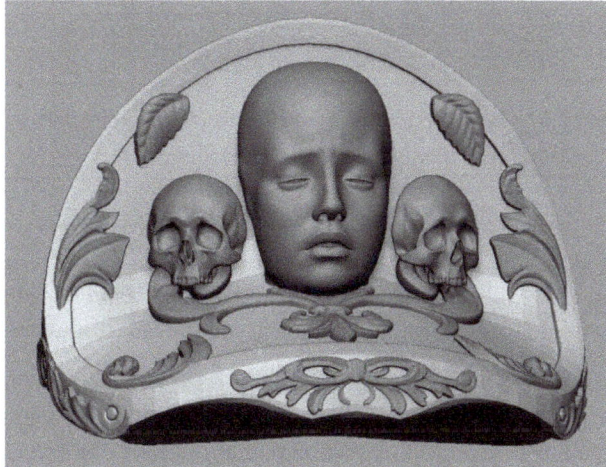

Figure 9.66 – Assembling the different 3D models on the armor base

The only missing piece of this Medusa armor is the snakes. Since multiple snakes are needed, and each snake has a unique pose, an effective workflow is necessary to do this efficiently. Luckily, ZBrush has a great tool for this scenario called an **IMM brush**.

Creating a tri-parts IMM Curve brush

Different **InsertMesh** brushes play a role throughout this book, as they are powerful tools for adding unique and complex details to a mesh. **IMM Curve** brushes function the same way as the **CurveStrapSnap** brush that was used in the first section to create leather straps. Here are some of the **IMM Curve** brushes in ZBrush's library:

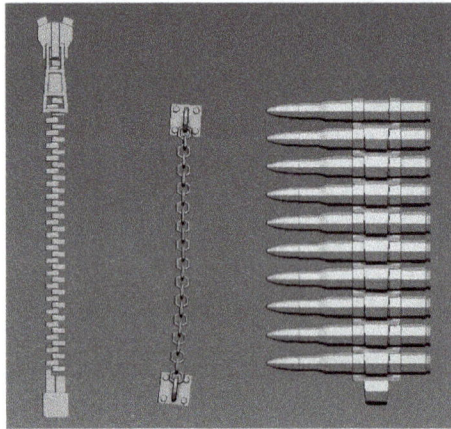

Figure 9.67 – IMM Curve brushes in ZBrush: IMM ZipperM (left),
IMM Curve (middle), and IMM Army Curve (right)

For this particular brush, you will be enabling a special option, called **Tri Parts**. This option provides a model with a start, middle, and end part, which ensures that you draw the mesh correctly. For example, each snake you draw will have only one head and one tail regardless of its length, whereas without this option, you would toggle through multiple heads and tails throughout the brush stroke.

To make a snake **IMM Curve** brush, you first need a snake model. Whether you buy one or make one from scratch, make sure that its polycount is not unnecessarily high – a polycount below **10,000** is a good value. Then, as mentioned, you need to make sure that you have one PolyGroup for each part of your model: the starting point, middle point, and end point:

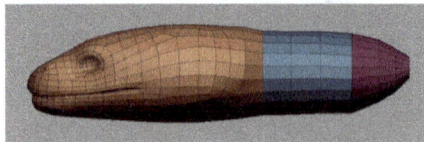

Figure 9.68 – A mesh with three PolyGroups, suited for the Tri Parts curve brush option

Try to keep the middle part of your brush as short as possible. This will make it easier to create curved brush strokes without having a long middle part that prevents a curved shape from forming.

If you've made a snake according to these requirements, proceed to create the curve brush in the following steps:

1. Make sure that your camera faces the snake so that it points upward, to ensure the brush will use the correct direction:

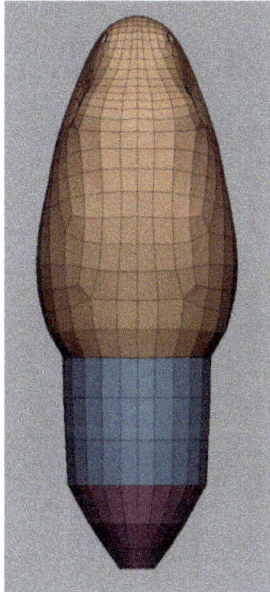

Figure 9.69 – Correct viewing angle of the mesh before creating the curve brush

2. Open the **Brush** library menu and select any **IMM Brush** type.

3. Open the **Brush** library menu again, and select **Create InsertMesh**:

Figure 9.70 – Create InsertMesh option in the Brush library menu

4. A pop-up window appears, asking if you would like to append your mesh or create a new brush. Select the second option.

5.   Open the **Stroke** palette, then its **Curve** submenu, and enable **Curve** mode:

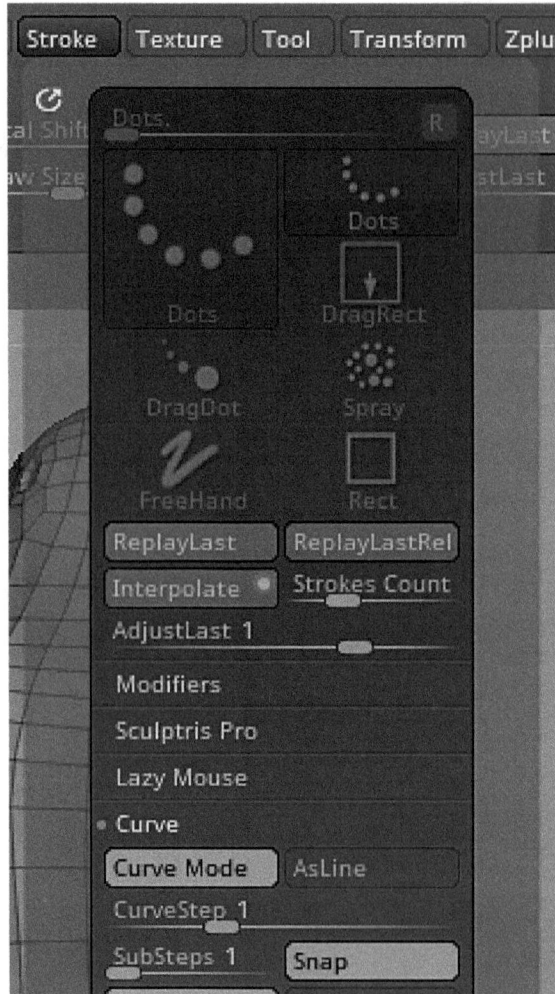

Figure 9.71 – Enabling Curve mode in the Stroke palette

6.   Still in the **Brush** palette, but in the **Modifier** submenu this time, enable **Tri Parts** and **Weld Points**. **Tri Parts** will ensure that the start and end points (the head and tail of the snake) will be used, while **Weld Points** will weld the points between PolyGroups of your mesh so that the result is a continuous mesh instead of many separate meshes:

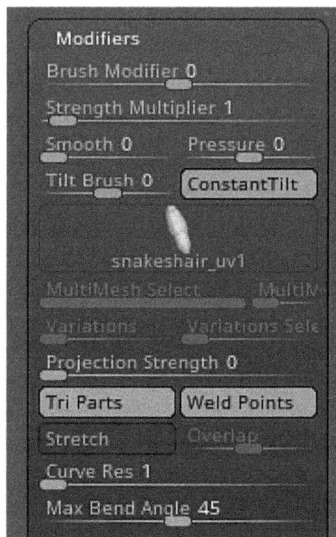

Figure 9.72 – Tri Parts enabled in the Brush palette

You have now successfully created a tri-part curve brush for drawing snakes. You could now draw snakes on the head model to create Medusa's snake hair. You could also add all the snakes on the shoulder armor piece, and assemble them to your liking. The result should look similar to this:

Figure 9.73 – Medusa shoulder armor piece

Since this model will ultimately be prepared for 3D printing, you will want to merge all the pieces of the shoulder armor so that it is one watertight, printable piece.

Merging the pieces with DynaMesh

Now that you have the design established and the pieces assembled, you can merge everything into one piece using DynaMesh. However, when merging complex objects with many tight spaces, as was the case with the snakes, you will find that DynaMesh will merge the mesh in places that you don't intend it to. This means you might have to go through some iterations, adjusting parts of the model until DynaMesh gives you a good result. Sometimes, you might need to move objects further apart to avoid merging close parts that you want to keep separate. Other times, you might get better results if you move objects closer and get a cleaner, merged result.

You can take a look at *Chapter 2* to learn more about DynaMesh; however, as a reminder, you can access the tool through **Tool | Geometry | DynaMesh**. In the following screenshot, on the left, you can see how DynaMesh creates a bad result when the objects are too close to each other, and on the right, extra space was added so that DynaMesh does not merge the objects:

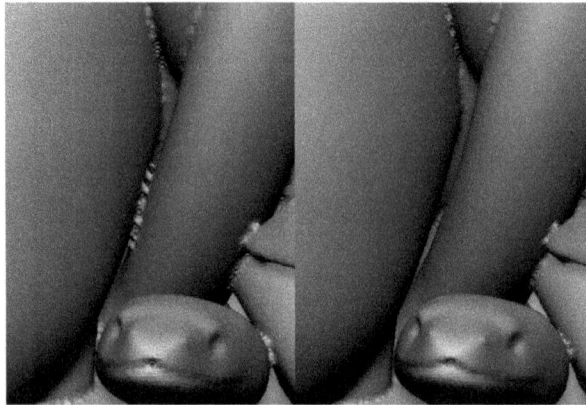

Figure 9.74 – Unintended merging of two parts (left) through DynaMesh

You may decide to ZRemesh this piece; however, it is not necessary for creating a 3D-print-ready object.

Creating a helmet

The helmet consists of two main elements: a basic helmet shape and a lion head, which will be added as an interesting detail that draws the viewer's attention:

Figure 9.75 – Creating a gladiator helmet

Sculpting a lion head

There are multiple ways of approaching this part. You can use a lion-head model or an animal model as a starting point, or you can start from scratch. If you do decide to start from scratch, you may start with any basic shape, such as a sphere.

Here is the process of turning a sphere into a lion's head:

1. Before you start sculpting, gather some references to ancient lion-head ornaments. This will help create a simplified design that looks authentic:

Figure 9.76 – Lion-head ornaments reference

2. Next, open the **Tool** palette, go to the **Subtool** menu, and choose **Append**. Then, pick the **Sphere3D** shape:

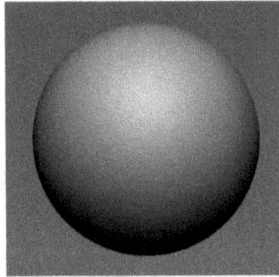

Figure 9.77 – Sphere3D mesh

3. Make sure to enable **Symmetry** mode by pressing *X*.

4. You can increase the resolution of your mesh by subdiving it, or alternatively entering **DynaMesh** mode with an accordingly high resolution.

5. To sculpt the lion's head, the **Claybuildup** and **DamStandard** brushes are great choices (but, of course, you may choose any brush you are comfortable with):

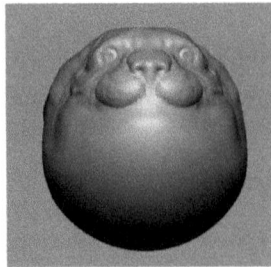

Figure 9.78 – Sculpting a lion head with ClayBuildup and DamStandard brushes

Continue to refine the sculpture until you are happy with its shape and level of detail:

Figure 9.79 – Final lion-head sculpture on a sphere

With this ornamental sculpture done, you can move on to create the basic shape of the helmet.

Creating the helmet base and adding the lion head

You could approach this modeling task from many angles, using various tools. You could start with a polymodeling workflow to get a precise result; however, since the helmet will be made of a rougher metal, you don't need as much precision, so you can use sculpting brushes to create it. Let's get started:

1. In the **Tool** palette, go to the **Subtool** menu and choose **Append**. Then, pick the **Sphere3D** shape.

2. Add some resolution to the sphere, since the default polycount is too low to create the shape you want. This could be done through subdividing the mesh or using **DynaMesh** mode with a high **Resolution** value.

3. Next, to block out helmet shape from a sphere, use the **MaskPen** brush to create a mask, then use the Gizmo and/or **Move** brush to pull out new forms:

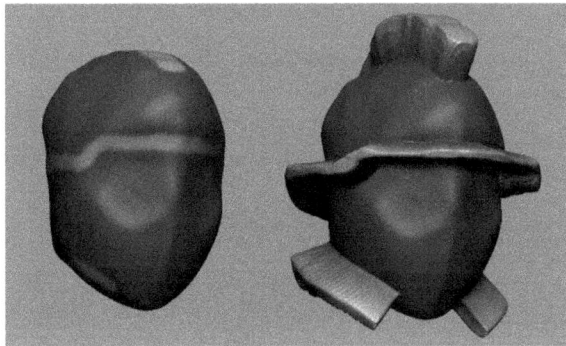

Figure 9.80 – Refining a shape using masking

Continue to sculpt the shape until it is as refined as you like it to be.

4. Once happy, switch to the **Lion Head** subtool you created before and merge it with the helmet by going to **Tool | Subtool | Merge | MergeDown**:

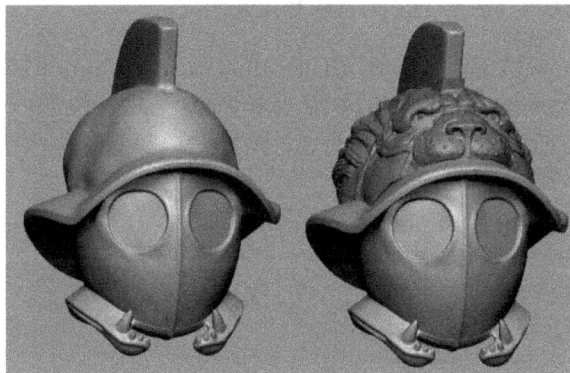

Figure 9.81 – Combining helmet and lion head

5. To merge the new subtool into one continuous mesh, perform a DynaMesh operation again, making sure to use enough resolution to keep all the detail from the lion head.

6. At this point, you can duplicate the helmet, go to **Tool | Geometry**, and use the **ZRemesher** tool to retopologize the model.

7. Finally, you can add subdivision levels and reproject the detail from the original DynaMesh model that you duplicated, until all the detail is restored on your new model (you can read about the detail reprojection process in *Chapter 5*):

Figure 9.82 – Helmet

Now, you have all the accessories blocked out and modeled, ready for detailing and the finishing touches. In the last section of the book, you will explore detailing techniques for the different materials of the gladiator's costume, including metal, leather, fabric, and wood.

Adding detail to the costume pieces

In this section, you will explore techniques for detailing the costume, using various brushes and tools. The goal is to create textures that enhance the realism and storytelling of your character and get the viewers' attention. This includes capturing the surface structure of the respective material, as well as creating small imperfections, damage, and wear and tear.

Let's take a look at the different materials present in the gladiator's costume, and which tools and techniques you can use to get the most out of these assets.

Metal

The metal parts of the character design can be enhanced in several ways. For the rough metal of the gladiator's armor, we focus on adding (ornamental) detail and adding damage.

Adding snake scales

ZBrush has a small library of **Scales** brushes. These brushes are not included in the main **Brush** library. Instead, you can access these brushes by pressing , to open the LightBox library, then navigating to **Brushes** and opening the **Scales** folder:

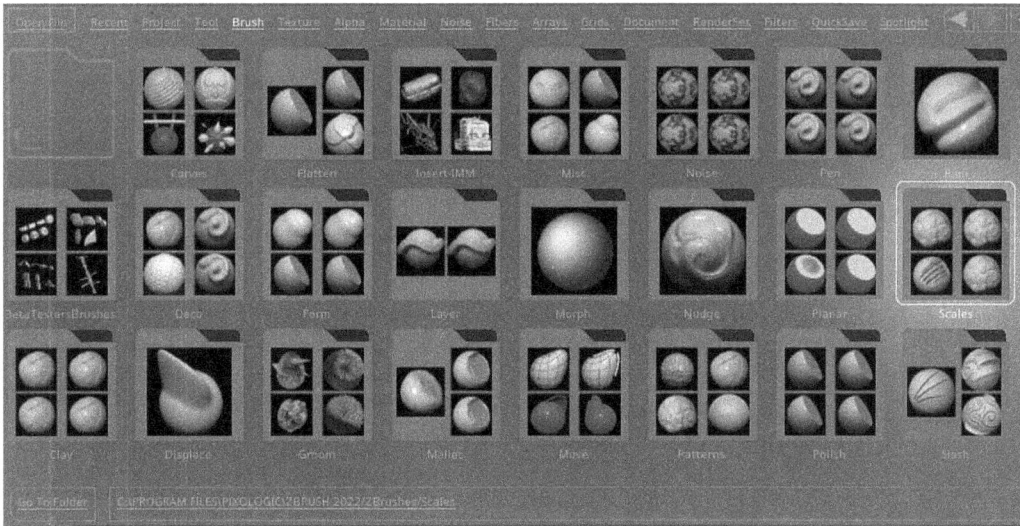

Figure 9.83 – LightBox's extra brushes

There's a brush called **ScalesSnake1** that you can use to add detail to the snakes on the shoulder armor, like so:

Figure 9.84 – Snake detail applied to the head

Adding ornamental detail

We've already used some baroque Alphas with the **Make 3D** option to add detail to the shoulder piece and arm armor. Now, you can use the same Alphas with a regular **DragRect** brush to add finer detail to the helmet. If you don't have any Alphas to use, you can create your own Alpha from a 3D model (take a look at *Chapter 4* to learn how to do this).

It is up to you whether you want to turn an Alpha into a mesh or apply it with a **DragRect** brush. Generally, when the detail is small, applying it to the model is more efficient. The biggest benefit of turning detail into geometry is that you can change its shape, while this is not possible in the same way when you apply details on a target mesh.

To apply an ornament Alpha on your 3D model, follow these steps:

1.   Make sure you have a grayscale image of an ornament pattern.

2.   Press *B* to open the **Brush** menu and select the **Standard** brush.

3.   Open the **Brush** menu again and select **Clone** on the bottom bar.

4.   In the **Brush** attributes on the left side of the canvas, change the **Stroke** mode to **DragRect**:

Figure 9.85 – Changing the Stroke mode to DragRect

5.   Click on the **Alpha** icon below **Stroke**, then select **Import** to load in your ornament Alpha:

Figure 9.86 – Using a grayscale ornament picture

6. You may need to adjust the **Z Intensity** slider, located above the canvas, in order to set the right brush strength to create your desired effect.

7. Now, you can apply the Alpha on your mesh:

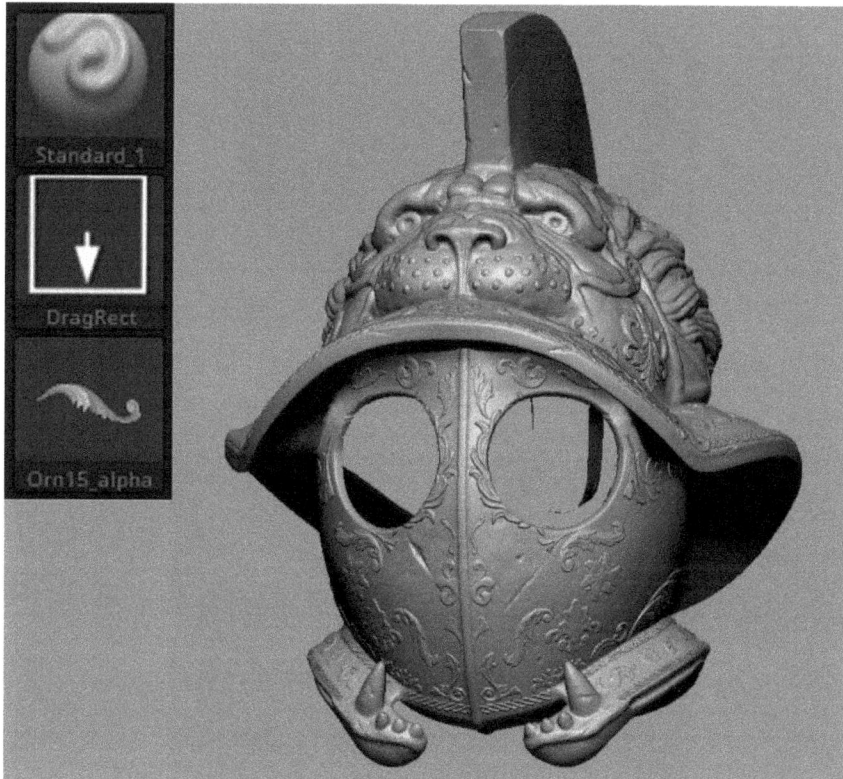

Figure 9.87 – The result after adding baroque-style ornaments

Creating detail using radial symmetry

With the option to apply brush strokes with radial symmetry, creating detail for circular elements in ZBrush is easy. The design of the gladiators' shoes includes coin-shaped elements with ornamental surfaces. In order to create one of those elements, do the following:

1. Go to **Tool | Subtool | Append** and pick a **Sphere3D** shape.

2. Rotate it 90 degrees so that the converging lines face the camera:

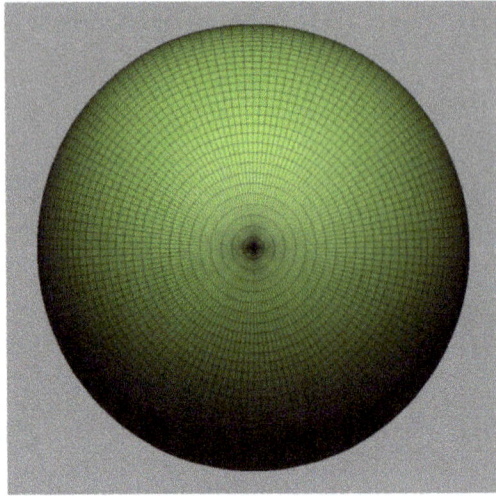

Figure 9.88 – Sphere3D rotated 90 degrees

3. Isolate the front half and delete the hidden part by going to **Tool | Geometry | Modify Topology** and selecting **Del Hidden**.

4. Use the Gizmo to scale the model, making it flat:

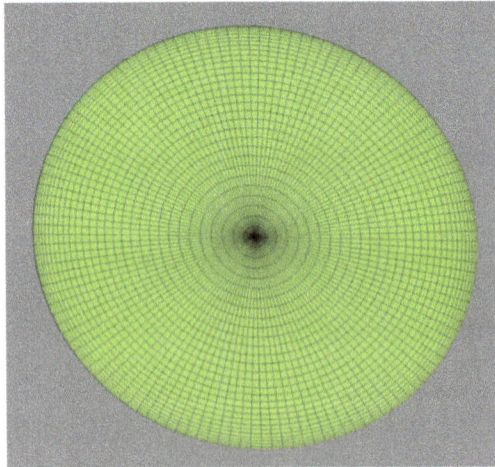

Figure 9.89 – Half a sphere, flattened

5. Open the **Transform** palette, then do the following:

    C. Enable **Activate Symmetry**.

    D. Change the affected axis from >**X**< to >**Z**<.

E. Enable **(R)** and change the **RadialCount** value based on how many points you want to sculpt on simultaneously (the finer you want the detail to be, the higher the value you should use):

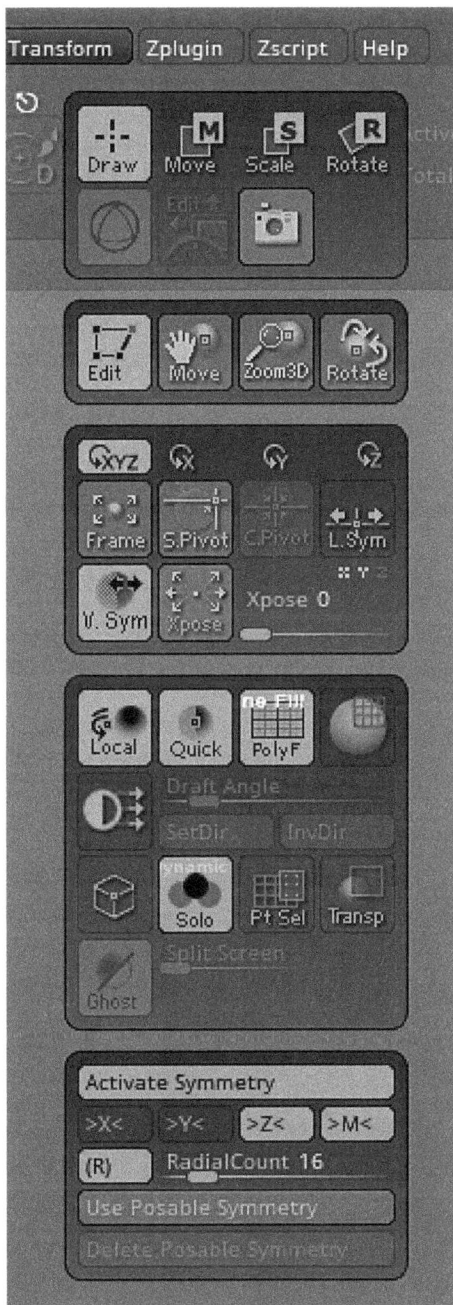

Figure 9.90 – Enabling radial symmetry in the Transform tab

6. You can now add as many subdivisions as you need and sculpt patterns in a radial fashion. You will find that **Symmetry** mode makes it very easy to sculpt interesting patterns without much effort:

Figure 9.91 – Creating radial detail using ZBrush's Symmetry mode

7. Of course, you can combine this detailing technique with other tools, such as **DragRect** brushes and ornamental meshes, to create elements for your shoe design.

You might end up with different ornament designs that you can duplicate and assemble on the shoe model, as is done in the following screenshot:

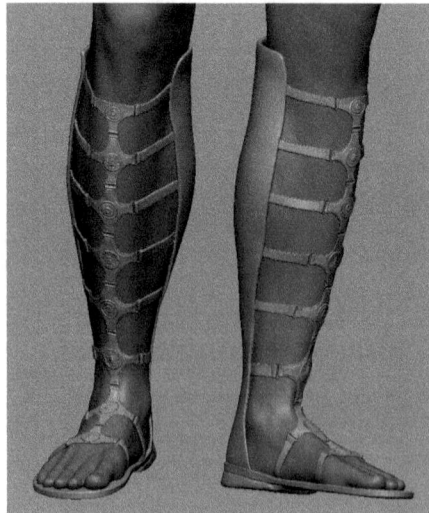

Figure 9.92 – The final shoe model

Adding metal damage and dents

If you create a character that has been in battles, you can design armor that shows this through dents, cuts, and imperfections. Depending on the state of the armor, you could include rust as well. Here is how you can achieve all those effects.

Dents

In theory, you could use any brush to sculpt dents into metal armor; however, just make sure to avoid creating a soft, muddy look. Using the **hPolish** brush can help re-establish the hard look of metal after deforming it. In the following screenshot, you can see a soft-looking dent on top and a dent with flat planes on the bottom:

Figure 9.93 – Soft (top) dents versus hard dents (bottom)

Cuts

There are several brushes that can create cuts, but the **DamStandard** brush is the classic choice. With its default Alpha, it creates sharp cuts, but you may choose another Alpha to make your cuts softer or even sharper:

Figure 9.94 – Sculpting in cuts using the DamStandard brush

Imperfections

Imperfections can include all kinds of forms that could realistically occur; for example, small holes or protrusions and large deformations to the shape and silhouette of the metal. All the basic brushes such as **ClayBuildup**, **Standard**, **Move**, and **DamStandard** will help you do this. Of course, as with any kind of detail, you have to be careful not to overdo it and make it too repetitive.

Here are just some examples of different imperfections that can be added to metal:

Figure 9.95 – Metal imperfections sculpted with basic brushes

Rust

Rust can be a nice detail, contrasting polished and shiny metal with rough and grainy details. You could look for a displacement map of rust or use the **NoiseMaker** tool, which applies a texture to the whole mesh so that you can hide or mask areas you do not want to be affected. However, there is an even better solution for this – a **Morph Target**. A **Morph Target** saves the mesh so that you can reload it at a later point or just reload parts of the mesh. Let's try it:

1. Go to **Tool | Morph Target** and select **StoreMT**.

2. Open the **Surface** menu, also located in the **Tool** palette, and click on **Noise**.

3. A window opens, in which you can adjust the **Noise Scale** and **Strength** values to your liking. Then, hit **OK**:

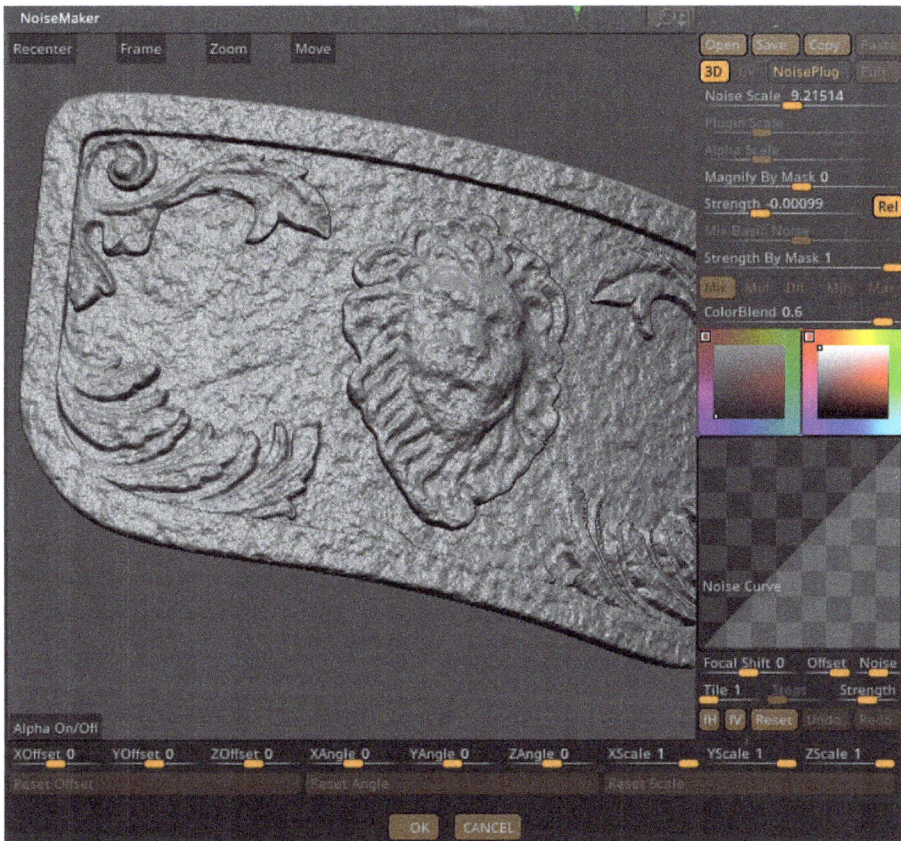

Figure 9.96 – Adding basic noise with NoiseMaker

4. In the **Surface** menu, click **Apply to Mesh**.

5. Go back to the **Morph Target** menu and select **Switch**, which will load the mesh at the state in which you saved the **Morph Target** with no noise applied:

Figure 9.97 – Switching the Morph Target to restore the state of the mesh without rust detail

6.   Open the **Brush** menu and select the **Morph** brush.

7.   Wherever you apply the brush on the mesh, rust/noise will appear. This gives you manual control over areas you want your noise/rust to cover:

Figure 9.98 – Bringing back rust detail with the Morph brush

You should now have some ideas for detailing the metal parts of your models. Next, you will explore various ways to refine leather and fabric parts.

Leather and fabrics

Leather and fabrics materials are part of most costumes, and they come in a variety of styles and patterns. Materials can look great when they are created with attention to detail, so let's take a look at how to create and enhance them.

Adding a base pattern

Unfortunately, ZBrush does not come with high-quality fabric patterns, and it does not contain a leather pattern at all. To get the best quality possible, you will have to buy these textures in a 3D marketplace such as `artstation.com` or `cgtrader.com`. Make sure that the texture is some sort of grayscale bump or displacement map, and that it's seamless so that it can be tiled on your models.

When you have access to a tileable leather texture, you can proceed with the following steps:

1.   First, you need to ensure that your model has UVs in order to apply a texture in **NoiseMaker** (to learn more about the UV creation process, take a look at *Chapter 6*).

2.   Next, make sure that your model has enough resolution. To apply the fine leather or fabric texture, you need a fairly high-poly model.

3.   With the highest subdivision of your model active, go to **Tool | Surface | NoiseMaker** and import your texture. Also, make sure to switch to **UV** mode, and set **Mix Basic Noise** to **0** if you don't want the texture to be mixed with the default noise:

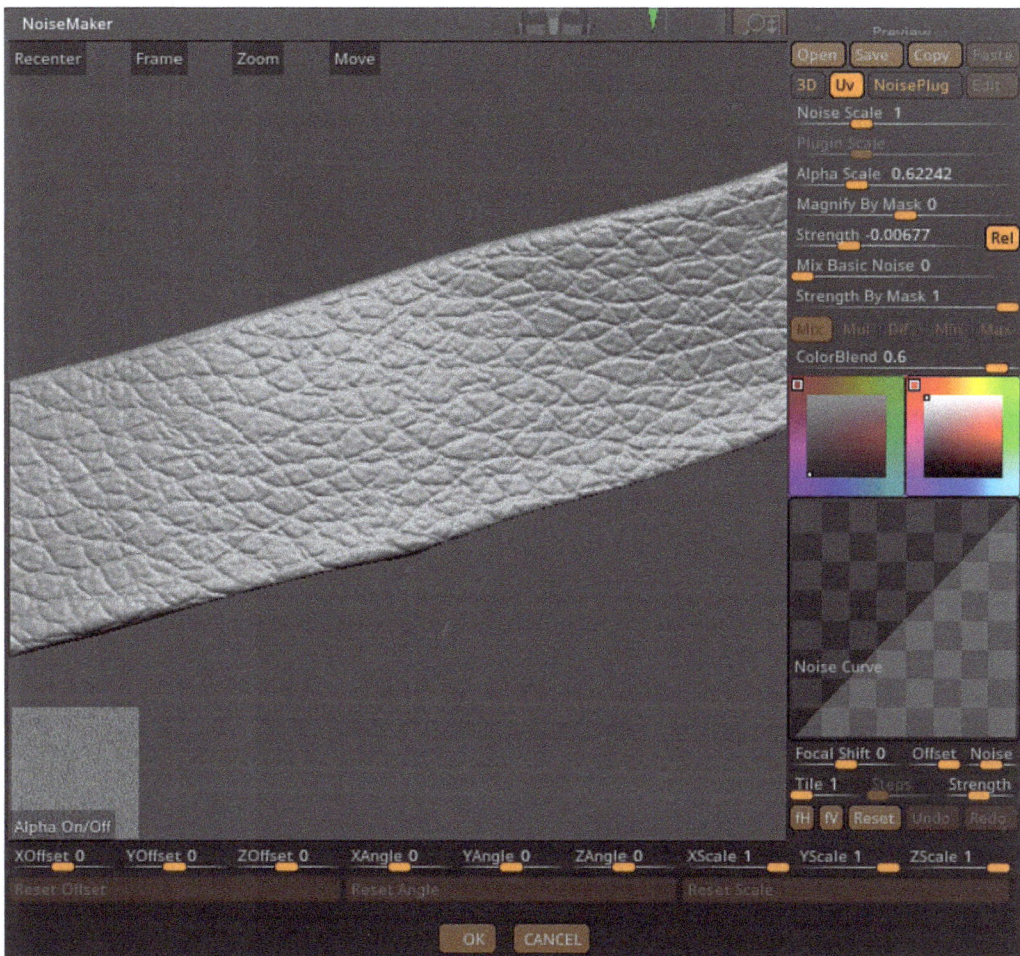

Figure 9.99 – Using NoiseMaker with a leather Alpha

4. Adjust the **Noise Scale** and **Strength** values and hit **OK**.

5. In the **Surface** menu, click **Apply to Mesh**.

At this point, a leather pattern is applied to your mesh, and you can proceed to further refine the surface with basic sculpting brushes.

Adding wear and tear

Creating a base texture is just the beginning. To really make an accessory interesting and appealing, you need to give it some extra attention through other Alphas and sculpting. Here are some tweaks you could make.

Creating irregularities

Since you tile your displacement textures in high frequency on your model, the detail will look very repetitive and, therefore, unnatural. You can break up this look by applying some light sculpting on top of it, using the **Standard** brush to enhance certain areas. This creates more contrast overall:

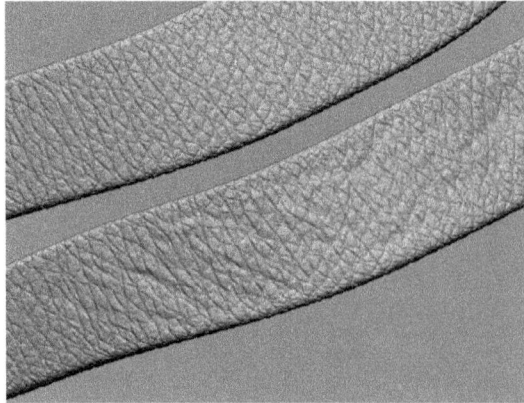

Figure 9.100 – Belt without sculpting (upper) versus belt with sculpting (lower)

Depending on the costume design, you may want to sculpt a little more subtly, and just a little bit of sculpting will already help to achieve a more organic, natural look.

Adding cuts

If you want your character to look like he has been in fights, you could add cuts from blades or claws. The **DamStandard** brush is perfect for this:

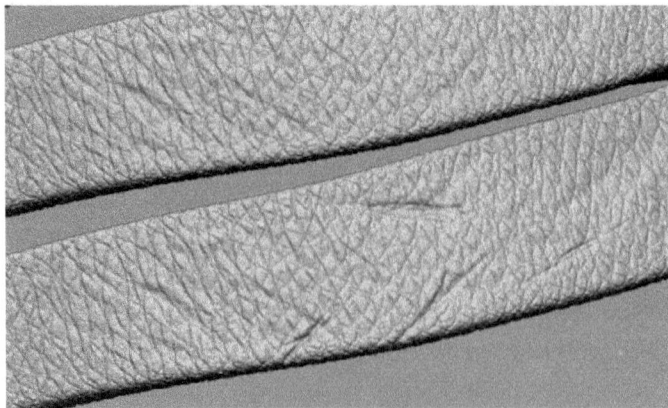

Figure 9.101 – Before (upper) and after (lower) using the DamStandard brush to add cuts

Breaking up edges

When you're working with 3D assets, it's a good idea to break up their edges a bit, or they may look too perfect and lack realism. A great way to do this is by using the **SnakeHook** brush with **Alpha 08**. You can use it to pull edges outward, creating a frayed look that feels more realistic:

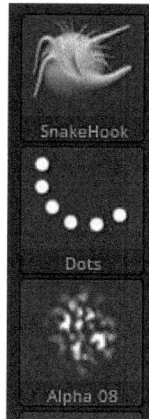

Figure 9.102 – The SnakeHook brush with Alpha 08 equipped

Here is the result:

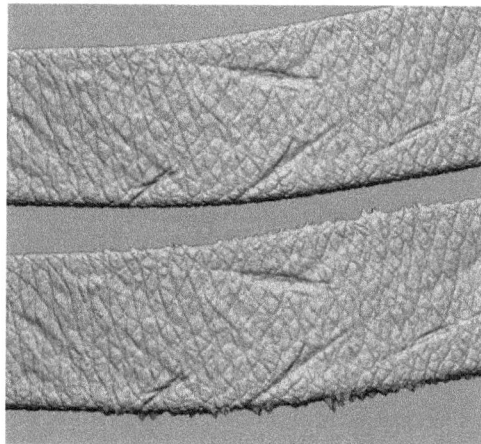

Figure 9.103 – Before (upper) and after (lower) using the SnakeHook brush

Adding holes and damage

Finally, you can use the **DamStandard** brush and other sculpting brushes to add small holes and larger damage to the belt. Adding larger imperfections to a surface such as this has the benefit of creating contrast, which is visible from a distance:

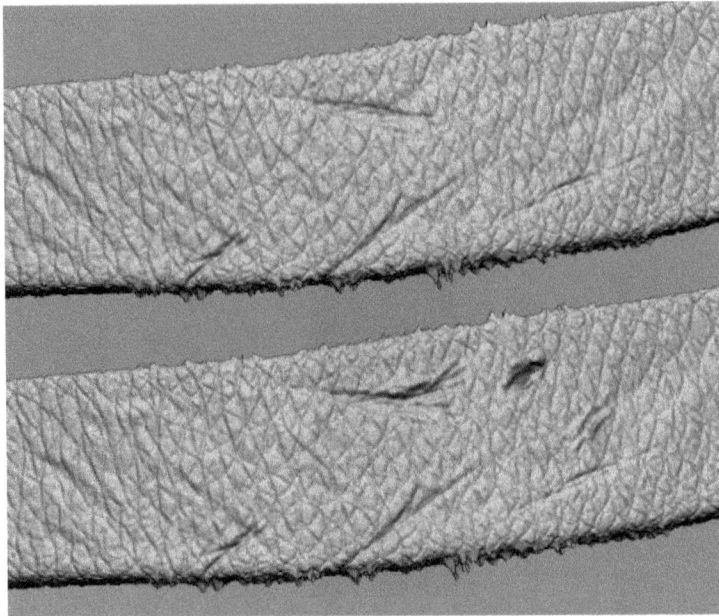

Figure 9.104 – Before (upper) and after (lower) adding damage

Wood

The last material worth looking at in the gladiator costume is the wooden spear. Again, let's add the base pattern and then start sculpting.

Adding the base pattern

NoiseMaker is a great way to add surface texture to a wood part. Unfortunately, ZBrush does not have a high-quality wood displacement in its content library, so you will need to use an external source for that – you can either look for a wood displacement texture or, alternatively, any texture that has long warped lines, simulating a wood-grain effect.

When you have such a texture, you can proceed with the following steps:

1. After making sure that your model has UVs and enough resolution to hold the detail, go to **Tool | Surface | NoiseMaker** and import your texture.

2. Make sure to switch to **UV** mode, and set **Mix Basic Noise** to **0** if you don't want the texture to be mixed with the default noise in **NoiseMaker**.

3. Adjust the **Noise Scale** and **Strength** values and hit **OK**.

4. In the **Surface** menu, click **Apply to Mesh**.

Here is what the model could look like after applying the texture with **NoiseMaker**:

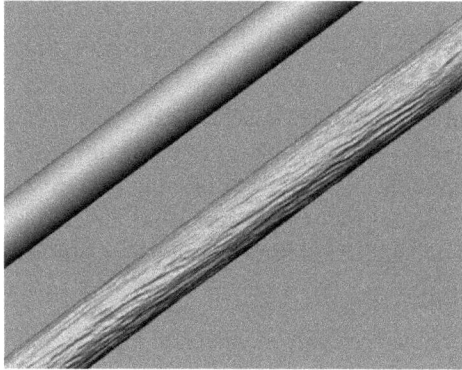

Figure 9.105 – Wood detail

Applying surface noise is a very effective way to add fine surface detail, which makes the texture look nice and sharp when looking at it close up. However, oftentimes, it lacks the strength and contrast in the detail to work well from a distance. To improve that aspect, you can enhance the surface with your basic sculpting brushes.

Sculpting wood

If you want to enhance the contrast of the wood and add some more intense detail, classic sculpting brushes can do a great job of enhancing surface texture. You can add some deeper cuts with the **DamStandard** brush or enhance the volume of some of the detail with the **Standard** brush. This is the before and after of using sculpting brushes on the wooden spear:

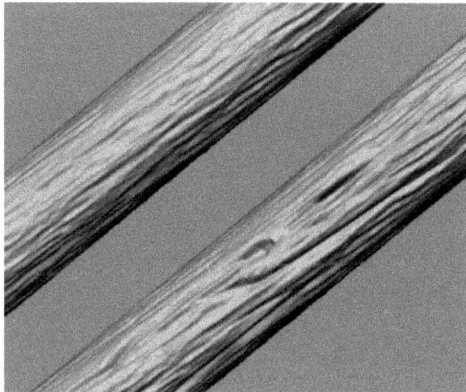

Figure 9.106 – Before (above) and after (below) sculpting on the wooden spear

This concludes the material detailing section. Of course, you can apply these methods to any material, but this should have covered some of the more common ones.

Summary

In this chapter, you learned about different tools and workflows for creating armor, clothing, and accessories, demonstrated on our gladiator model.

First, we focused on using primitive shapes and the **ZModeler** brush to construct parts of a gladiator's costume. This way, you created several costume parts in an easy and precise way. Then, we continued the creation process by incorporating DynaMesh, kitbashing, and the **IMM Tri Part Curve** brush into our workflow. Finally, you learned how to add surface texture, imperfections, and some minor damage to improve the quality and realism of these meshes. You did this by using a combination of brushes and the **NoiseMaker** tool.

In the next chapter, we will learn how to prepare the gladiator for 3D printing by making it watertight, cutting it into pieces, and keying each piece.

10

Preparing and Exporting Our Model for 3D Printing

To successfully 3D print your 3D models, they need to meet several requirements that ensure that the printing works well and that no issues occur. In this chapter, you will learn how to prepare your digital sculpture for 3D printing, based on the gladiator example created in the previous chapters.

First, you will learn how to test the mesh topology for holes and errors so that you can create a watertight model with the right properties for printing.

Next, you will learn about good practices for cutting your model, which will save print time and material costs. Based on this information, you will merge parts of the gladiator model and split it into multiple pieces. After that, you will create a key model that you can use to create matching male and female keys. This will allow you to assemble your printed figure out of multiple pieces.

Finally, you will learn how to scale and export the model. Your 3D-printing software of choice then takes over and translates your digital sculpture into a physical one.

So, this chapter will cover the following topics:

- Creating watertight models and troubleshooting errors
- Merging and cutting your model
- Adding keys, scaling, and exporting your model

Technical requirements

For the best experience, it is recommended that you have a strong PC that meets the minimum requirements described in the first chapter's *Technical requirements* section. However, you can work on this chapter with just a mouse, a functional PC setup, and a ZBrush license.

If you completed *Chapters 8* and *9* and created a gladiator (or your own character model), you can apply the lessons in this chapter to your character, but of course, you can use the workflows and tools on any other 3D model.

Creating watertight models and troubleshooting errors

When you prepare 3D models for printing, the first thing to do is reduce their polycount by decimating them. Then, you need to ensure that your model has the proper topology and is watertight, giving it a volume. Errors and inconsistencies in the topology, such as non-manifold geometry or overlapping parts, will not be translated properly, leading to faulty or failed printing. In this section, you will learn how to prepare your model accordingly.

Using Decimation Master to reduce your meshes' polycount

If you have highly detailed models with a high polycount, **decimating** them allows you to significantly reduce their polycount while still keeping their details. This is done through a topology with varying densities of polygons based on where the detail is – so, the polycount is low in smooth areas while it has a higher poly density in highly detailed areas, capturing the detail there. You can see what the decimation looks like in the following screenshot, where the polycount is reduced by 90%:

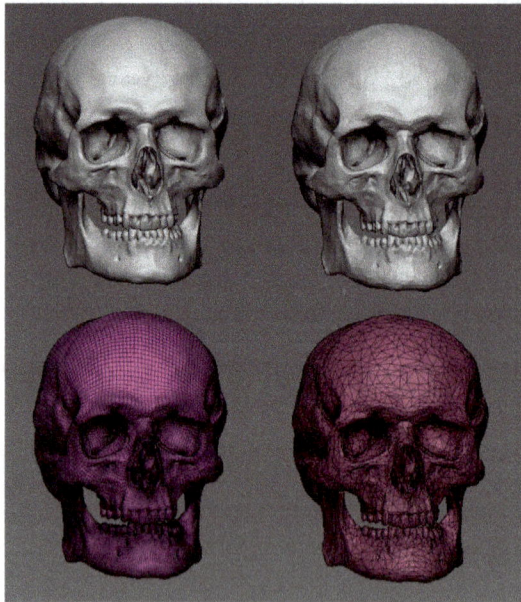

Figure 10.1 – Decimating a skull model by 90%

ZBrush has a plugin called **Decimation Master** that does just this. In order to decimate your model, follow these steps:

1. With your subtool selected, switch to its highest subdivision level. You can switch to the next higher subdivision level with the *D* key.

2. Navigate to the **Zplugin** palette, then open the **Decimation Master** menu:

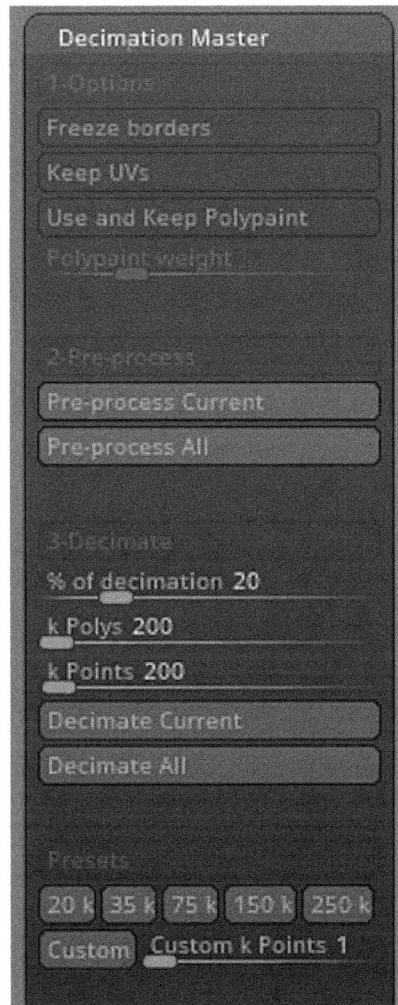

Figure 10.2 – ZBrush's Decimation Master decimation tool

3. Select **Pre-process Current** (without letting ZBrush run this preprocessing, the decimation will not work).

4. Now, you have several options to determine the polycount of the decimation result – you can put a polycount target in the **Custom k Points** field or enter a value in the **% of decimation** field, which will result in a polycount based on your meshes' polycount. **20%** decimation usually gives a result that keeps the maximum amount of detail, though you might try out lower values from **5%** to **15%**, which can also work well depending on the topology and detail of the mesh:

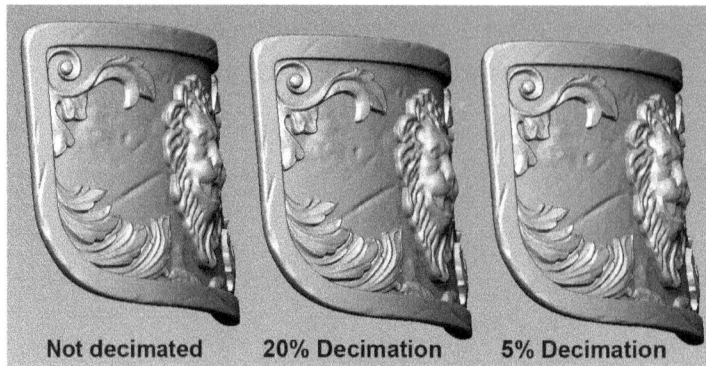

Figure 10.3 – 20% and 5% decimation

You can see how the 20% decimation looks almost the same, while at 5% (the low polycount), the mesh starts to lose the shape of its detail.

5.   Next, hit **Decimate Current** to decimate the mesh.

As an optional step, readjust the % value, repeating *steps 4–5*, until you have a result that reduces the polycount the most while keeping enough of the detail for the printing process.

Here is a comparison of a mesh before (with 3.7 million points) and after decimating (now with 186,000 points) with a decimation value of **5%**:

Figure 10.4 – Before and after decimating a mesh

As you can see, there is almost no visible difference in the mesh even after 95% of polygons have been removed. Having lower-poly meshes means that all subsequent operations such as merging, cutting, and adding keys will take less computation time.

Additionally, you'll be able to save space due to smaller file sizes, and the whole process will be smoother because the printing software won't have to deal with unnecessary 3D information.

You can repeat this process with the other subtools of your character, which will leave you with a faster-performing scene, ready for the next steps.

Checking your mesh for holes and errors

In this stage, you should check if there are any holes in your mesh or other issues, such as shared vertices or non-manifold geometry.

There are three kinds of **non-manifold geometry**. The most common one, which can often occur in ZBrush, is an edge that is shared by more than two faces, which can occur during an extrusion operation. This is problematic because it prevents several tools from functioning, and is especially problematic in the context of 3D printing, where printing software will struggle to properly process a 3D model with this type of topology.

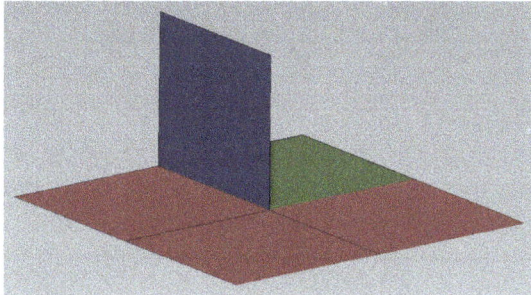

Figure 10.5 – An edge sharing more than two faces: non-manifold geometry

First, let's check for holes by opening the **Tool** palette, navigating to **Geometry | Modify Topology**, and selecting **Close Holes**. This closes any holes in the mesh that you might have missed.

Next, go to **Geometry | MeshIntegrity** and select **Check Mesh Integrity**. If there are no issues, you will see the following message: **Mesh integrity test completed successfully.**

If there is an issue in the mesh integrity, you will get an error message instead, such as **The mesh contains [Number] edges that are shared by more than two polygons**, or **The mesh contains [Number] faces with multiple references to the same vertex**.

In that case, simply select **Fix Mesh**, and you should see a success message saying the mesh is fixed:

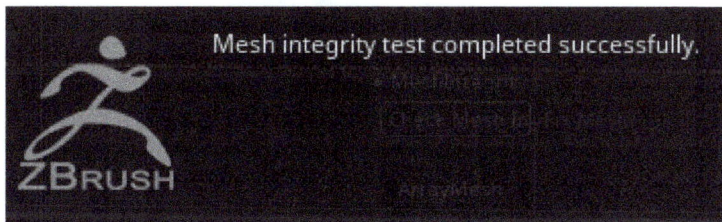

Figure 10.6 – Successful mesh integrity test message

This tool will fix various issues, including non-manifold geometry. With holes in your mesh closed and the mesh integrity test successful, you can proceed and check the volume of the mesh.

Checking the mesh volume

In order to test whether your mesh is actually watertight and ready for printing, you need to check whether the mesh volume can be calculated.

To do this, navigate to **Tool | Zplugin | 3D Print Hub** and select **Check Mesh Volume**. If the volume can be calculated, you will see a message like this:

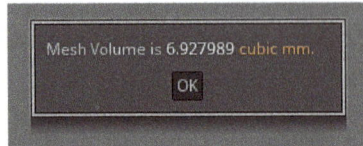

Figure 10.7 – Calculated mesh volume message

If the mesh is not watertight, the volume will not be calculated and you will get an error message:

Figure 10.8 – Failed volume computation message

You might get this message even after performing the **Close Holes** operation and checking for mesh integrity issues. In that case, you have to go through some additional steps to create a watertight mesh.

Fixing a tool that is not watertight

If you have closed the holes of your model and checked for mesh integrity issues, and ZBrush still cannot compute the mesh volume, you need to do some extra fixing. A common problem is that there are one or more meshes inside of the mesh:

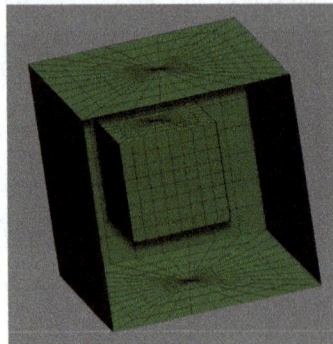

Figure 10.9 – A mesh hidden inside the main mesh might be the issue for a non-computable mesh volume

Here are the steps to fixing (almost) any mesh and making it watertight, allowing for a successful mesh volume computing test:

1. First, make sure to close any holes the mesh might have by going to **Tool | Geometry | Modify Topology** and selecting **Close Holes**.

2. Then, open the **Gizmo** menu by pressing *W*.

3. Click on the gear icon on the left and select **Remesh By Union**:

Figure 10.10 – Remesh By Union Gizmo operation

You might not get a message if the operation completes without issues, or you might get a notification about some inaccuracies in the result:

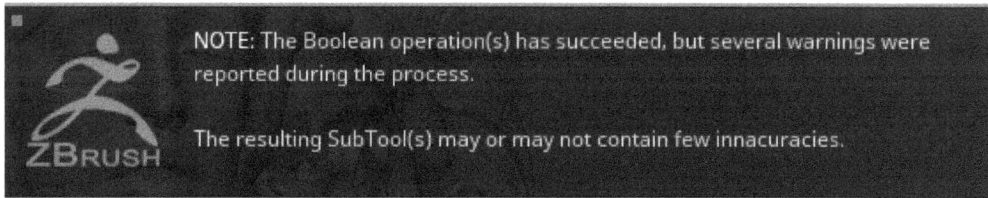

Figure 10.11 – Potential warning message

Either way, this isn't a problem, and you can continue with the instructions.

4. Use the **Selection Rectangle** tool to isolate a small section of your model on the outer border, making sure to only make the section at the very edge of your model so that you do not accidentally include part of a hidden interior model.

5. Go to **Tool | Visibility** and select **Grow All**. This will grow the selection, including every poly that is attached to it. If there are separate meshes inside your model that are floating without attachment to the outer shell, they will not be included in your selection.

6. Invert your selection by holding *Ctrl + Shift*, and left-click by dragging on a space on the canvas. Now, you can see all the separate meshes that are not part of your "main" model, if there are any:

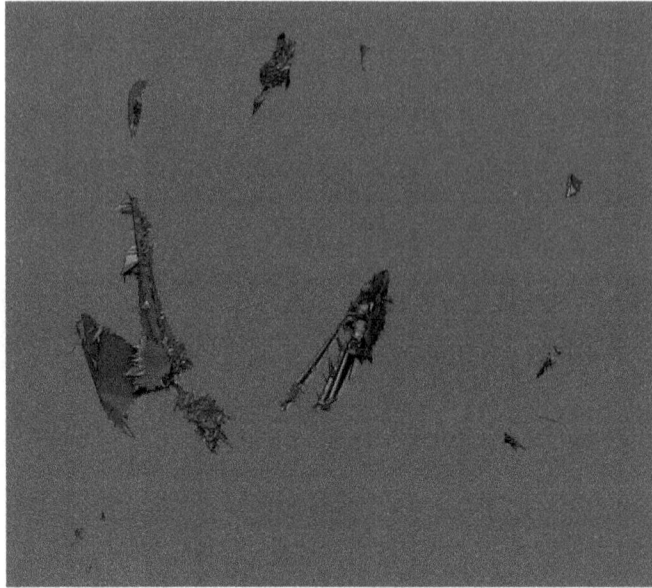

Figure 10.12 – Fragments that can hide inside the model

7. Invert the selection again so that you have only the main mesh left that you want to keep.

8. Go to **Tool | Geometry | Modify Topology**. Select **Delete Hidden**, then **Close Holes**.

9. Go to **Tool | Geometry | Mesh Integrity**. Select **Check Mesh Integrity**, then **Fix Mesh** if an issue was found.

10. Repeat *steps 1–11* until the **Mesh Integrity Check** functionality finds no issues.

11. Go to **Zplugin | 3D Print Hub** and select **Check Mesh Volume**. If you get an error message, repeat *steps 1–12*.

Most meshes can be fixed through these steps. However, if you have a mesh for which the volume is still not calculated, you have to do some extra work and create a new model using **DynaMesh** and then reproject the detail (covered in *Chapter 5*). Let's take a look at how that could work using the gladiator's complex shoulder piece, which is prone to errors due to its varying structure:

1. Go to **Tool | Subtool | Duplicate** so that you can keep the original highly detailed model, which is not watertight.

2. Switch to the highest subdivision level, then go to **Tool | Geometry** and select **Del Lower** to delete the lower subdivision levels.

3. Next, go to **Tool | Geometry | DynaMesh** and enter a **Resolution** value that is high enough to capture the general shape of your mesh. Then, hit the **DynaMesh** button:

Figure 10.13 – The mesh after entering DynaMesh mode

4. Close any holes the mesh might have by going to **Tool | Geometry | Modify Topology** and selecting **Close Holes**.

5. Then, go to **Zplugin | 3D Print Hub** and select **Check Mesh Volume**. It should be able to calculate a volume, but if an error message appears, repeat *steps 3–4*. If the mesh has a mesh volume, it means that it is watertight and ready for 3D printing.

6. Enable the visibility of the current DynaMesh mesh, as well as the duplicate that you made earlier, and disable the visibility of any other subtool.

7. Subdivide the mesh by pressing *Ctrl + D*.

8. Go to **Tool | Subtool | Project** and select **ProjectAll**.

9. Repeat *steps 7–8* until your DynaMesh mesh has the same level of detail as the original model.

Now, you have a watertight model that you can further modify, merge, and cut in the following sections. You may continue to prepare the remaining subtools in your scene until your whole character is without mesh integrity issues and ZBrush can calculate a volume for every piece.

So, in this section, you learned how to fix errors in a mesh, which is necessary to enable the **3D Print Hub** plugin to compute the mesh volume. This means that you have a watertight model that can be 3D printed without issues.

Next, you will learn about guidelines for splitting a model for 3D printing so that you can create a blueprint for your own character. Then, you will begin to prepare your model by merging certain parts and cutting them into multiple pieces for optimum printing.

Merging and cutting your model

Before starting with the actual process, you need to decide what the best way to merge and split your model will be. These are some things to take into consideration:

- Consider the size of the printer that you will be using and the size of the object that you want to print. Splitting the model into smaller pieces will allow you to print larger objects, even with a smaller printer.

- Think about overhangs in your model. Some parts will require the printer to create significant support structures, which wastes a lot of material and slows down the printing process. Splitting a model in these critical areas helps to reduce this:

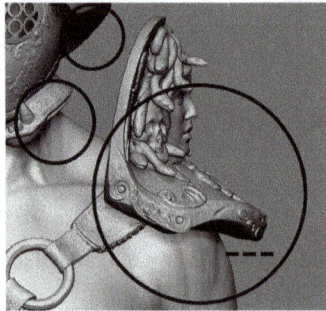

Figure 10.14 – Areas that have a lot of overhangs

- Following on from the previous point, you can split the model where the cut is hidden or less obvious. This could be in an area where different elements meet, such as the beginning of a clothing piece or a piece of armor covering the body. For example, the split gladiator model could look like this:

Figure 10.15 – Possible split of the gladiator model

- Finally, if you're planning to paint your model, it might make the painting process easier if you separate the model based on different elements, materials, and colors so that each piece can be painted separately, without accidentally applying paint in nearby areas.

With these things in mind, you can begin preparing your model accordingly, starting by merging pieces that should be part of the final 3D print part.

Merging meshes

Let's take a look at the easiest way of merging two separate meshes into one mesh that shares a continuous surface and has one volume.

It is important to note here that this does not mean merging two subtools into one, which does not affect the topology and meshes themselves but simply combines them into one subtool. Instead, you'll use the Gizmo **Remesh By Union** operation, which merges meshes like a **DynaMesh** operation, changing the topology of the meshes to create a new continuous surface that gives the merged meshes a shared volume:

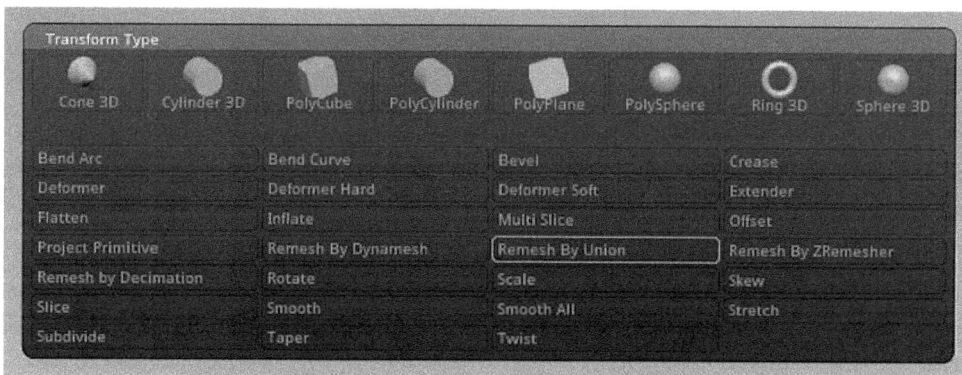

Figure 10.16 – Remesh By Union tool of the Gizmo

However, while **DynaMesh** merges vertices within a radius, determined by the **Dymanesh Resolution** value, the **Remesh By Union** operation changes only areas that intersect while the rest of the topology stays the same. This has the advantage that you can merge high-poly, highly detailed objects without losing any detail, which frequently happens with DynaMeshing.

Here, you can see the difference between a simple merging of subtools, the merging with **DynaMesh**, and finally the merging through the **Remesh By Union** operation:

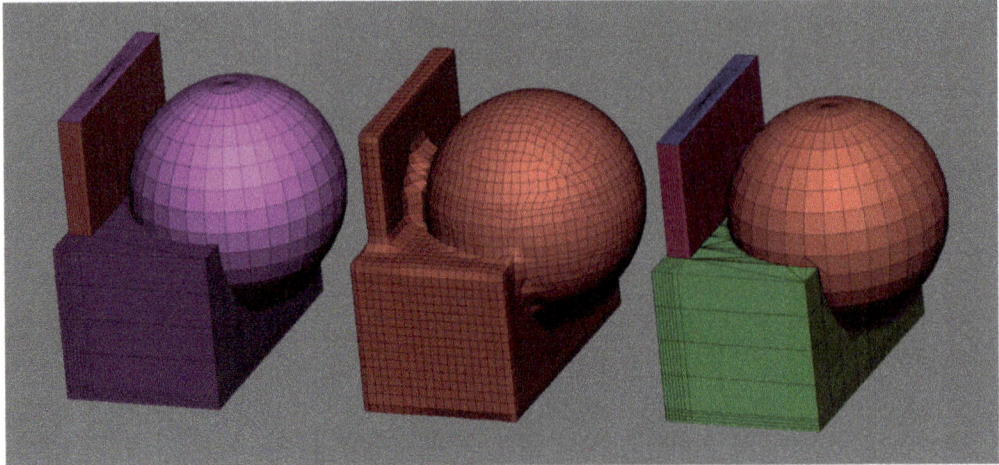

Figure 10.17 – Merging subtools (left), merging with DynaMesh
(middle), merging with Remesh By Union (right)

The model on the left shows that while the two parts are combined into one subtool, they are not combined through topology. Instead, the sphere intersects and passes through the other mesh.

In the middle, you can see how the **DynaMesh** operation combines the model, but the sphere also merges with the wall, due to their close proximity. This is not a desirable result if you want to keep the wall intact. While this can be avoided with a high-resolution value, this stops working when you have highly detailed meshes with close shapes.

On the right, the **Remesh By Union** tool creates a new topology just for the polygons where both shapes intersect, while the remaining topology stays the same. The sphere and wall are also not merging, as is the case with the **DynaMesh** operation.

After merging the shapes, you can test if you successfully created a new object with a continuous topology by navigating to **Zplugin | 3D Print Hub** and selecting **Check Mesh Volume**. If you get an error message instead of the mesh volume, go through the troubleshooting process as described earlier.

Now, you can continue to merge all parts of your character that you need to combine, based on how you ultimately want to split your character.

Cutting a model

With your subtools watertight and meshes merged where needed, you can start cutting the model based on the recommendations from the beginning of the section.

Let's use the left arm of the gladiator as an example. Splitting the arm from the body will save printing time and material, and the bands on the arm are a great place to hide seams:

Figure 10.18 – Splitting the arm

There are several ways to cut your model, but they all come down to splitting the model based on a selection of polygroups.

First, make sure that each piece – the body, wraps, and armor – is decimated. That way, the polycount is not unnecessarily high, printing will be easier, and computation times in ZBrush will be shorter. Also, make sure to check for mesh volume and mesh integrity issues. While you need to check for these again in the end, it is a good practice to ensure these things for the individual subtools already, because it will make the merging and cutting easier for ZBrush. These are the pieces we will be using here:

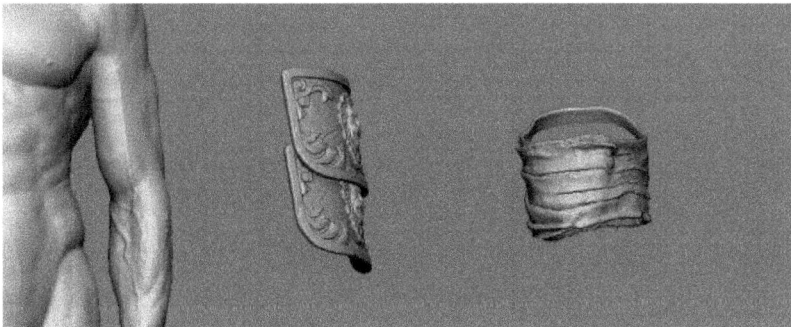

Figure 10.19 – Merging and splitting the arm

Here is the process for preparing the arm part, in which you will be merging and splitting the model:

1. Assign a new PolyGroup to the arm by using the **Masking** or **Selection** tools. You will want to create the new PolyGroup based on where you want to split the model. In this case, a good place for splitting the arm is right where the wraps begin:

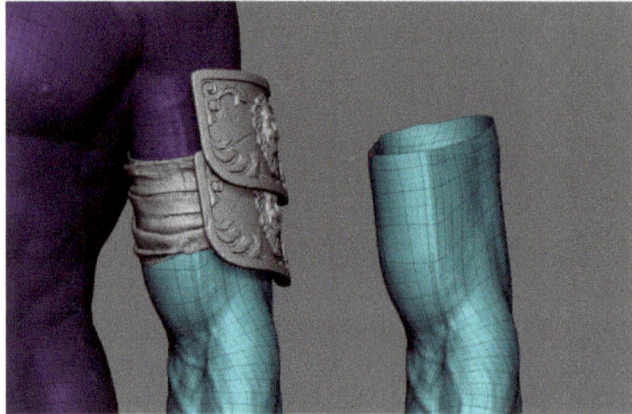

Figure 10.20 – Creating a PolyGroup based on the planned cut

2. If the border of the PolyGroup is jagged, isolate the PolyGroup, then go to **Tool | Masking** and select **MaskByFeatures** with only **Border** enabled:

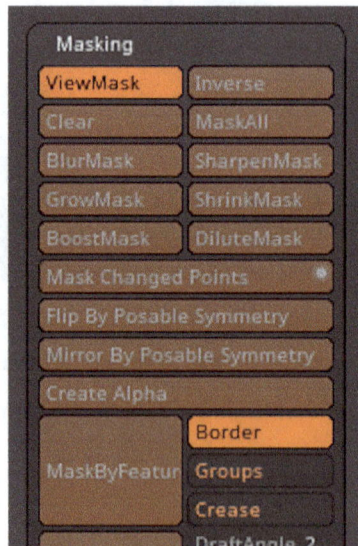

Figure 10.21 – MaskByFeatures option

3. Now, invert the mask by *Ctrl* + left-clicking on a space in the canvas.

4. Then, navigate to **Tool** | **Deformations** and use the **Polish By Features** slider to soften the edge:

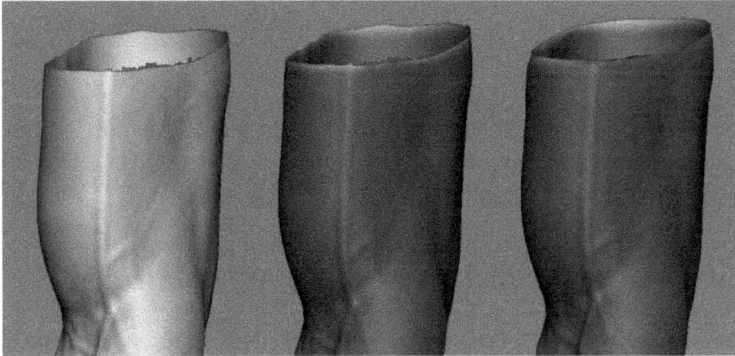

Figure 10.22 – From jagged edge (left) to smooth edge using
MaskByFeatures (middle) and Polish By Features (right)

5. Next, with your arm part isolated, go to **Tool** | **Subtool** | **Split** and select **Split Hidden**.

6. Go to **Tool** | **Geometry** | **Modify Topology** and select **Close Holes**. Repeat the steps for the body so that both arm and body subtools have no holes.

7. Merge the arm subtool with the wraps and armor:

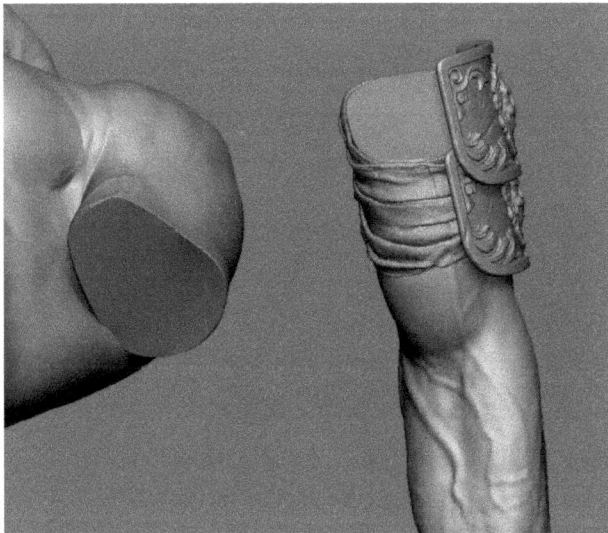

Figure 10.23 – The resulting elements after splitting and closing holes

8. With the arm selected, press *W* to open the **Gizmo** menu, click on the gear icon, and select **Remesh By Union**. From here, continue to go through the steps described in the earlier section, *Fixing a tool that is not watertight*, until the mesh integrity test is successful, and ZBrush can compute the mesh volume.

9. Repeat the steps for the body too.

In this section, you merged and cut your character based on the pieces you want to print. In the final section of this chapter, you will learn how to add keys, scale your model, and export it so that you can proceed with any 3D printing software and bring your character to life.

Adding keys, scaling, and exporting your model

In this section, you will add keys that let you combine individual pieces to assemble the full character. Then, you will learn how to scale and export your model so that it can be properly interpreted and processed by the printing software. Let's get started.

Creating a key model

There are many options for the shape of the key, but a simple tapered cube is a popular choice that gets the job done well:

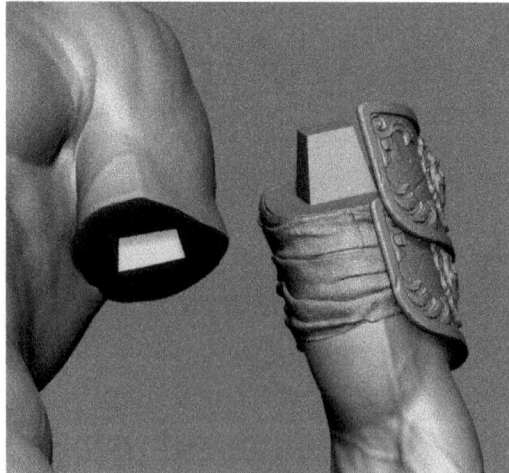

Figure 10.24 – Male key as a tapered cube (on arm) and matching female key (on torso)

To create this shape, follow these steps:

1. Go to **Tool | Subtool | Append** and pick any shape (for example, **Sphere3D**).

2. Navigate to **Tool | Initialize**. Set **X Res**, **Y Res**, and **Z Res** to **6**, and select **QCube**:

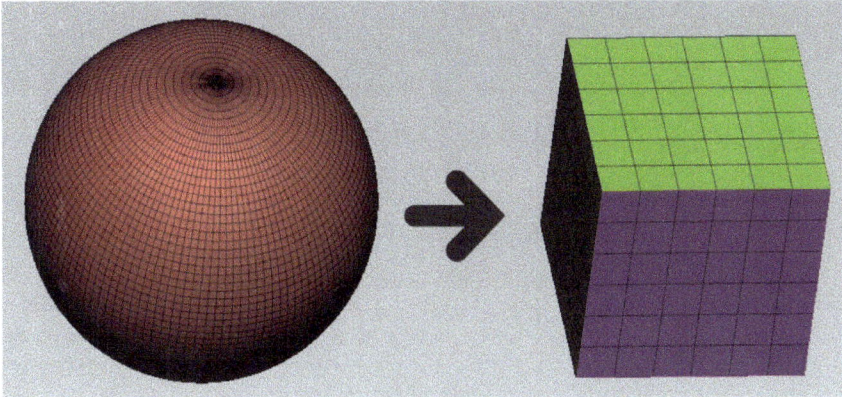

Figure 10.25 – Turning a mesh into a primitive shape with the Initialize function

3. Now, open the **Gizmo** menu, click on the gear icon, and select **Taper**. Taper the cube by dragging on the orange arrows. There is also a white arrow saying **Exponent**, which lets you adjust the profile of the taper to get a more curved or straight result.

Once done, the key is finished:

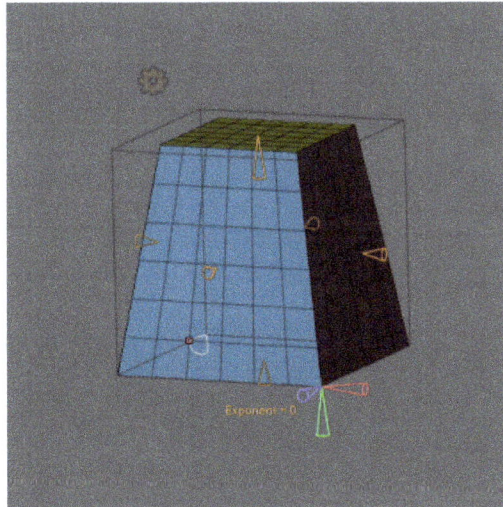

Figure 10.26 – A simple tapered cube

Now, let's use this model to create a male key on the arm and a female key on the body of the gladiator. In this screenshot, you can see what a male and a female key are:

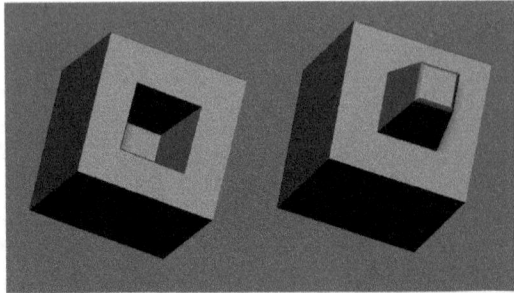

Figure 10.27 – Female (left) and male (right) keys

We need these to fit the two parts together. So, let's start creating them.

Adding a male key

To create a male key, follow these steps:

1. Position the cube on the arm, where you want it to attach to the body:

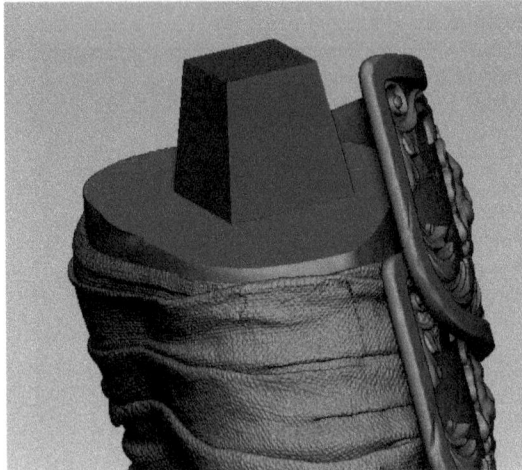

Figure 10.28 – Positioning the cube on the arm

2. Duplicate the cube and scale the duplicate cube up ever so slightly. This key will be used to subtract it from the body and create a hole. It has to be slightly larger so that the fit is not so tight. Depending on the printer, material, and desired tightness, you should aim to leave .1 to .5 mm of a gap. The most accurate way to get the scale of the gap right is to scale the cube up in an external 3D software such as Maya or Blender and then import it back into ZBrush. But scaling up the cube carefully in ZBrush so that it is slightly larger than the other one will be good enough in most cases:

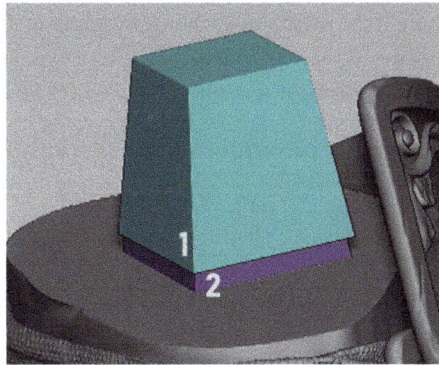

Figure 10.29 – Duplicating the key for adding a hole later

3. Switch to the arm subtool, then merge it with the original smaller cube subtool by going to **Tool | Subtool | Merge** and selecting **MergeDown** (assuming that you positioned the small cube subtool in the subtool list below the arm subtool).

4. Open the **Gizmo** menu, click on the gear icon, and select **Remesh By Union** to properly merge the key with the model:

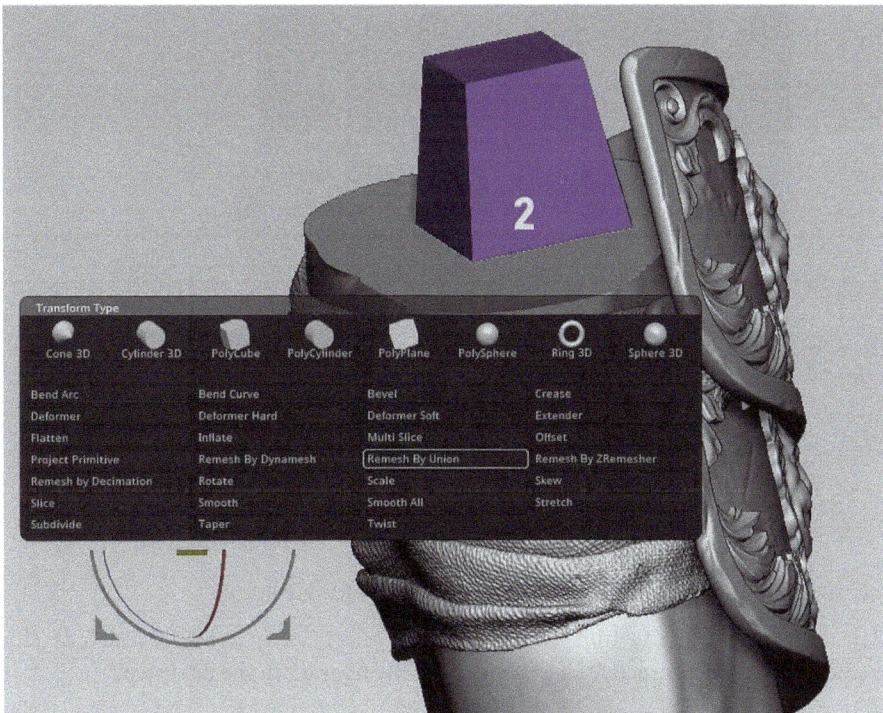

Figure 10.30 – Merging the arm with the smaller key with Remesh By Union

5. You need to check once more if the mesh has a volume by going to **Zplugin | 3D Print Hub** and selecting **Check Mesh Volume**. If there is an issue, continue to go through the steps described in the *Fixing a tool that is not watertight* section.

Now, you have successfully created an arm part with a key that is ready to be printed. Next, you have to add a matching hole on the body.

Adding a female key

The steps for adding a female key are very similar to the process of adding a male key:

1. The cube for the female key should already be positioned correctly since it was duplicated during the creation of the male key. It is important that it is not moved anymore so that both parts – the arm and body – will fit nicely:

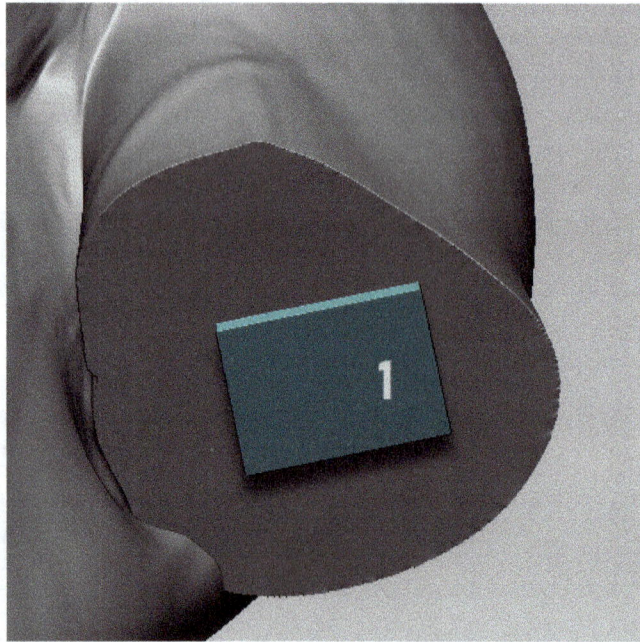

Figure 10.31 – The female key position

2. Then, switch to the arm subtool and merge it with the female key.

3. Isolate the cube, navigate to **Tool | Polygroups**, and select **Group as DynaMesh Sub**. This will turn the cube into a subtractive shape, which will be relevant in the next step:

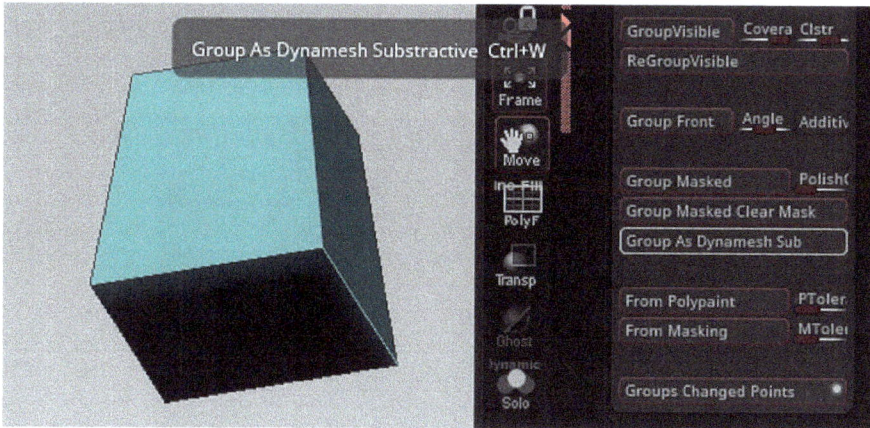

Figure 10.32 – Group as DynaMesh Sub function

4.   Open the **Gizmo** menu, then click on the gear icon and select **Remesh By Union**:

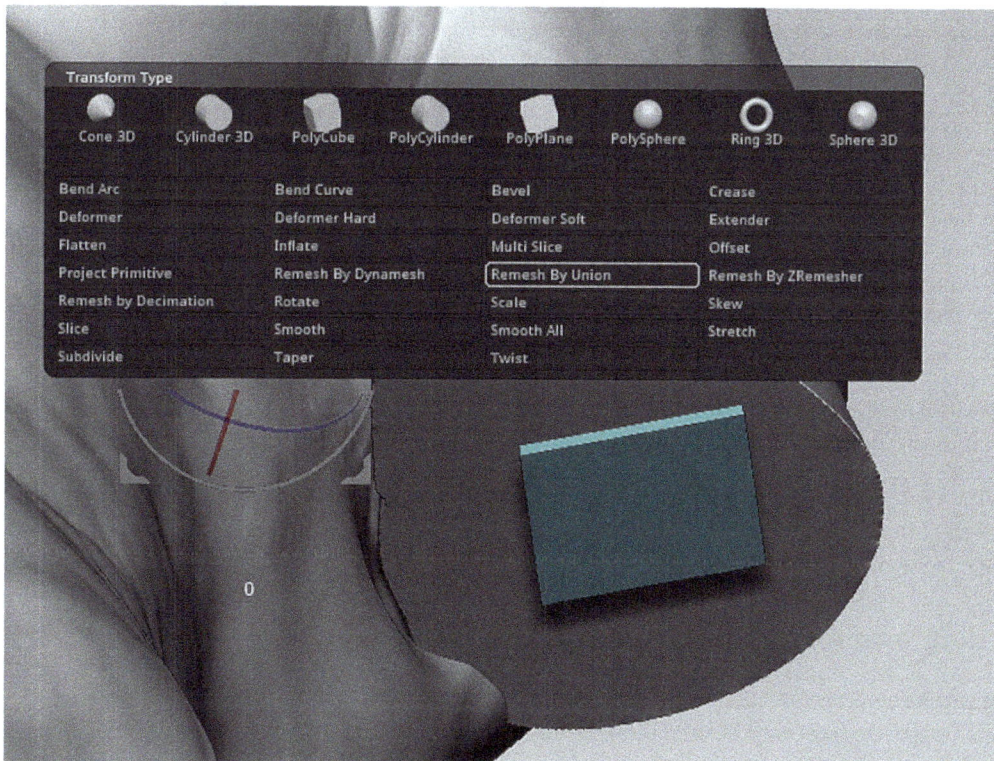

Figure 10.33 – Remesh By Union function

This will result in the cube being subtracted from the body since you used the **Group as DynaMesh Sub** operation. Now, you have the body model with a hole that matches the arm part with the male key perfectly:

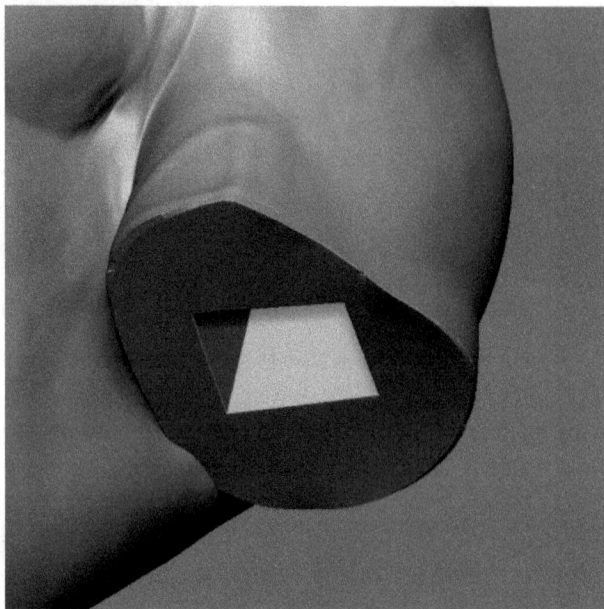

Figure 10.34 – The body after using a Remesh By Union operation with a DynaMesh Sub shape

5.   Finally, you need to check once more if the mesh has a volume by going to **Zplugin | 3D Print Hub** and selecting **Check Mesh Volume**. Again, if there is an issue, continue to go through the steps described in the *Fixing a tool that is not watertight* section.

Now, you have the 3D print-ready body piece in which you can insert the arm part. You can continue with this workflow, preparing the remaining parts of your character.

At this point, you are ready to scale your model using the **Scale Master** plugin, which you can also use to export your model and import it into your 3D printing software to turn your character into a physical model.

Scaling and exporting your model

Scaling your model is essential since you want to print the model in that size later on. In order to scale your character to have your desired dimensions, you can use ZBrush's **Scale Master** plugin, which you can find in the **Zplugin** palette:

Figure 10.35 – Scale Master plugin

Here are the steps to scaling your model with **Scale Master**:

1. In the **Scale Master** plugin menu, select **New Bounding Box Subtool**. This will create a box that fits all subtools of the scene perfectly, giving you the complete dimensions for the *x*, *y*, and *z* axes:

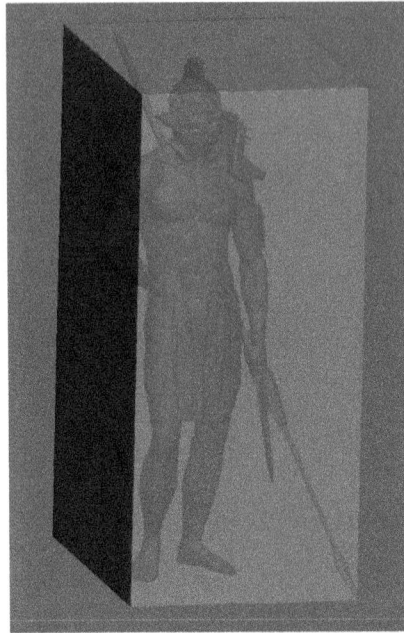

Figure 10.36 – Bounding box showing the character's dimensions

2. Click on **Set Scene Scale**. Now, you can pick between different units, such as **mm**, **cm**, or **in**. Don't worry about the exact numbers for now:

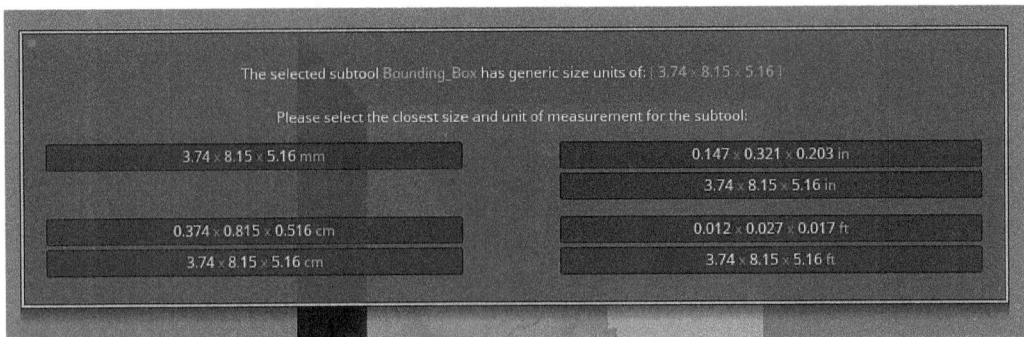

The selected subtool Bounding_Box has generic size units of: (3.74 × 8.15 × 5.16)

Please select the closest size and unit of measurement for the subtool:

3.74 × 8.15 × 5.16 mm	0.147 × 0.321 × 0.203 in
	3.74 × 8.15 × 5.16 in
0.374 × 0.815 × 0.516 cm	0.012 × 0.027 × 0.017 ft
3.74 × 8.15 × 5.16 cm	3.74 × 8.15 × 5.16 ft

Figure 10.37 – Set Scene Scale options

3. Now, you can set the scale with any of the sliders. Make sure **All** is enabled, which it is by default. Then, click on **Resize Subtool**. At this point, ZBrush will scale every subtool in your scene accordingly.

4. Delete the **Bounding Box** subtool since you don't need it anymore.

5. You can export your model as a .obj file in the same menu. Make sure **All** is enabled if you want to export every subtool, then click on **Export to Unit Scale** (which you can see in *Figure 10.35*).

Alternatively, while .obj is a valid format for 3D printing, it stores unnecessary information, such as color, texture, and material. The standard format for printing is .stl, and it only stores the object's geometry, which gives it lighter file sizes.

To export your model as .stl, simply change the file format in the **Export** window that pops up after clicking **Export to Unit Scale**:

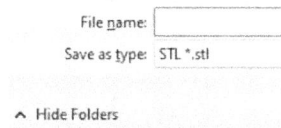

Figure 10.38 – Exporting your model as .stl

You now know how to scale your model and export it in two common 3D print file types, .obj and .stl.

This concludes the last section of this chapter, in which you learned how to add male and female keys to your models, as well as scale and export the finished pieces.

Summary

In this chapter, you learned how to turn your digital sculpture into a 3D model that can be 3D printed effectively.

First, you explored potential issues with your 3D models that prevent them from being 3D printed flawlessly, and you learned how to troubleshoot these problems to create watertight, print-ready meshes. Next, you learned how to merge parts and make clean cuts to separate your model into multiple pieces that can be printed individually, saving time and material costs.

In the last section of the chapter, you created a key model that you used to add a male key and a female key to have two matching pieces. Then, finally, you learned how to scale your character and export it so that it can be processed by any 3D-printing software.

In the next chapter, we will dive into head sculpting, which will cover everything from facial anatomy and blockout to likeness tips and hair creation with **FiberMesh**.

Part 3:
Sculpting a Female Head:
Tips and Techniques

Part 3 of this book teaches you how to sculpt a female head, including realistic skin detail, sculpted hair, and FiberMesh. The last chapter gives you tips for building a portfolio and using social media to promote your artwork, allowing you to showcase your ZBrush skills and create job opportunities.

This part includes the following chapters:

- *Chapter 11, Sculpting a Female Head*
- *Chapter 12, Adding Skin Detail, Sculpting Hair, and Using FiberMesh*
- *Chapter 13, Building a Portfolio and Leveraging Social Media*

11

Sculpting a Female Head

Portraits and busts are among the favorite subjects of artists because they allow you to connect to the artwork in a unique way. However, portraits are as challenging as they are popular because they allow for fewer imperfections, with even subtle flaws making a head look "off." This makes it essential to get a thorough understanding of the shapes and proportions, as well as the anatomical features, of a face.

This chapter will begin with a brief comparison of three potential starting points for blocking out a head: using a sphere, using a base mesh, and using ZBrush's Head planes model. You will learn about the pros and cons of each method, after which we will focus on the workflow with the Head planes model.

After that, our focus will turn to facial anatomy and individual facial features, including how to sculpt them properly. This way, you get to refine the head until it is missing only skin detail and hair, which will be added in the next chapter.

Then, you will learn about likeness sculpting and get some tips on how to make this difficult task as easy as possible. This section covers different criteria for picking a useful reference and how to use that reference material to create the likeness in ZBrush.

Finally, the last section contains a list of mistakes beginners tend to make when sculpting a head in ZBrush and how to avoid them.

So, this chapter will cover the following topics:

- Blocking out the head
- Refining the head
- Reviewing tips for sculpting likeness
- Avoiding beginner sculpting mistakes

Technical requirements

For the best experience, it is recommended that you have a strong PC that meets the minimum requirements described in the first chapter's *Technical requirements* section. However, you can work on this chapter with just a mouse, a functional PC setup, and a ZBrush license.

Blocking out the head

In this section, you will learn about different starting points for sculpting a head and what their pros and cons are. Then, you will learn how to block out the head from the **HeadPlanes_Female_256** model, from ZBrush's content library.

> **Important note**
>
> Before you start to sculpt the head, make sure you have collected enough pictures and reference material so you do not have to rely on memory and imagination. A great reference software is PureRef. You can find more information about it in *Chapter 2.*

Starting points for blocking out the head

As so often is the case in ZBrush, there are many ways to start modeling something. When it comes to sculpting a head, there are three common ways – using a sphere, using a base mesh, and using Head planes. Let's look at each of these options.

Sphere

Sculpting a head out of a sphere is one of the most popular methods, and many artists swear by it.

To load a sphere into your scene, simply go to the **Tool** palette, select **Append**, and pick the **Sphere3D** shape. Now, you can start sculpting and adding subdivision levels by pressing *Ctrl + D*, as you need to add resolution for sculpting more detail.

There are pros and cons to this method. Here are the pros:

- It is great for anatomy practice because you are required to establish the overall head shape first (good knowledge of the skull and realistic proportions and measurements are required for this).

- It leads to a unique-looking head because every facial feature will be modeled from scratch.

- It is a very flexible way to work, especially when combined with DynaMesh. It will allow you to experiment freely with proportions in a way that is not possible with a base mesh, where you usually want to commit to its topology.

Here are the cons:

- It is the most time-consuming method as you are sculpting everything from scratch
- There is also the need to retopologize the model when you want to refine it and add fine detail

Base mesh

The most effective and least time-consuming method is to use a base mesh. In some cases, the base mesh might already look similar to how you envision your model, so you just have to adjust the proportions.

You can import your base mesh by opening the **Tool** palette and then selecting **Import**. Alternatively, you can load a head from LightBox, which you can open by pressing ,. From there, you can navigate to the **Project** tab and select **DemoHeadFemale.Zpr**.

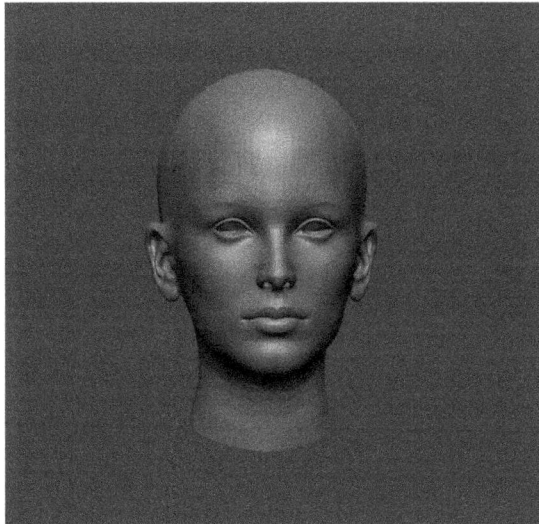

Figure 11.1 – DemoHeadFemale

Again, there are pros and cons to using a base mesh. The pros are as follows:

- Since all the facial features are already established, it is a very efficient method when quick results are needed (that's why this option is often used in commercial work)
- A good base mesh should have good animation- and sculpting-friendly topology, as well as UVs, which are another time-saver
- Some base meshes and head scan products come with texture maps, which let you view your head sculpture with skin color, without having to paint it yourself

The cons are as follows:

- There is little anatomy practice involved since the proportions and individual facial features are already established. The artist does not need as much knowledge of facial anatomy as they would with the other two methods.

- There is a risk of sticking to the shape of the base mesh. This often leads to a head model in which you can recognize the original base mesh.

- The topology can be restrictive if you want to sculpt a head with different proportions. Retoplogizing the base mesh would then cost more time.

Head planes

The third starting point for sculpting a head is the Head planes model, provided in ZBrush's LightBox.

In some ways, it is a balance between starting from a sphere and starting with a base mesh, as it allows you to sculpt a unique head, learn anatomy, and be flexible in adjusting the proportions. At the same time, it saves a bit of time, compared to the sphere, as the overall head proportions are already in place.

To load the **HeadPlanes** tool, open the LightBox by pressing , and navigate to the **Project** menu. Then, open the **Head planes** folder and pick any of the male or female ZProjects.

Figure 11.2 – Head planes

> **Important note**
>
> You will notice **128** and **256** at the end of the **HeadPlanes** model names – they signify the DynaMesh resolution value. **256** models will have a higher resolution, although this can be changed later.

Like before, there are pros and cons to using Head planes. Here are the pros:

- You have a good starting point, saving you time

- It allows you to practice sculpting facial features such as eyes, nose, and mouth

- You can experiment and change proportions at any point during the process since you are using DynaMesh with it

Here are the cons:

- You don't have good topology, UVs, or texture maps like you tend to have with base meshes

- As is the case with the sphere, eventually you need to do a retoplogy for a high-fidelity model

Now, you have three methods to choose from, based on your situation and goals. In order to cover more of the anatomy and sculpting process, this book uses the Head planes method, but you are free to use a base mesh and apply the lessons to it.

Using the Head planes blocking-out method

To start the head sculpting process, as mentioned, make sure you have some reference images of the face you want to sculpt. Then, open LightBox, navigate to the **Project** menu, open the **Head planes** folder, and pick the **HeadPlanes_Female_256** ZProject.

> **Important note**
>
> Make sure to enable **Symmetry** mode by pressing X. Having a symmetrical model will not only save time but also help you sculpt a prettier face. You may choose to break the symmetry toward the end of the project, but you should try to keep the model symmetrical until you are completely satisfied with the proportions and overall shapes of the facial features.

The first thing you should do is to add the eyelids. For that, you can use the **IMM Primitives** brush and pick the sphere from the shapes that are displayed above the canvas. Then, place the spheres where the eyes will be, and apply a **DynaMesh** operation to merge them with the head. The result should look similar to this:

Figure 11.3 – Adding spheres to create the eyelids

Next, you can create holes between the eyelids – so you can add the eyeballs later – using any sculpting brush, such as the **ClayBuildup** brush:

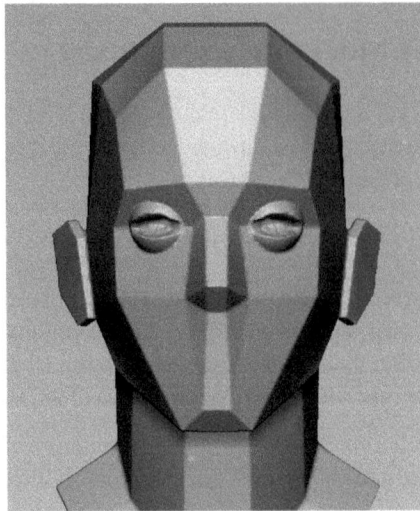

Figure 11.4 – Adding holes for the eyeballs

At this point, you can use basic sculpting brushes, such as **ClayBuildup**, **DamStandard**, **Standard**, and **Move**, to sculpt the face. You can soften up the harsh lines and copy what you see on your reference pictures. The result could look like this:

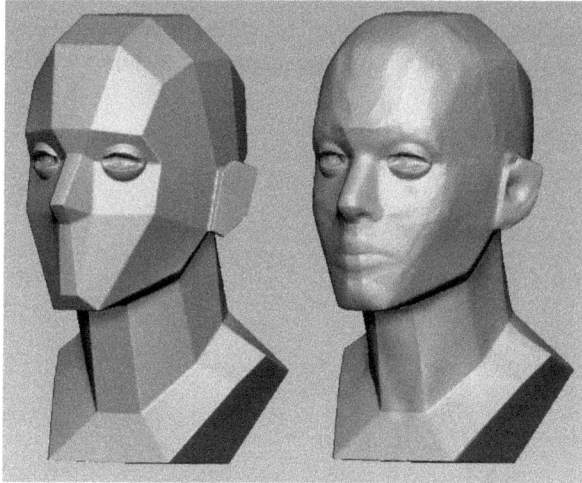

Figure 11.5 – Softening up the edges of the model

When using the **ClayBuildup** brush, you can also alternate between **Zadd** and **Zsub** mode (using *Alt*) to add/remove forms and carve away at your model). Don't worry if your model does not look great to begin with or if you are struggling to sculpt some of the facial features because the next section covers the anatomy of each of them.

As a tip, when you are sculpting a head, it can be beneficial to look at the face in terms of shapes and planes. You can look for reference pictures of *a simplified head* or *planes of the head*. These could be very rough, to be used for the blockout phase, or slightly more defined, to help with later stages of the sculpture.

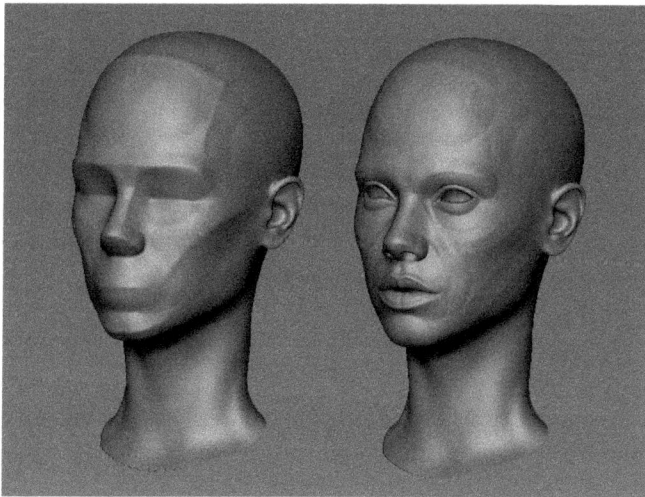

Figure 11.6 – Simplified Head models highlighting shapes and planes of the face

Also, if you are new to sculpting heads, it can be hard to get the proportions of the head to look correct. Here, it will help to look at the most important measurements. A simple way to get the proportions of the face right is to break it up into thirds:

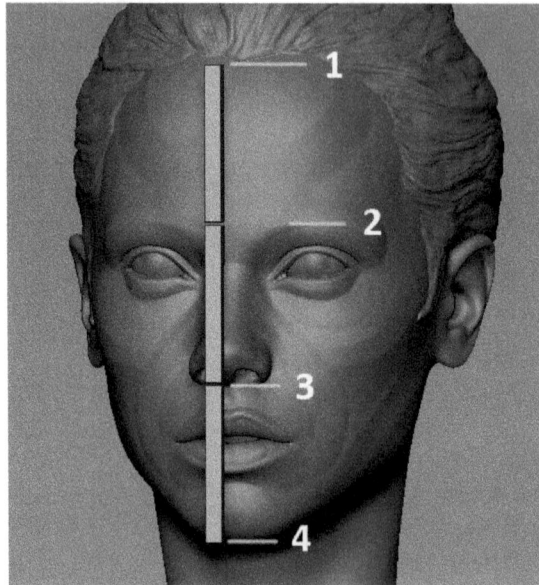

Figure 11.7 – Establishing the basic face proportions using thirds

Let's briefly explain this:

- The upper third (labels *1* to *2*) spans from the hairline to the eyebrows
- The middle third (labels *2* to *3*) continues to the bottom of the nose
- The last third (labels *3* to *4*) measures to the bottom of the chin

This concludes this section, in which you learned about different ways to approach the sculpting of a head. Then you used one of these methods, loading the Head planes model from LightBox, to block out the head. In the next section, you will take a closer look at individual facial features, such as the nose, lips, and ears, to create a more detailed and accurate model.

Refining the head

In this section, you will continue to refine the face, plus you will learn about the characteristics of the various facial features and the role that the skull, muscles, and fat deposits play in the shapes of these features.

Again, studying reference material and keeping the facial anatomy in mind will help you create a more realistic and visually appealing head.

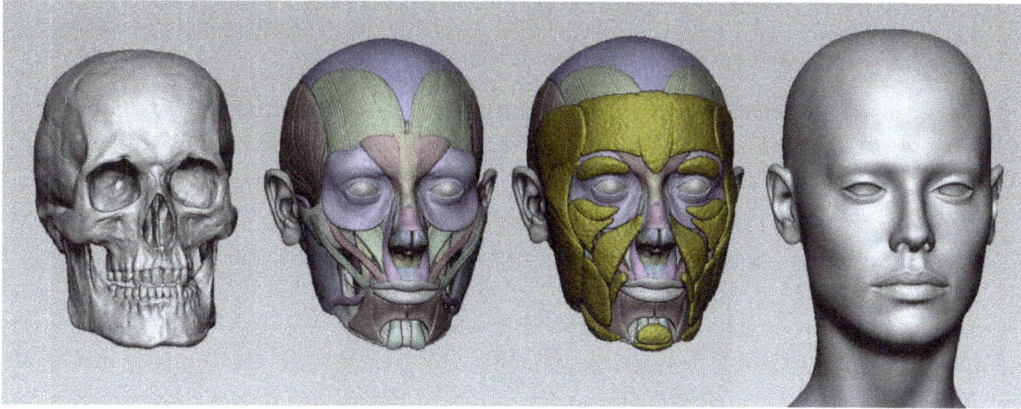

Figure 11.8 – Skull, muscles, and fat of the face

Also note that this project is based on a Caucasian woman. Differences in the proportions and shapes of skulls in other ethnicities give their faces a different appearance, so you might have to find extra reference pictures and look at the specific development of each facial feature more closely. However, the underlying anatomy is similar, and the information can still be used to sculpt a more believable head.

Here, you can observe differences between a European, Asian, and African skull:

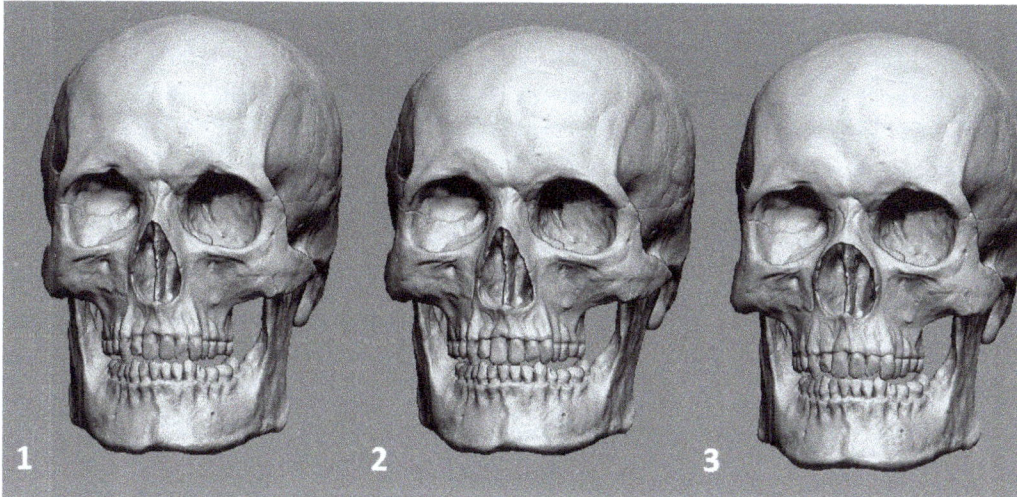

Figure 11.9 – Differences in European (1), Asian (2), and African skulls (3)

You can see how the Asian skull has more pronounced cheekbones, while the African skull has a bigger nasal cavity.

Now, before you proceed with the sculpting, you may want to increase the DynaMesh resolution so you have a high enough density of polygons to sculpt more defined forms.

In order to do that, go to **Tool | Geometry | DynaMesh** and increase the value of the **Resolution** slider. Try to experiment with lower values first, and see at what point you reach a resolution of around one million. This should give you a sufficient density to sculpt more defined forms.

Figure 11.10 – Increasing the mesh resolution with the DynaMesh Resolution slider

Then, press *Ctrl*, left-click, and drag any empty space on the canvas to apply the new resolution to the mesh.

At this point, you can proceed to add the eyes and refine the eyelids that we have already modeled.

Eyes and eyelids

The eyes are possibly the most important facial feature, as they are naturally a focal point, and therefore you have to treat them with attention to detail. Here is what the areas around the eyes look like as simplified shapes:

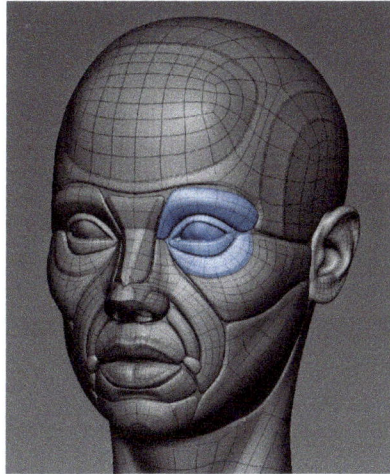

Figure 11.11 – The eye area as simple shapes

First, you need to add the eyeballs, because it will guide how you sculpt the eyelids.

Adding the eyeballs

The eyes can be simple spheres, but it is important to get the size and position right.

To add a sphere, go to **Tool | Subtool | Append** and pick the **Sphere3D** shape. Position the sphere on one side of the head where the eye would be, then go to **Tool | Geometry | Modify Topology** and select **Mirror and Weld**. After that, enable **Symmetry** mode by pressing *X*, so you work on both eyes simultaneously.

In order to scale the eyes to a realistic size, you can follow the rule that there should be one eye length of distance between the eyeballs. As long as you keep the position of the eyeballs centered behind the eyelids, this will give you a good estimate of the scale.

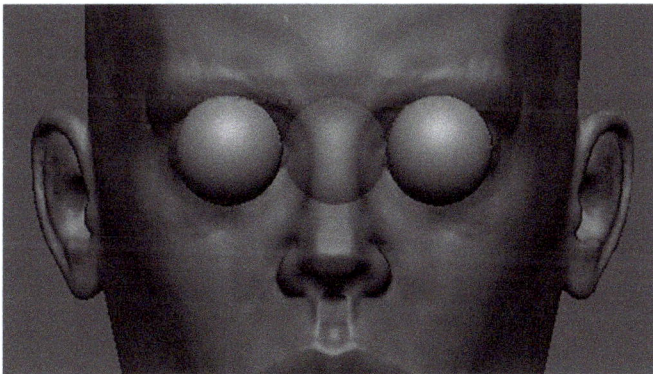

Figure 11.12 – Eye distance

To get a more accurate measurement, you may scale them down slightly or measure them to ensure a diameter of roughly 2.5 cm.

For the depth of the eyeballs, make sure that they touch an imaginary line from the brow to the lower part of the orbital bone. You can use **Transparency** mode to better see the placement.

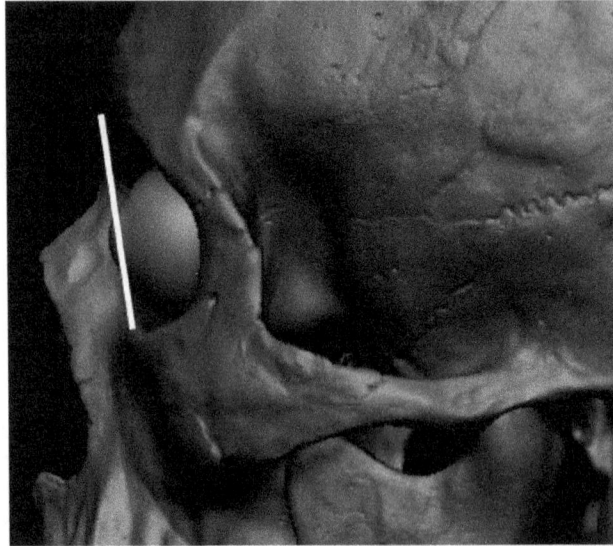

Figure 11.13 – The depth of eyeballs

Now let's move on to refining the eyeballs.

Refining the eyeballs

To create a more realistic effect on a sculpted eye, you can add the effect of the iris:

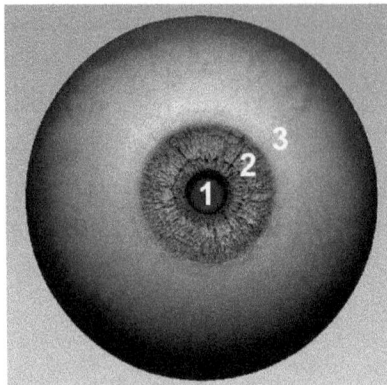

Figure 11.14 – Pupil (1), iris (2), and sclera (3)

To do so, go through the following steps:

1. Mask off the front of the eyeballs based on the size of the eyes' iris.
2. Invert the mask by *Ctrl* + left-clicking on the canvas.
3. Then, *Ctrl* + left-click on the mask to blur it.
4. Press *W* and use the Gizmo to move back the unmasked part.
5. Use the **Smooth** brush to create a flat surface.
6. Use the **ClayBuildup** brush on **Zsub** mode to chip in a little hole where the pupil is.

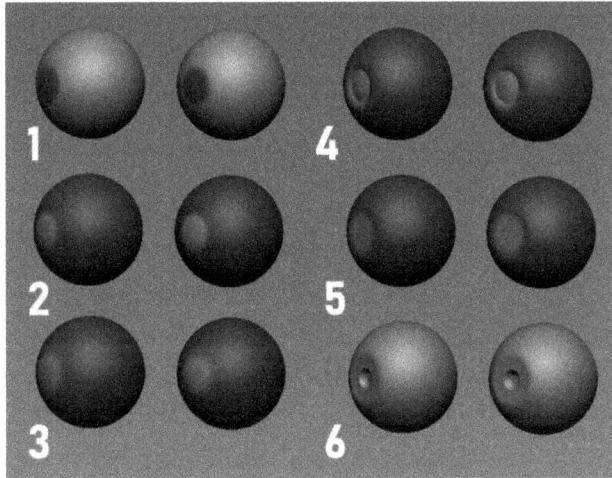

Figure 11.15 – Creating simple eyes

> **Important note**
>
> If you want to learn how to sculpt a realistic eye in ZBrush, there is a fantastic tutorial on J Hill's YouTube channel called *Making an Eye in Zbrush and Rendering*: https://www.youtube.com/watch?v=Qj5uK6RSdUo&t=1s.

With the eyeballs done, let's move on to refining the eyelids (we already started creating them earlier in the chapter).

Refining the eyelids

The eyelids can come in various shapes, and it is important to follow the reference closely. Make sure to avoid making them too round and indicate the change in direction of the eyelids' silhouette, as illustrated in the following figure:

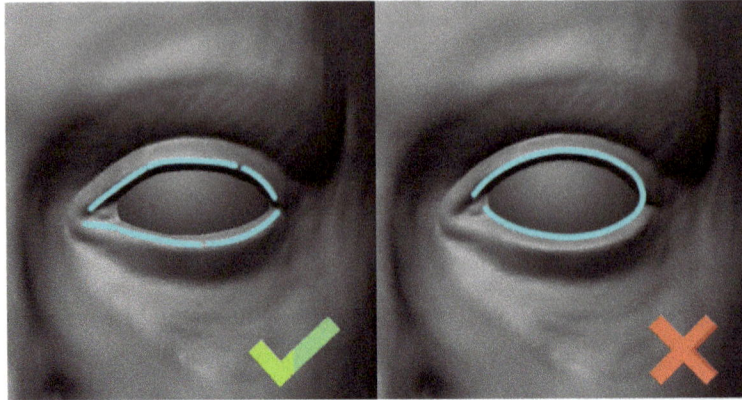

Figure 11.16 – Avoid the mistake of making the eyelid shape too round

When you sculpt the upper eyelid fold that sits above the upper eyelid and the forms around the eye, there are a few points to keep in mind:

- Pay attention to the shape of the eyelids from every angle. It is useful to look from a downward angle, where you can see the unique curvature of the eyelids, wrapping around the eyeballs, and then changing direction toward the inside of the eye (see label *1* in *Figure 11.17*).

- Make sure to also look at the side profile of the head and ensure that the silhouette of the face and brows, transitioning to the forehead, looks anatomically correct. A good reference collection that includes those profile shots will be essential (see label *2* in *Figure 11.17*).

- You can visualize the upper eyelid fold and brow region as a simple bent cylinder shape, with the most protruding part being in the middle, where the eyebrows are (see label *3* in *Figure 11.17*).

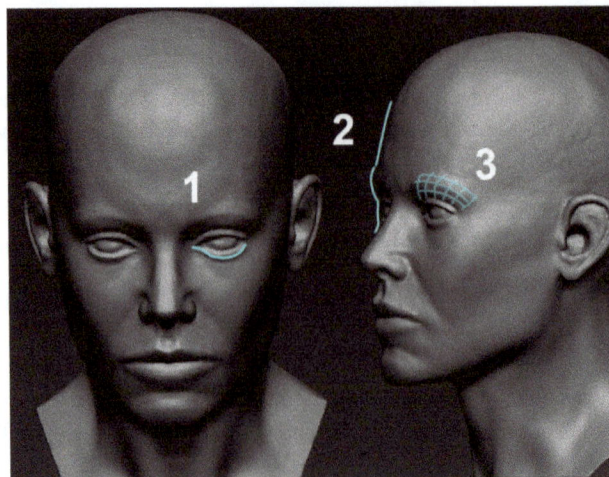

Figure 11.17 – Important shapes and silhouettes of the eye region

- When placing the eyeballs, make sure to give the eyelids enough thickness. It is a common beginner mistake to make the eyelids too thin, which results in a lack of shadows cast from the upper eyelid and will make the eyes look less believable.

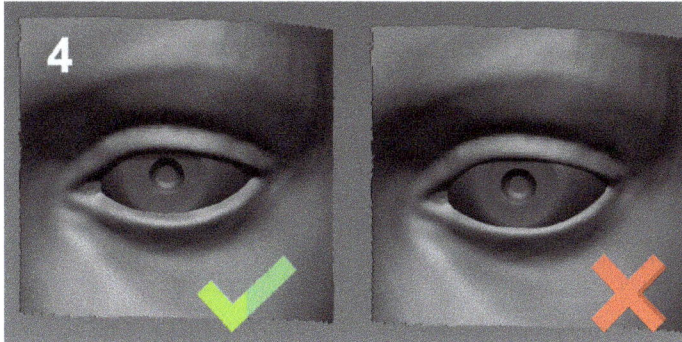

Figure 11.18 – Ensuring proper eyelid thickness

- Don't forget to add the caruncle. It is a small, fleshy protuberance that sits in the inner corner of the eyes. It is important to add this detail when aiming to sculpt a realistic eye.

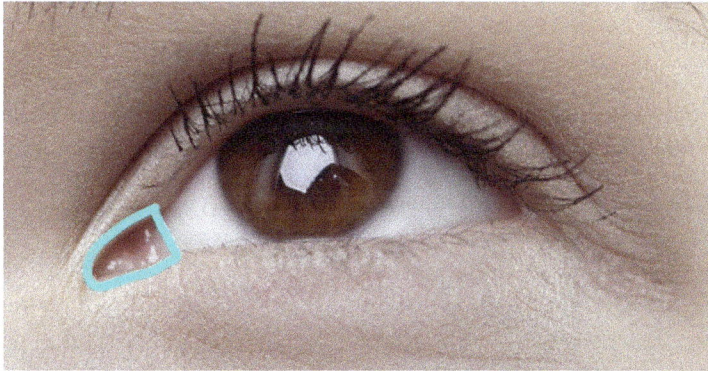

Figure 11.19 – The caruncle of the eye

Now that you have made some essential modifications to the eyelids, you can refine the eyeballs by adding the iris.

Here is what the result could look like after the refinement of the eyes and eyelids:

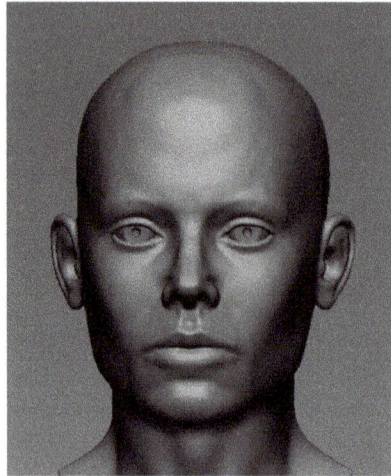

Figure 11.20 – The head model after working on the eyes and the surrounding areas

Next, let's look at another important and complex facial feature: the nose.

Nose

The nose is one of the more challenging facial features to sculpt accurately. There is a lot of variation in the shape of the nose, but some characteristics are always similar due to the underlying anatomy.

Let's take a look at the nose in a simplified model:

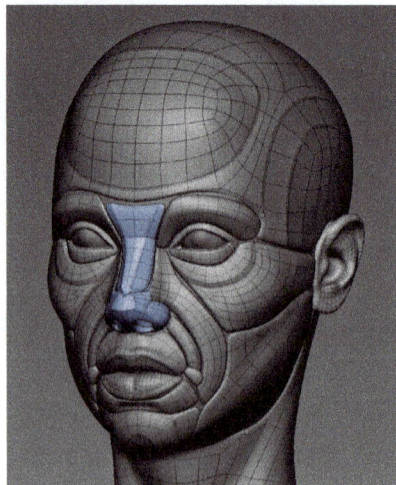

Figure 11.21 – The nose in a simplified model

Dissimilar to many parts of the face that get their appearance mostly through the bone structure, facial muscles, and fat deposits, the nose shape is also determined by cartilage, which makes it essential to take a closer look at these cartilage parts before starting to sculpt.

There are four main parts of the nose: the nasal bone (*1*), lateral cartilage (*2*), major alar cartilage (*3*), and fibrofatty tissue (*4*):

Figure 11.22 – Anatomy of the nose

When sculpting the nose, it is especially important to also take a look at the side profile. Since we are sculpting a woman here, the nose silhouette will be a bit more concave and shallow. This type of shape creates a more feminine, as well as younger, appearance:

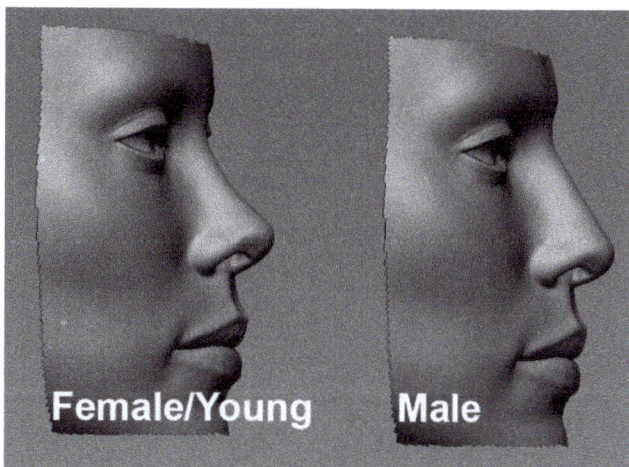

Figure 11.23 – Nose shape for a more female versus male nose

After taking a look at the side profile, you can change the perspective to a lower angle and work on the shape of the nostrils. Here, it will increase the realism to pay attention to the shape of the nostrils, which will make the nose unique and interesting.

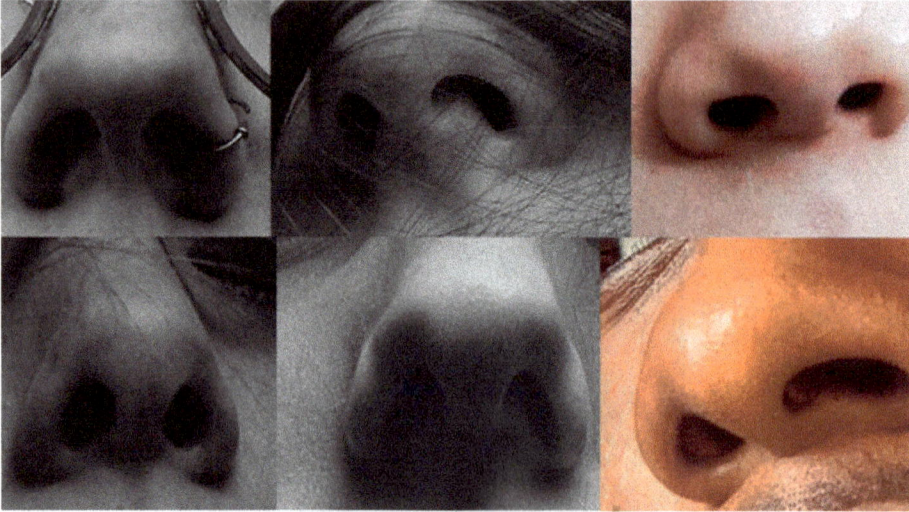

Figure 11.24 – The variety of nostril shapes

In this figure, you can also see the shape of the major alar cartilage and how big of a role it plays in the nose's appearance. Ideally, you should have many reference pictures, in good resolution, and from different angles, on which you can identify the nose's anatomy and translate it to your sculpture.

Combining those ideas, this is what a female nose could look like in ZBrush:

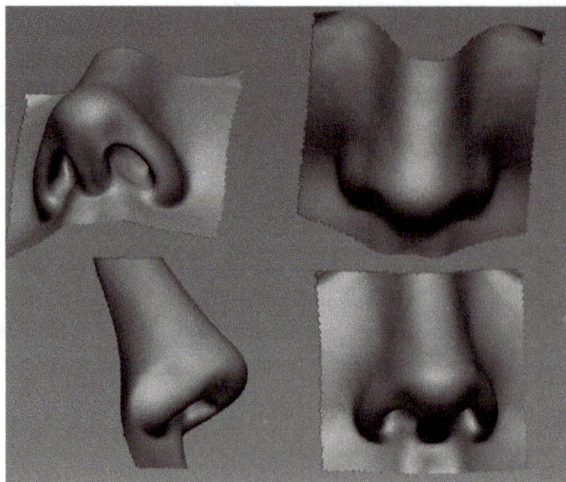

Figure 11.25 – Female nose

On a more general note, when you are sculpting a female face, you should generally create more rounded shapes and smoother transitions, avoiding harsh lines and intense, angular features:

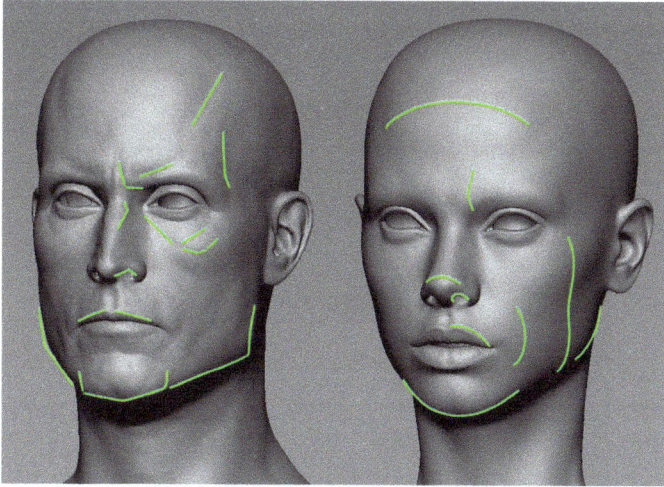

Figure 11.26 – More angular versus rounded shapes in a male and female face

Lips

The lips can be a very expressive facial feature – although they are relatively simple in shape, it is important to pay attention to the subtle details in silhouette and forms in order to get the best result. Here are the lips and surrounding area as simple shapes:

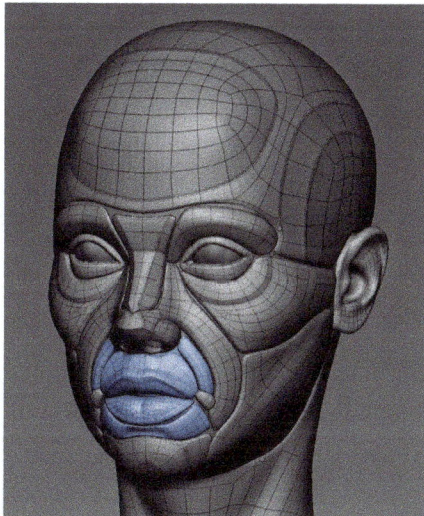

Figure 11.27 – Lips and surrounding area simplified

Try to keep these things in mind when working on the lips:

- Pay attention to the overall shape and silhouette of the lips. Capturing the right curvature is essential to sculpting good-looking lips, which may be a particular focus when modeling female characters.

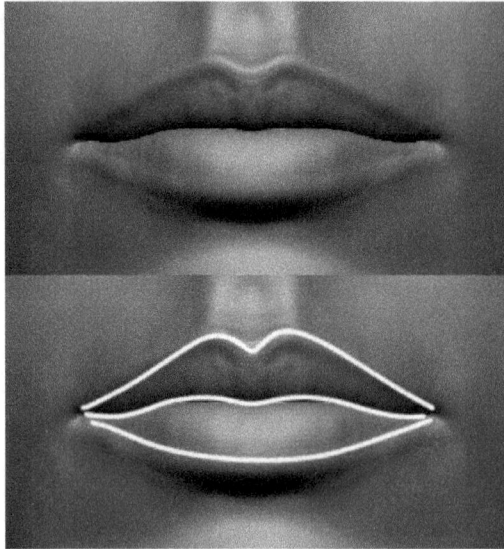

Figure 11.28 – Silhouettes and curves of female lips

- There are many convex and concave shapes around the lips, and it is essential to sculpt them with the right curvature. The following figure shows some of these shapes:

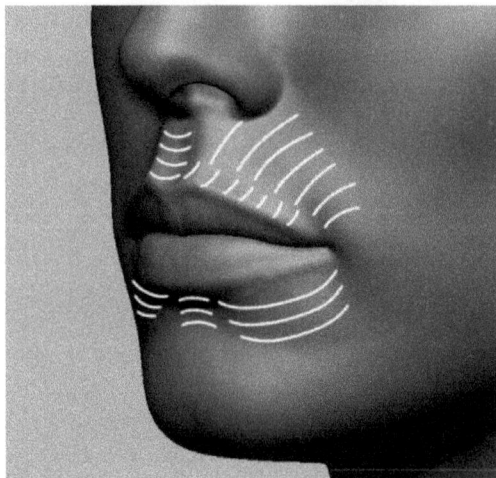

Figure 11.29 – Concave and convex shapes around the lips

- Take a look at the side profile of the face. Make sure that the upper lip protrudes further than the lower lip and that the concave curvature below and above the lips is present.

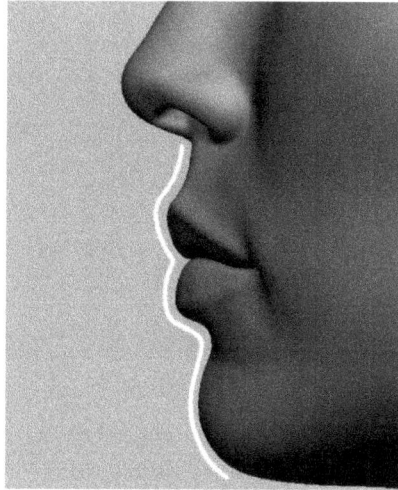

Figure 11.30 – Side profile of the lips

- When going into more detail, and working on the forms of the upper and lower lips, you can use spheres to help you, like so:

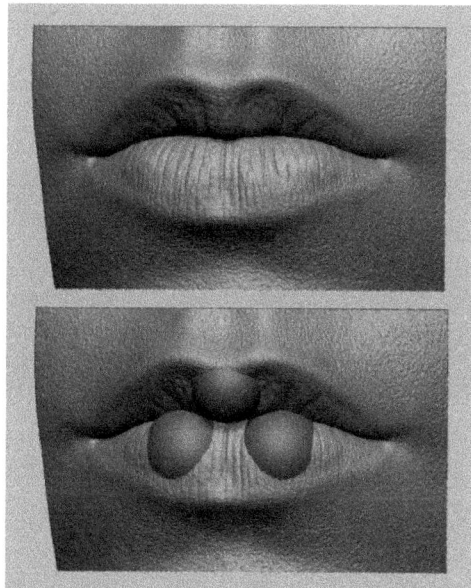

Figure 11.31 – The forms of the lips, visualized as spheres

- There is a small mass next to the corners of the mouth, which is essential for capturing an accurate sculpt. This mass comes from the facial muscles meeting in this spot, as well as a facial fat deposit being there:

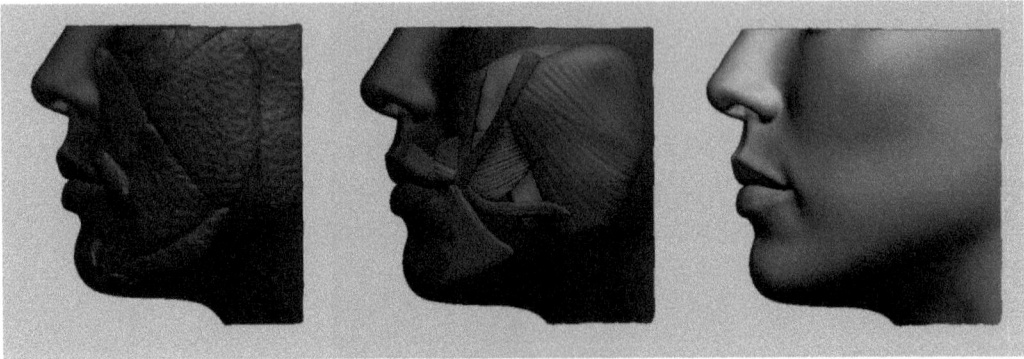

Figure 11.32 – Facial muscles and facial fat deposits creating a shape next to the corners of the mouth

- Finally, you need to ensure that the curvature of the lips from a low angle looks as it does here, sloping backward, toward the outside, so that the lips do not look too flat:

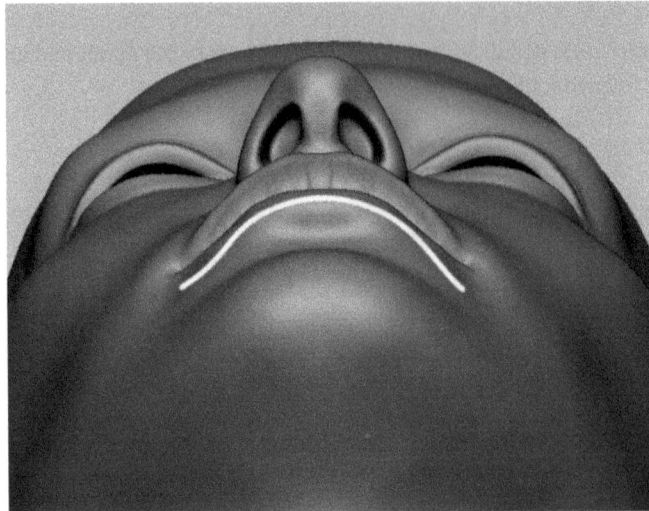

Figure 11.33 – The curvature of the lips, seen from a low viewing angle

Ears

The ears are one of the more complex facial features, and although they tend not to get the same attention as the more centered facial features, they are an important element. Well-sculpted ears can improve the quality of the portrait significantly.

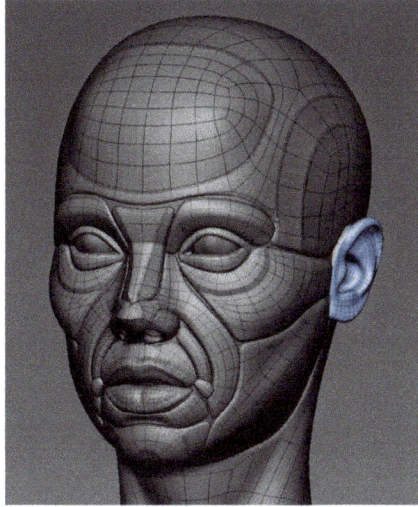

Figure 11.34 – The ears in a simplified model

As a beginner in sculpting and facial anatomy, it can help to break down the ears into shapes to make it more easily digestible, so that you can construct it from scratch. There are four main parts: the helix (*1*), the antihelix (*2*), the tragus (*3*), and the lobule (*4*):

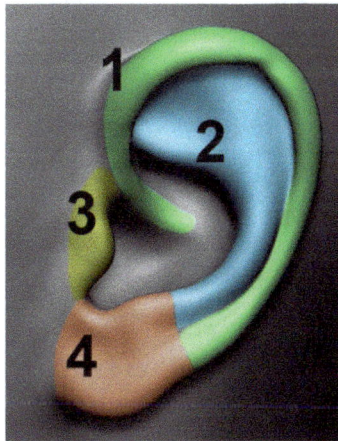

Figure 11.35 – The shapes of the ears

Pay attention to these parts when sculpting the ear, and make sure that you have reference images from several angles. Of course, the ears silhouette is especially important from the front, but another characteristic of the ear from the side view is also important – the ears should be oriented with a slight angle, and not be completely straight. This is a common beginner mistake and should be avoided for a realistic sculpture.

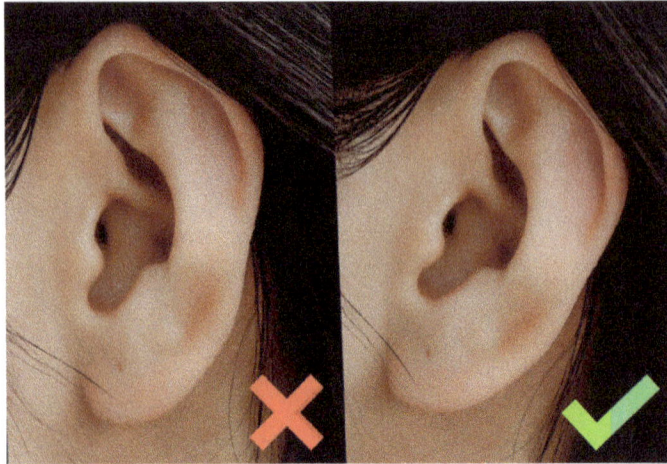

Figure 11.36 – Incorrect and correct tilt orientation of the ear

When it comes to the silhouette of the ear from a front view, there is not a clear shape, but rather the amount of variation makes it a matter of looking at a reference and matching the picture accordingly.

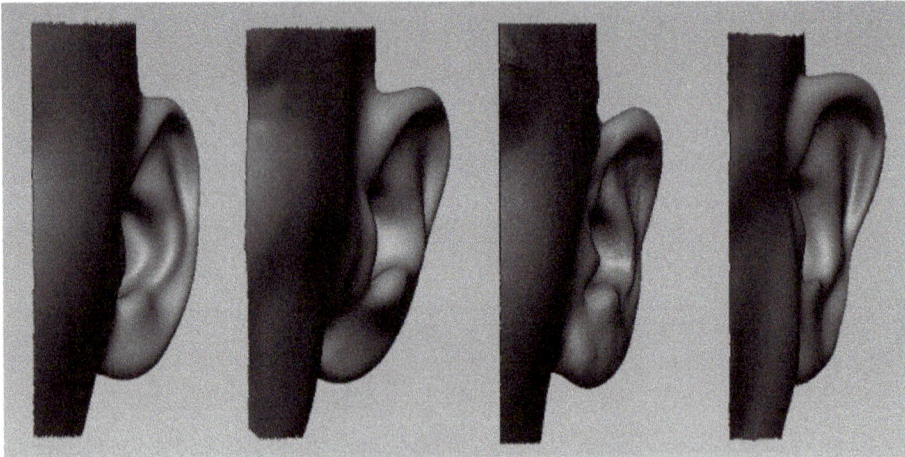

Figure 11.37 – Different silhouettes of ears

With these tips in mind, you can go ahead and sculpt the ears of your portrait, making them a strong part of the sculpture instead of just being an afterthought.

Forehead

When it comes to sculpting the forehead, the shape is mostly determined by the skull. Here is what the forehead looks like as simple shapes:

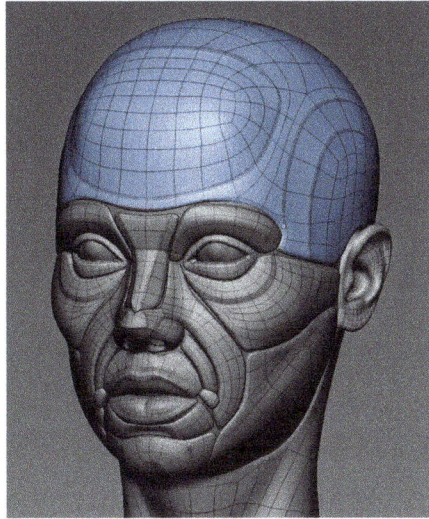

Figure 11.38 – The forehead in a simplified model

Before you start sculpting, it is worth taking a look at references of skulls, or even loading in a skull scan 3D model to your scene. With your reference collected, you can take a look at the shapes and anatomical features:

- **Brow bone**: The brow bone is much more pronounced in men and is a good feature to sculpt more pronounced for a masculine and fierce-looking portrait. Facial muscles are attached to the brow region and can exaggerate the brows when they are contracted in a fierce expression, but the overall size is mostly determined by the skull, which makes it a useful subject to study.

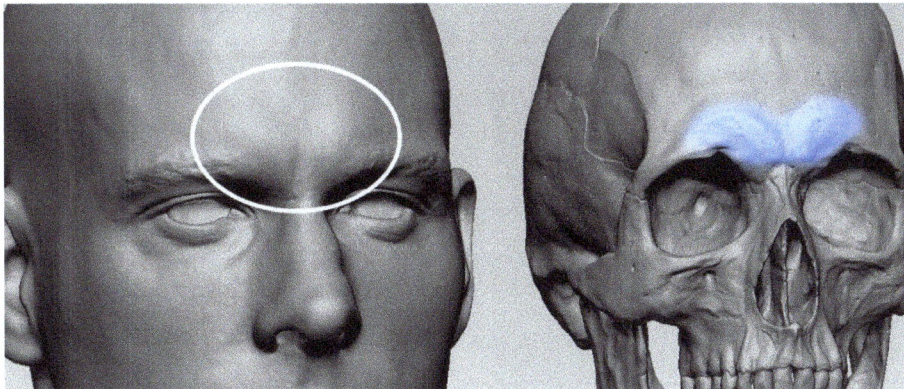

Figure 11.39 – Brow bone

- **Temporal line**: This is another common bony landmark that you need to consider when sculpting the forehead. Like the brow bone, this is mostly visible in men and becomes more pronounced with increasing age. As we are sculpting a female, make sure to sculpt this subtly and not like a harsh line – the **Standard** brush on a low intensity is a good tool to accomplish this.

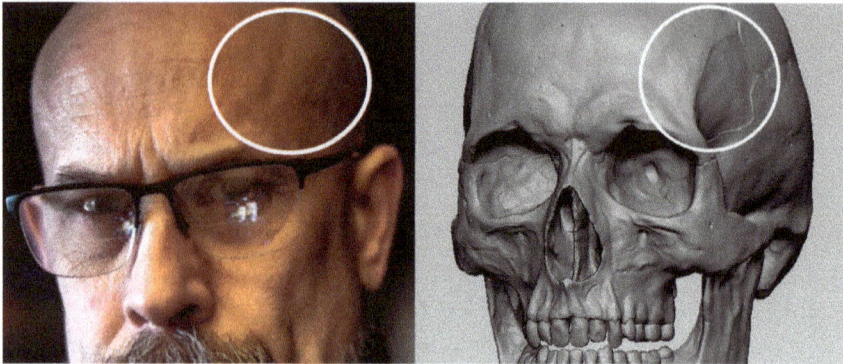

Figure 11.40 – Temporal line

- **Glabella**: This is a feature that is more apparent in women, which makes this an important shape for our female portrait. Don't sculpt it too pronounced as female brows should not look too strong.

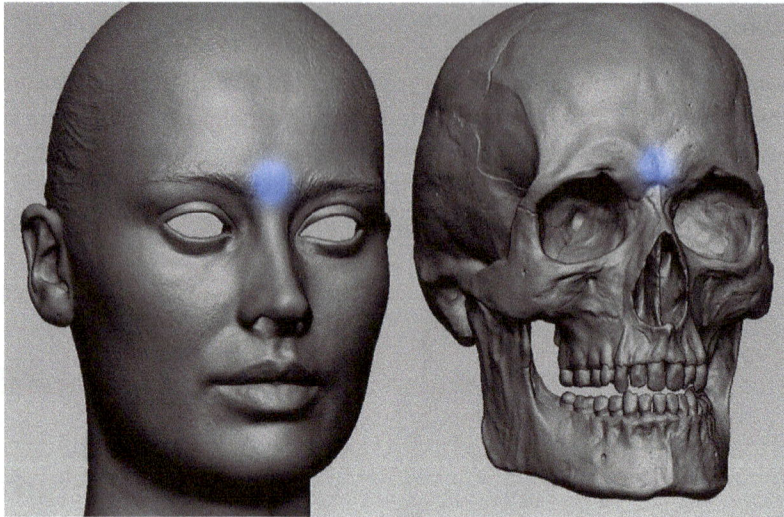

Figure 11.41 – Glabella

- **Frontal eminences**: This is another important shape on the forehead that is more visible in women. Incorporating this shape in your sculpture can help break up the uniform, spherical shape of the forehead that beginners tend to sculpt, with a lack of forms and shapes.

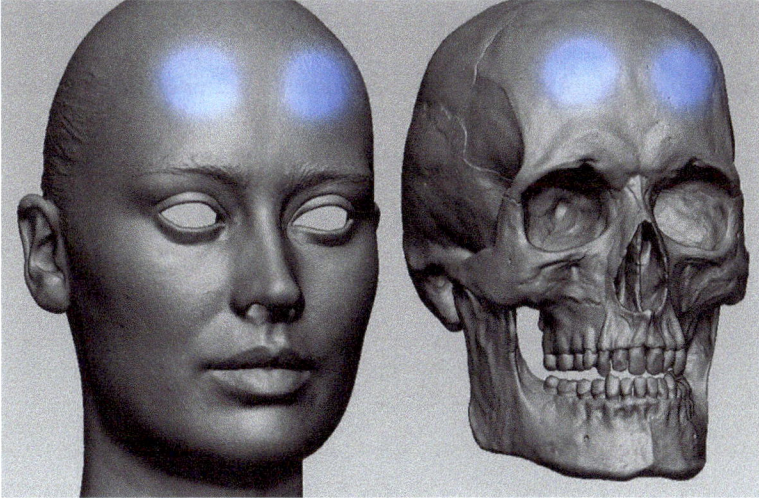

Figure 11.42 – Frontal eminences

Some of these features could be considered somewhat more advanced sculpting, as they are more subtle and do not stand out as much as other areas. But as a final broader guideline, you should watch out for the most common mistake with sculpting the forehead, which is making it too flat, that is, not giving it enough of a spherical shape:

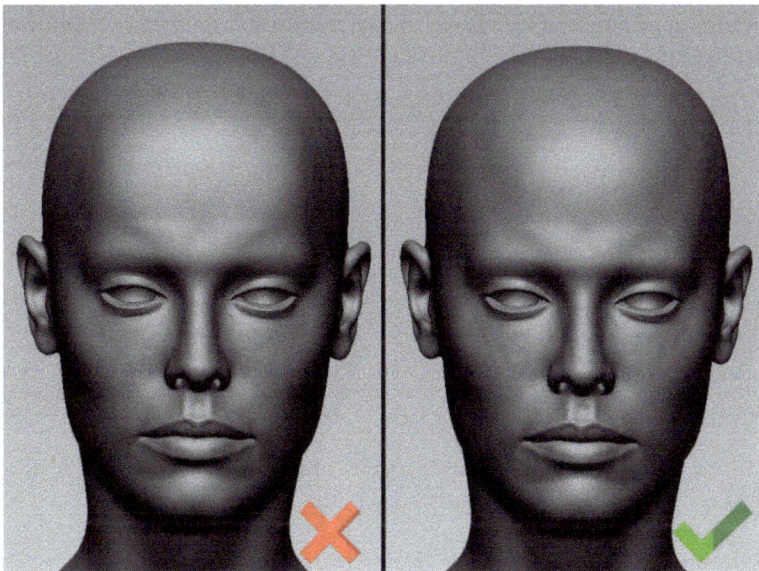

Figure 11.43 – Beginner mistake: Flat forehead

Cheeks and nasolabial fold

The cheeks and nasolabial fold are one of the most prominent features of the face. Here they are as simple forms:

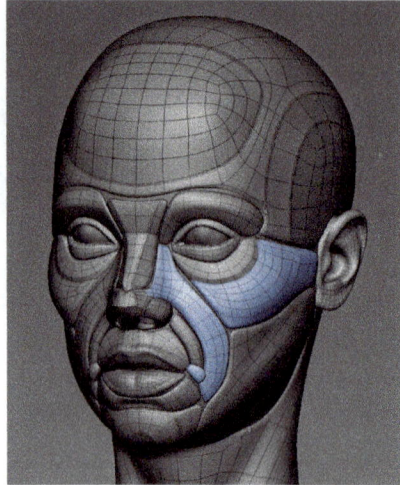

Figure 11.44 – Nasolabial fold and cheekbone

When it comes to the appearance of the cheeks, it comes mostly from the zygomatic bone, or what most people refer to as the cheekbone:

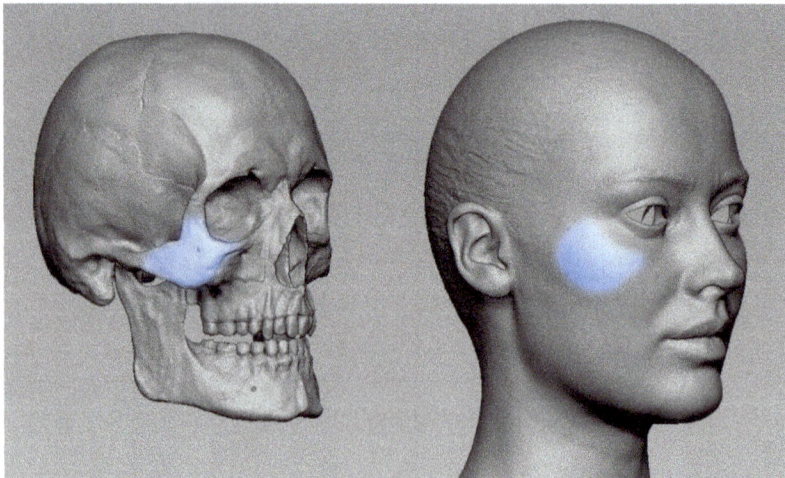

Figure 11.45 – Zygomatic bone/cheekbone

Below the cheekbone, toward the mouth, there are two important fat deposits: the nasolabial fat deposit and the superior jowl fat deposit. The crease line below the nasolabial fat is called the nasolabial fold, and it is an important landmark on the face that becomes more pronounced with age.

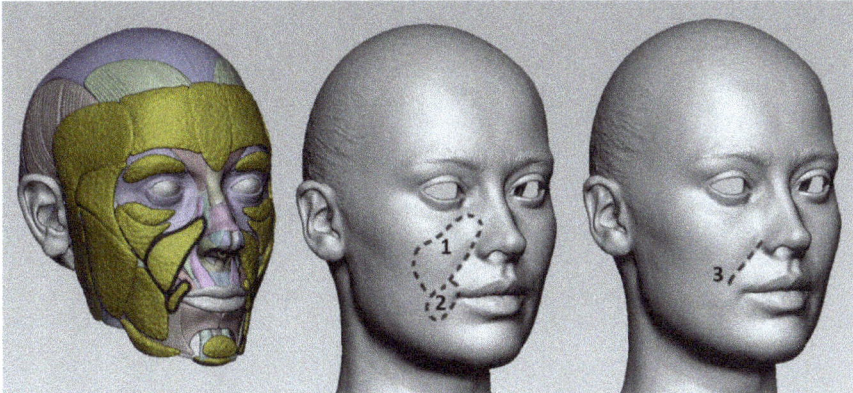

Figure 11.46 – Nasolabial fat deposit (1), superior jowl fat deposit (2), and the nasolabial Ffold (3)

Keep these anatomy features in mind when you observe reference images and sculpt your head.

Chin and jaw

When it comes to the face's silhouette, the chin and jaw can play an important part in giving a face a strong, youthful, and pretty appearance. This is how both features can be simplified:

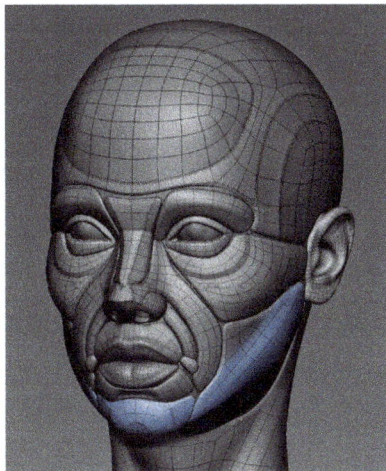

Figure 11.47 – Chin and jaw as simple shapes

From the front, the chin can be smaller and pointier, giving it a more female and youthful look, or more broad and strong for a more masculine effect. Looking at the chin from the side profile, you can see how the most protruding part tends to be around the center of the chin.

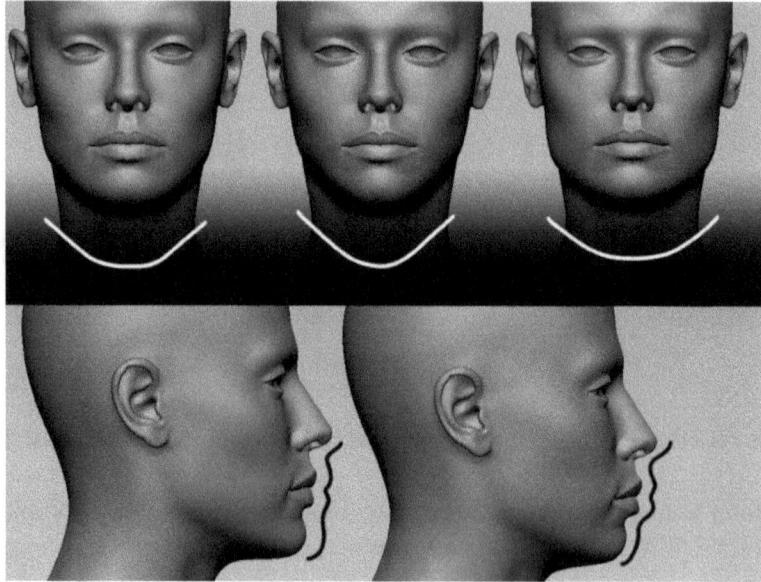

Figure 11.48 – Different profiles and shapes of the chin

Besides the silhouette and profile of the chin and jaw, it is also worth taking a closer look at the shapes that make up the chin, as they are indicated in the following figure. The facial muscles, fat, and skull determine these shapes.

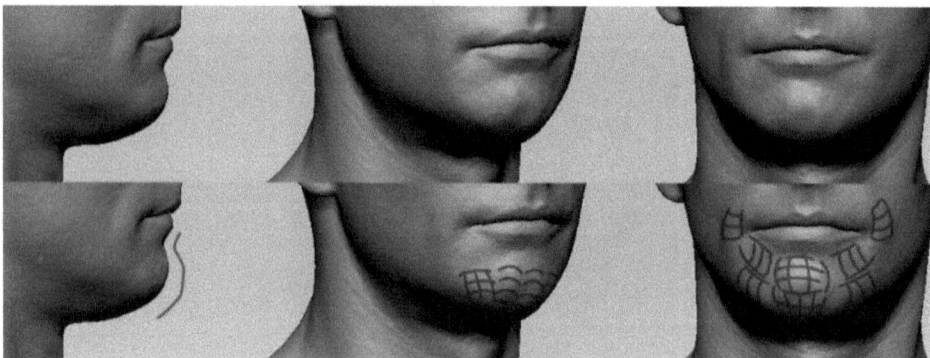

Figure 11.49 – Shapes that make up the chin

As with the previous facial features, the chin comes in a variety of shapes and forms, but it is good to study some of the common characteristics in terms of convex and concave shapes.

The skull largely determines the shape of the chin, and especially from a side profile, it is important to pay attention to the angle of the mandible and the profile the chin has.

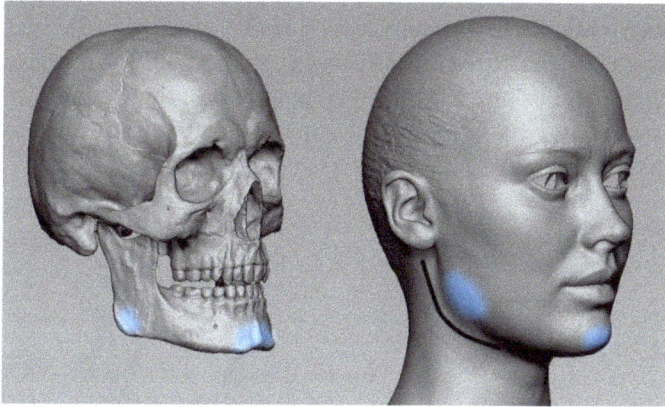

Figure 11.50 – Bony landmarks of the mandible

> **Important note**
>
> If you would like to study facial anatomy in more depth, you can study the subject with the *Anatomy of Facial Expressions* e-book by Uldis Zarins.

Neck

Most likely, you will not just need to model a head, as a portrait typically also includes the neck. Here, the neck muscles are depicted as simple shapes:

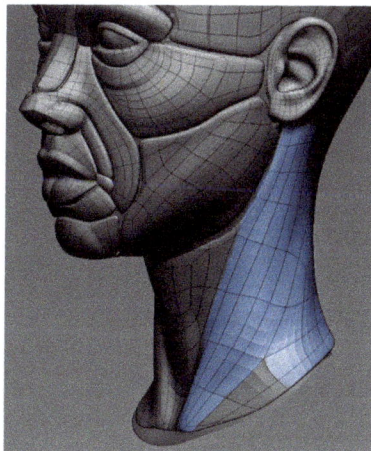

Figure 11.51 – The neck muscles in a simplified model

When it comes to sculpting the neck, there are two main muscles that affect the shape of the neck the most, which we will focus on here:

- **Sternocleidomastoid muscle**: This muscle is especially apparent in a tensed or turned head position and is a major element of the frontal view of a portrait. The main part of this muscle originates on the manubrium, which is the upper part of the sternum, and another part originates on part of the clavicle. Make sure to not sculpt this muscle too pronounced, unless the pose you pick requires it.

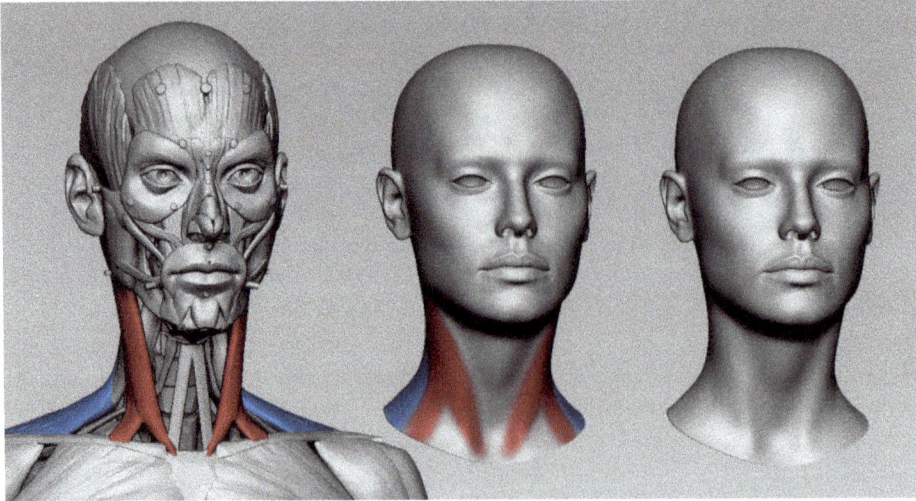

Figure 11.52 – Sternocleidomastoid (red) and trapezius (blue) muscles

- **Trapezius muscle**: This is an especially important muscle for a powerful, strong appearance. It originates in the lower part of the back of the skull and attaches to the spine, scapulae, and clavicles. The way you sculpt this muscle depends on the subject, but to create a more feminine and elegant head, you should definitely keep this muscle slim.

Now, you have some insight into how to approach these important areas of the head (and neck), which hopefully translates to a better sculpt.

You can spend any amount of time you like on this stage, improving the model until you are satisfied. Before you call the modeling stage done, though, the last essential step is to give the head an expression and even add a slight pose by adjusting the neck.

Adding a pose and facial expression

When it comes to creating a pose or facial expression for the portrait, it only needs a subtle change to make the face much more interesting. The main point is to not have a completely neutral and symmetrical head.

> **Important note**
>
> If you decide to open the mouth of your head model, you can follow the workflow in the *Adding teeth* section of *Chapter 2*.

Here are the steps for creating a simple pose with a basic head tilt and a change in eye orientation:

1. Mask the lower part of the neck, then blur the mask by *Ctrl* + left-clicking on the masked part.
2. Press *E* to use the Gizmo and rotate the head.

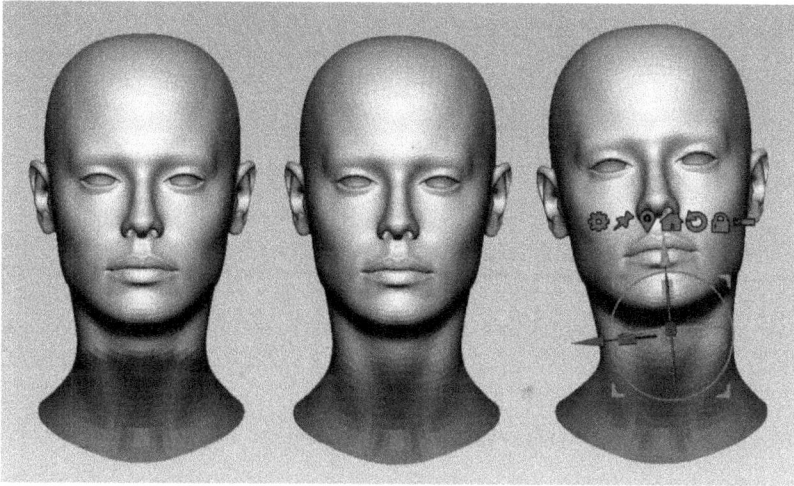

Figure 11.53 – Using masking and the Gizmo to create a head pose

3. Use the **Move** brush to change the eyelid shapes so that they match a glance to the side.
4. Use the **Move Topological** brush to open the mouth slightly.

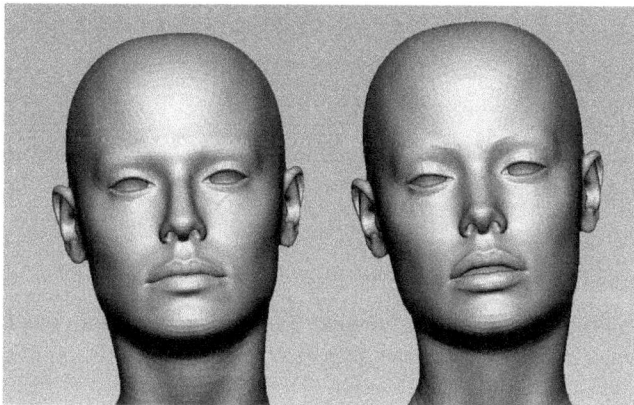

Figure 11.54 – Before (left) and after (right): Adjusting eyelids and lips shape

5.   Then, rotate the eyes to have the iris point to the side, matching the new eyelid shapes.

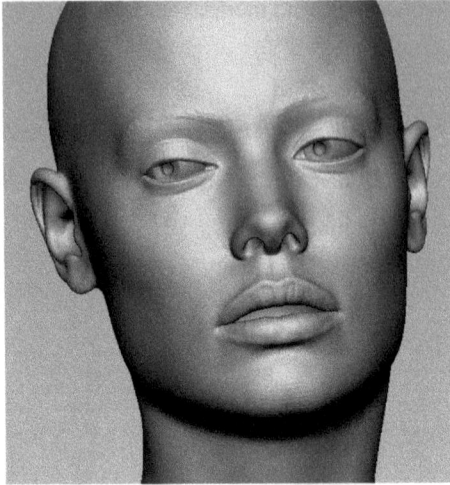

Figure 11.55 – The final result of the basic head sculpting process

These are all the steps for creating a subtle and slightly more interesting pose. The basic head shape is established now, and in the next chapter, you will finalize it by adding skin detail and hair.

Reviewing tips for sculpting likeness

Likeness sculpting is one of the most popular subjects of 3D artists, as it tends to be popular on social media and often gets more attention than other themes. Creating digi-doubles, which requires likeness sculpting, is also a high-in-demand task, so it is worth adding likeness sculpts to your portfolio to boost your chances of getting a job in this field.

Creating a great likeness is one of the more challenging projects an artist can undertake. While sculpting a realistic portrait is challenging in itself, likeness sculpting takes that to the next level, as even the smallest imperfections will throw off and even ruin the likeness.

To create a convincing likeness, the artist needs to have good knowledge of facial anatomy and high attention to detail; but above all, they need to have a lot of patience. That being said, here are some tips to help you get progress on your likeness sculpts just a little bit faster and easier.

Choosing a likeness subject

This point is more significant than it may seem because not every person is equally easy or hard to copy in a likeness sculpt. Here are some of the things to consider when picking the subject:

- **Reference material availability**: Actors, athletes, or any other famous person are usually a good pick when finding references because there are plenty of photos available, and many of them are great quality and have a high resolution. The more high-quality photos available, the easier achieving the likeness will be. (We will go into more detail about this topic in the next subsection.)

- **Gender**: Men are a bit easier to sculpt because they tend to have more pronounced facial features, whereas female faces are often more subtle and with softer transitioning shapes. The use of make-up can also make it harder to recognize the shapes and proportions, though older women can be an exception.

- **Age**: Young faces tend to have more subtle features and so can be more challenging, but with increasing age comes more skin detail and overall change in the faces. This makes older faces more unique and recognizable, and so often easier to sculpt.

- **Facial structure**: If you pick a subject with very uncommon characteristics, such as very narrow or distant eyes, or a big or long nose, it makes them more easily recognizable, so look out for these facial features when picking a subject.

At the end of the day, you want to pick a subject that you feel motivated and interested in working on, but in order to avoid some frustration with an already difficult task, you can make life easier for yourself by paying attention to these points.

Once you have decided which person you would like to sculpt, you can start gathering pictures. Next, you will learn about some tips for creating your reference board.

Finding references

The difficulty and success of a likeness depend, to a large degree, on the availability and quality of reference material. Luckily, most celebrities have a great number of photos available. Here are some tips.

Use Google Advanced Image Search

Look for pictures in very high resolution. The more details that are visible, the easier it is to understand the facial structure and detect the smaller and subtler forms of the face. Google Advanced Image Search is the perfect tool for this purpose:

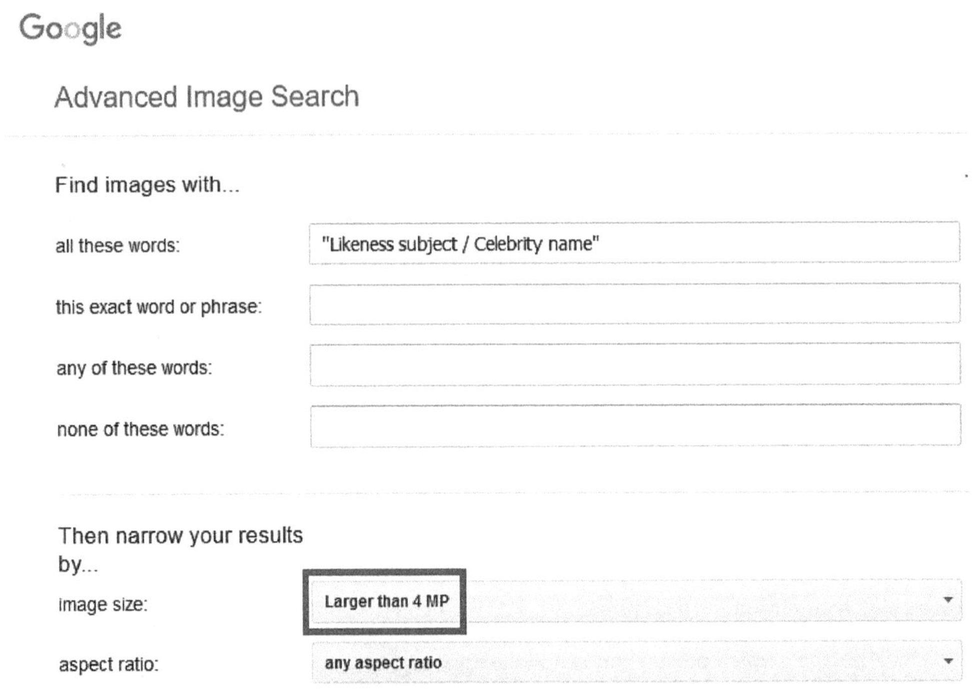

Google

Advanced Image Search

Find images with...

all these words: "Likeness subject / Celebrity name"

this exact word or phrase:

any of these words:

none of these words:

Then narrow your results
by...

image size: Larger than 4 MP   ▼

aspect ratio: any aspect ratio   ▼

Figure 11.56 – Google Advanced Image Search

In the **all these words** box, put the subject's name, then in **image size**, select something with large dimensions, such as **Larger than 4 MP**. Note that going too high in resolution might give fewer pictures, and you might miss out on great pictures that are strong in aspects other than the resolution, so it is worth checking 2 MP as well.

Pay attention to the focal length

When looking for reference pictures, make sure that they have been taken with a long focal length. This information is rarely accessible, but in general, just make sure that the head looks rather wide, the ears are largely visible, and the nose and middle of the face do not look too big compared to the side of the head. This reference ensures that there is little perspective distortion, and it makes it easier to match it when sculpting it in ZBrush.

Here is the difference in appearance between no perspective distortion enabled and perspective distortion turned on with a focal length of **50** mm:

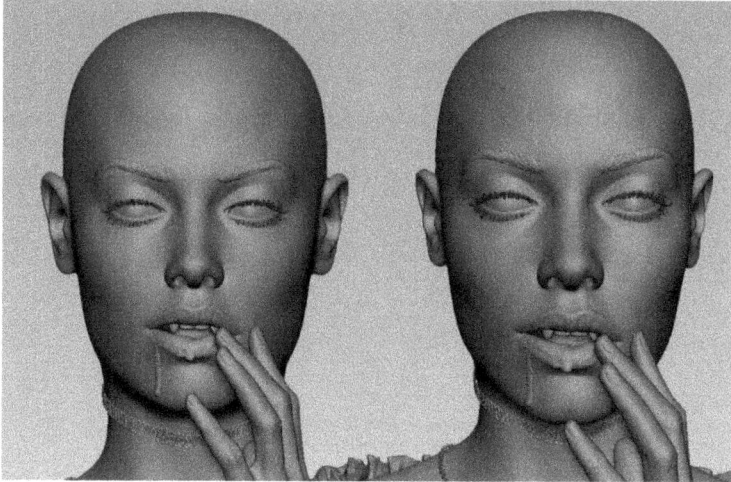

Figure 11.57 – No perspective distortion (left) and perspective with a 50 mm lens (right)

You can turn on perspective distortion by pressing *P*, or by navigating to the **Draw** palette and clicking on the **Dynamic Persp** button. Below this button, you can enter the **Focal length(mm)** value.

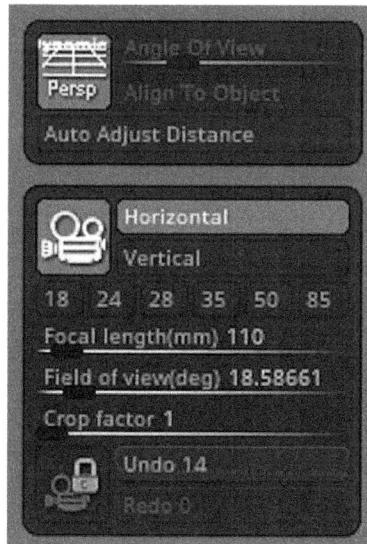

Figure 11.58 – Dynamic Persp mode and Focal length in the Draw palette

If you decide to sculpt with perspective distortion turned on, make sure to pick something higher than **110** mm, so you can match it to your reference photos more easily.

Prepare the main reference

When it comes to likeness sculpting, a straightforward and effective approach is to match your model as closely as possible to one main reference. Here are some of the criteria for this main reference picture:

- The picture should be a perfectly neutral front shot. Avoid even a slight head tilt in any direction, so the subject's face is oriented perfectly neutral toward the camera.

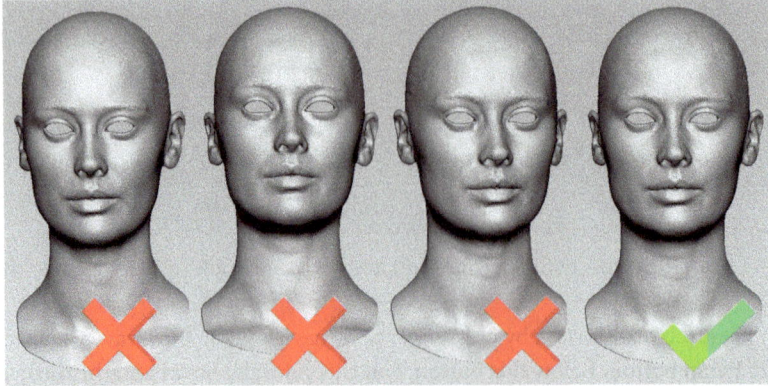

Figure 11.59 – Neutral head pose

- It should have a neutral, relaxed facial expression. This will make it easier to find similar pictures and make the process easier.

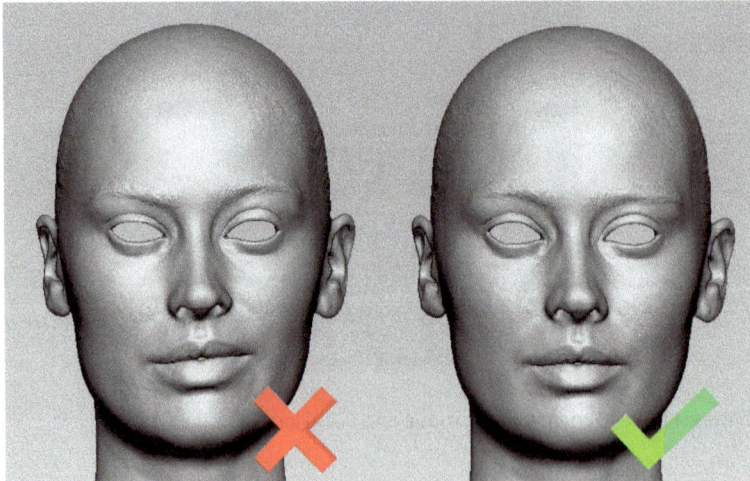

Figure 11.60 – Choose a neutral expression for your main reference picture

- It needs to be as high resolution as possible, ideally 4 MP+.

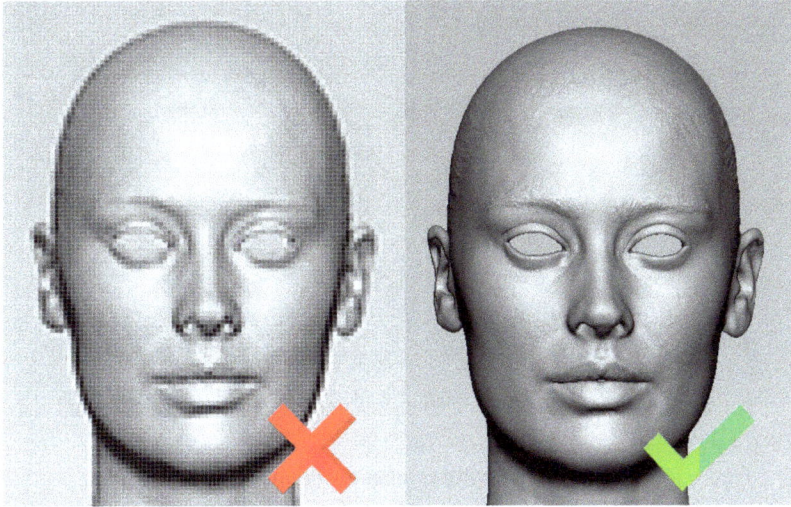

Figure 11.61 – Ensure your main reference picture has a high resolution

- Make sure to find an image from the front, one from the side, and one from a roughly 45-degree angle (all the previous tips apply to these different angle pictures too).

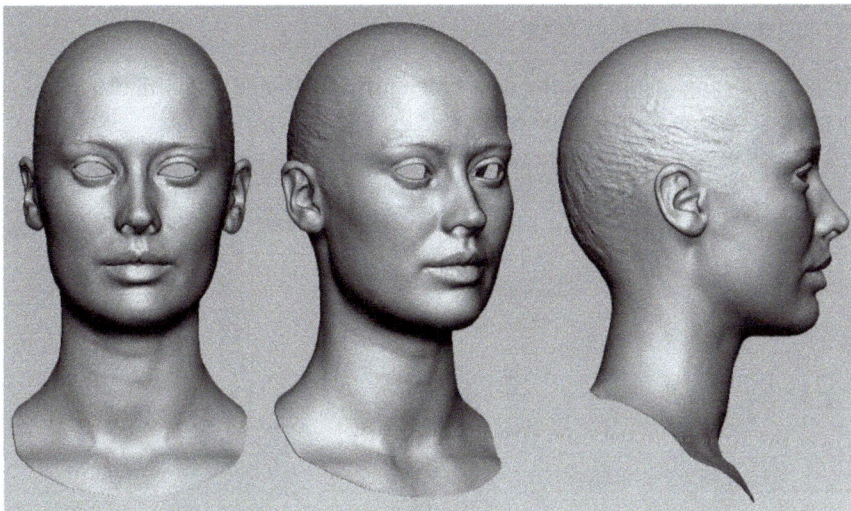

Figure 11.62 – The important angles for the main reference

- Finally, I recommend using these images to trace important lines of shapes and silhouettes in the face with a saturated color. This can be used inside of ZBrush as a blueprint for matching your sculpture to the reference. Make sure the picture has square dimensions, so you can apply it to a square plane in ZBrush with no distortion.

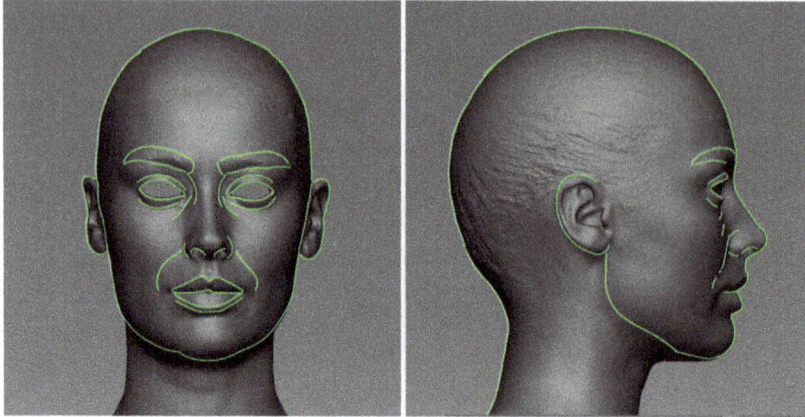

Figure 11.63 – Tracing lines with color as a blueprint for your likeness

Gather additional references

Besides the front view, side view, and 45-degree-angle view, you need to find additional reference images that will help fill the gaps and provide visual information for different angles and details in the face. Every photo has some unique advantages, so make sure to spend some time on this essential step of gathering references. Here are some things to consider:

- Try to get many angles, not just the ones previously mentioned. In particular, a low or high angle can be useful to complete the reference.

Figure 11.64 – Variety in camera angles will help with reading facial features better

- Collect images with a variety of lighting. Different light directions and intensities can emphasize certain facial features and make it easier to read the shape of them.

Figure 11.65 – Variety in lighting helps understand the forms of the face better

- Add plenty of pictures to your reference collection. While there is no need to have hundreds of pictures, something such as 10 is definitely not enough when you are trying to sculpt an accurate likeness.

- If your subject is a famous actor, there is a good chance that images of a life cast exist. These are extremely helpful as they show the forms of the face much more clearly than in photos, as there is no color or sheen on the face that could conceal the forms. Professional sculptor and life cast seller Joe Slockbower has a great selection of life casts on his Instagram account: `https://www.instagram.com/joeslockbower/`.

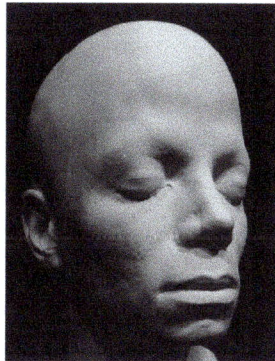

Figure 11.66 – One of Joe Slockblower's life casts, which can help you see the forms of the face

Sculpting a likeness

Once you have a strong reference collection prepared, including the main reference with draw-over, you can start the sculpting process in ZBrush.

First, you can load any model you prefer to start with (remember that the first section of this chapter gave an overview of different starting points).

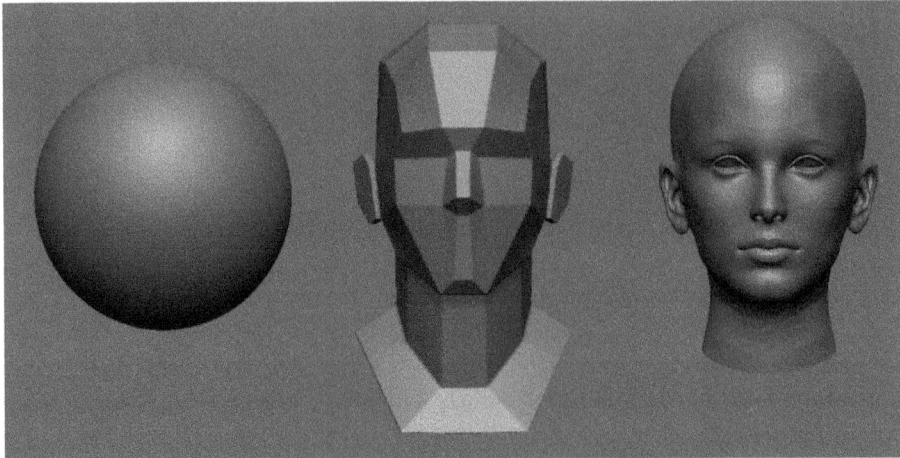

Figure 11.67 – Potential starting points for the portrait

Next, you can add your main reference by doing the following:

1. Go to **Tool | Subtool | Append** and pick the **Plane3D** model.

Figure 11.68 – Plane3D

2. With your **Plane3D** subtool selected, go to the **Tool** palette, then go to **Texture Map** and click on the empty, square field. Then, click on **Import** and load in the main reference.

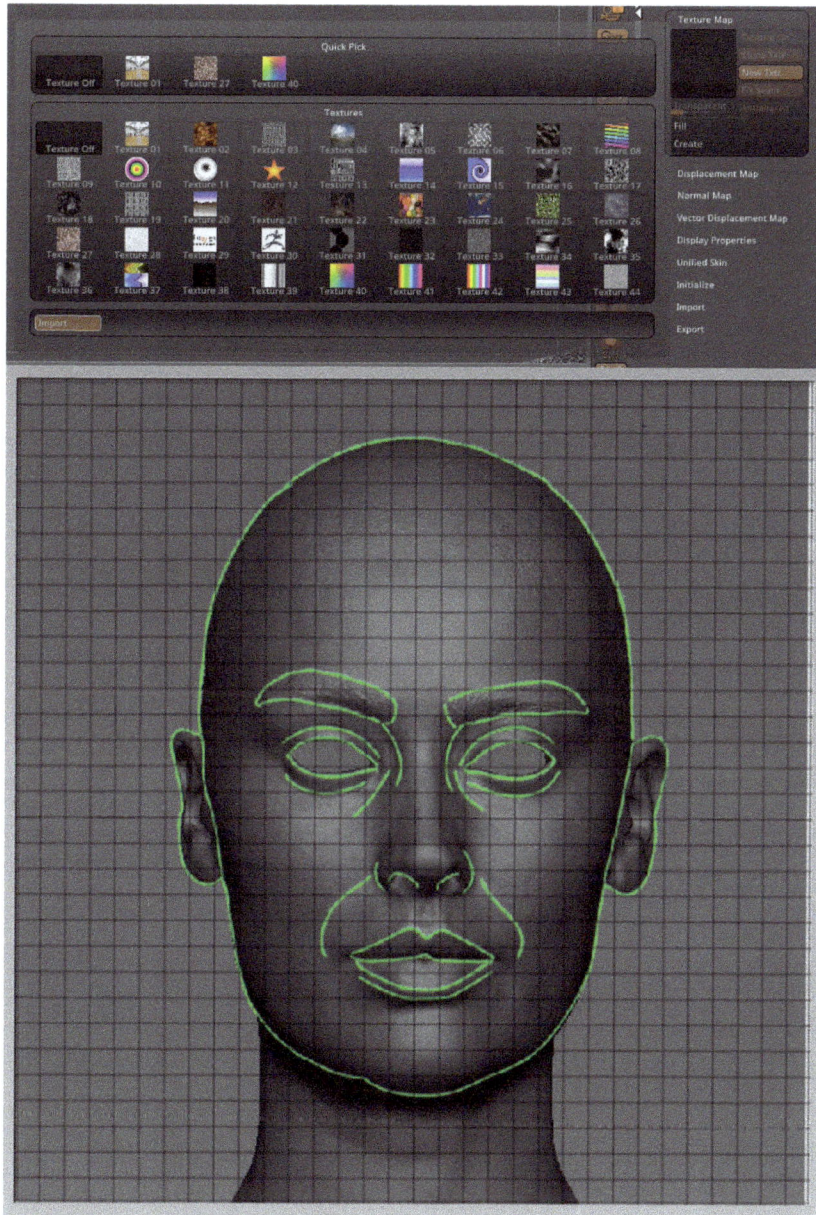

Figure 11.69 – Importing a texture

3. Now, enable **Transparency** mode (found on the toolbar on the right side of the canvas) and position the plane so that the lines are roughly aligned with your head model.

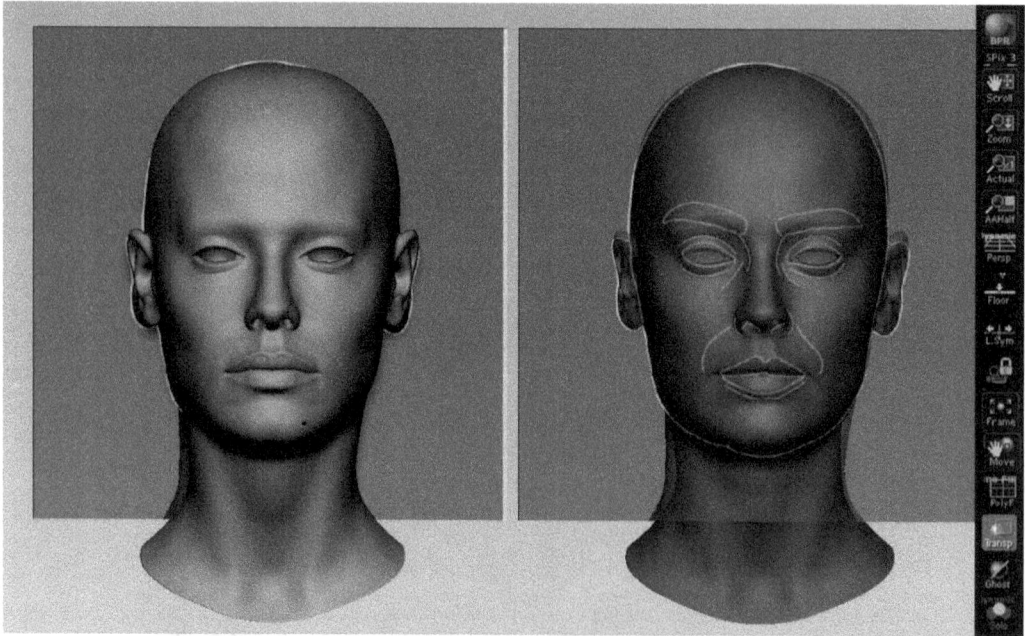

Figure 11.70 – Before (left) and after enabling Transparency mode (right)

4.   Start adjusting your model so that it matches the green lines.

5.   Don't forget about the side profile. Now, you can repeat *steps 1-4*, but using the side view reference.

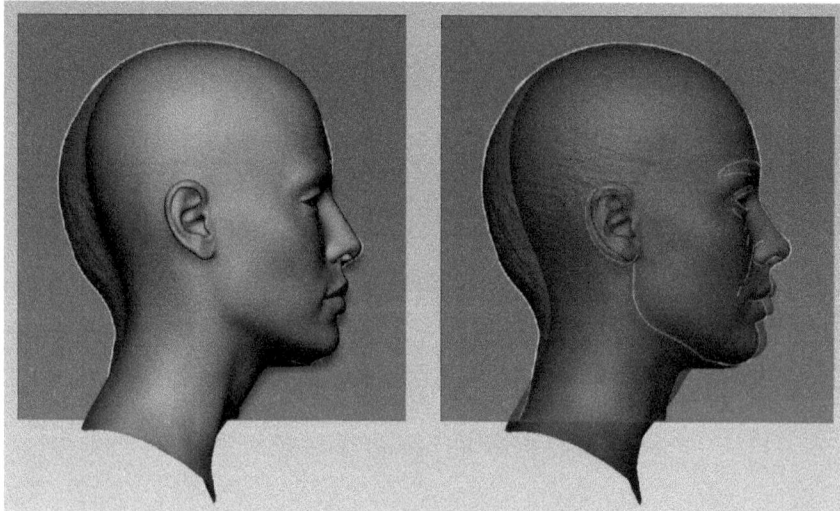

Figure 11.71 – Using the side view reference

At this point, you can continue to sculpt and refine your model, based on your reference images and the facial anatomy information that was mentioned in the first part of this chapter. Here are some more general tips for this sculpting process:

- Import a 3D scan and position it next to your model. This allows you to compare the facial structure. Big differences and inconsistencies between your model and the scan could indicate that something is off in your model, although the natural differences of skulls mean this is not always the case. Either way, it is worth looking at more references to investigate whether you made a mistake or not.

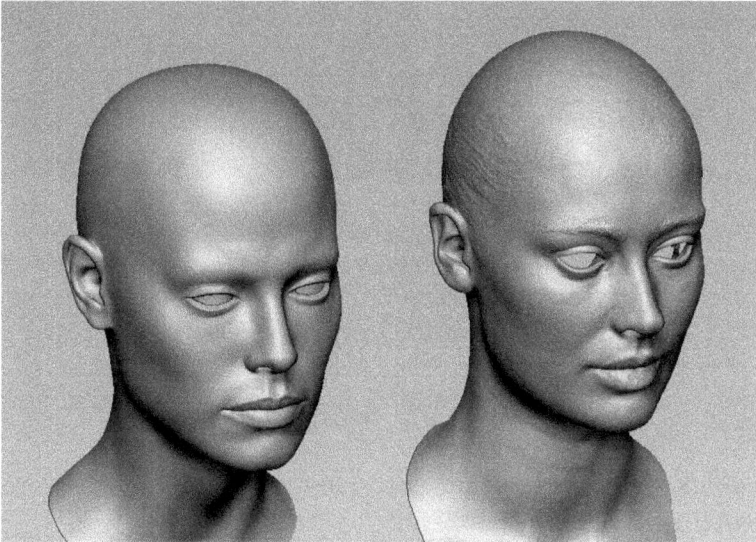

Figure 11.72 – Importing a head scan model to compare the facial structure

- Every once in a while, make sure to export a screengrab of your model and import it in Photoshop in a layer along with the main reference. You can toggle between the layers and see subtle differences more easily than in ZBrush.

- If you have access to 3D software such as Maya or Blender, you can create renders of the head model there. Since the lighting is more realistic than in ZBrush, it will highlight some flaws in the face more clearly, which makes it a great tool for when you feel stuck.

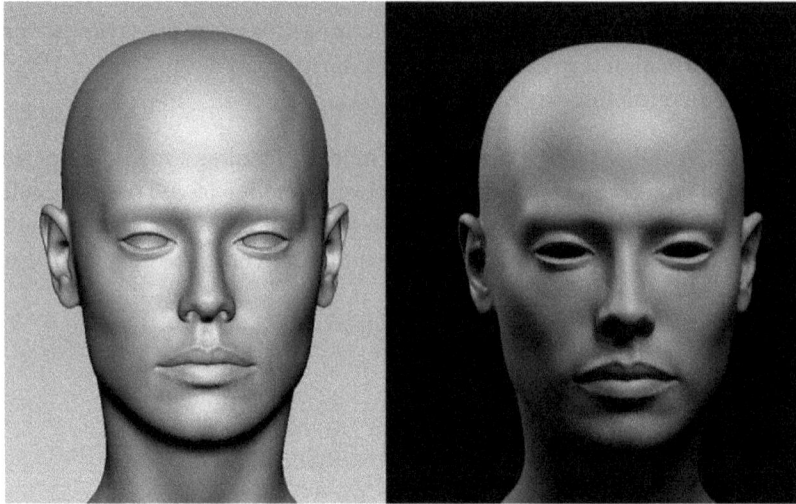

Figure 11.73 – Rendering in Maya with a different light setup gives a different perspective on the model

This wraps up the tips for creating a likeness sculpt. Ultimately, creating a successful likeness allows for very few flaws, and success is rather binary, meaning you either achieved the likeness or you did not. There is not really a partial success, or steps you could celebrate in between start and finish, as is the case with other types of artwork. Therefore, you need to have some faith and patience with yourself during the initial stages, which are not too rewarding.

However, when it comes to sculpting a head, there are some common mistakes that you could make, so let's take a look at those next.

Avoiding beginner sculpting mistakes

On your journey to master facial anatomy and likeness sculpting, you will likely have to go through some iterations until you see the results you hope for. Some skills cannot be learned in the short term by reading or even practicing, but they come through the experience of sculpting various faces. Through this process, you will get an understanding of proportions and shapes because you will begin to realize similarities and fix mistakes you previously made.

That being said, you can shorten the learning curve by checking your model for these common beginner mistakes:

- One of the more common mistakes is making the brow bone protrude too much. This creates an excessive shadow under the eyebrows, making it look wrong even in the front view. While the brow can be more pronounced in men, this is still often pushed too far.

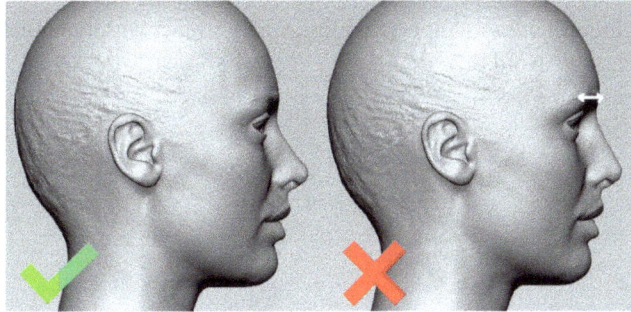

Figure 11.74 – Mistake: Excessive protrusion of brows and forehead

- Another common mistake is applying too much volume between the eyes and nose. Unless you are sculpting someone who is overweight, this area should not look inflated.

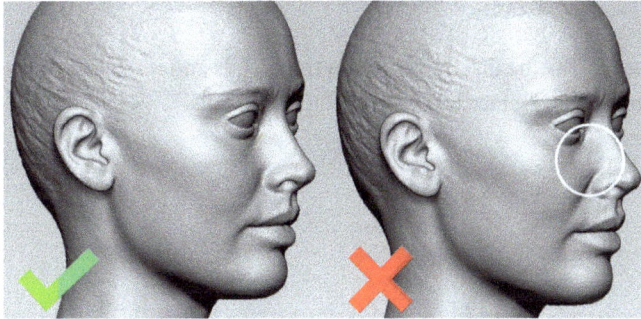

Figure 11.75 – Mistake: Too much mass below the eyes

- The forehead is not easy to sculpt correctly – beginners especially tend to make the forehead too flat, instead of giving it the curvature it should have.

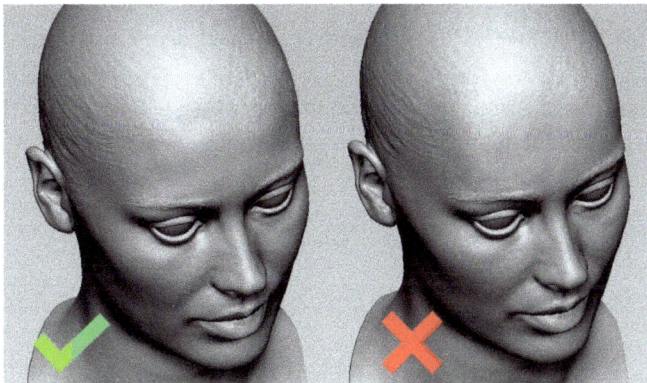

Figure 11.76 – Mistake: Lack of curvature in the forehead

- Sometimes, beginners tend to sculpt crease lines and wrinkles with too much intensity, which looks unnatural, so do avoid this.

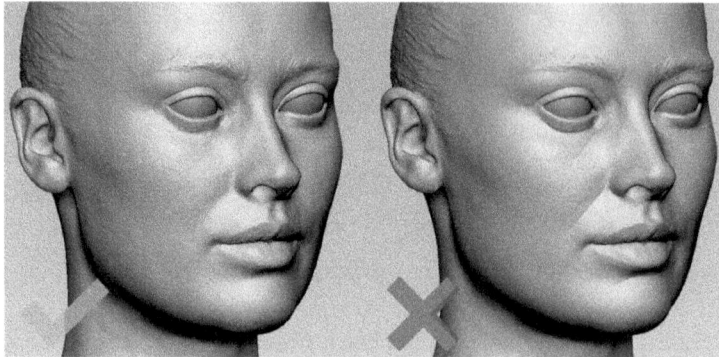

Figure 11.77 – Mistake: Sculpting lines too intensely

- The depth of the face, or the distance between the eyelids and nostrils, needs to be appropriate. Make sure to not make the face too protruding or too flat.

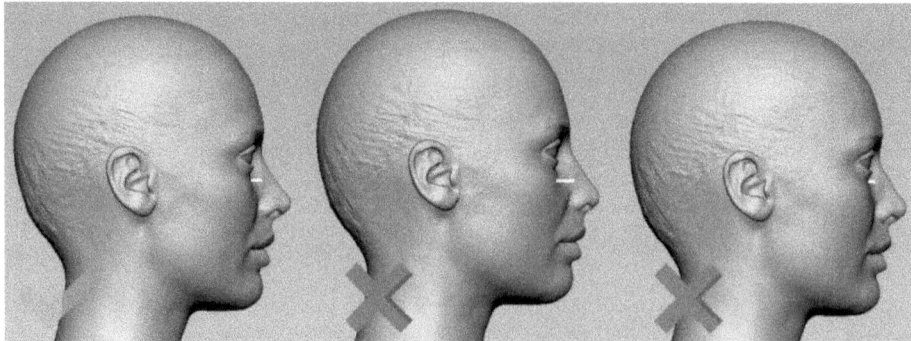

Figure 11.78 – Mistake: Lack of depth or too much depth of the face

- Another common mistake is making the face too flat, which happens when the corners of the eyes do not have sufficient depth.

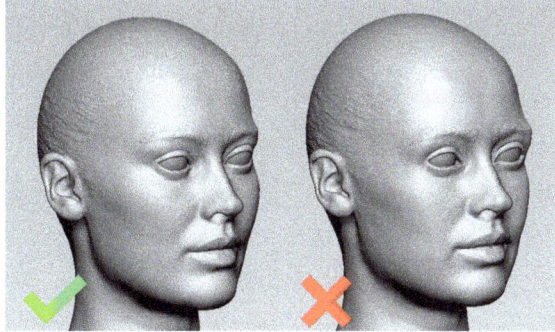

Figure 11.79 – Mistake: Lacking depth of the corners of the eyelids

- This mistake is extremely common – not giving the eyelids enough thickness. Especially on the upper eyelids, a lack of thickness will reduce realism, as it cannot cast as much of a shadow on the eyes. This is a small but essential detail for creating a believable eye in CGI, so make sure to give the eyelids enough thickness.

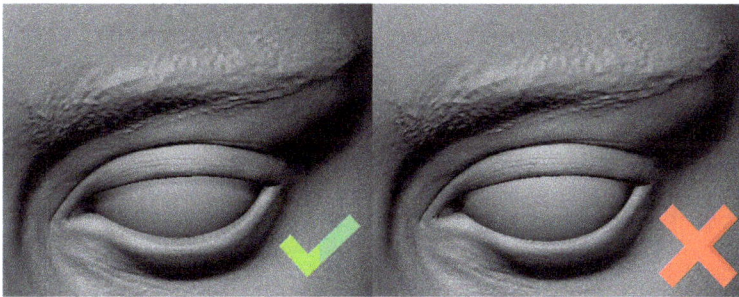

Figure 11.80 – Lack of thickness of the eyelids

- Sculpting the border of the lips too intensely and too defined is another common mistake that is easy to avoid. While there is a clear transition from lips to the skin around it, the transition needs to be fairly subtle. As always, good reference material and attention to detail will ensure a successful sculpt.

Figure 11.81 – Mistake: Defining the lip shape too intensely

- Don't make the lips too flat. Pay attention to the profile of the lips from the side view and observe the depth.

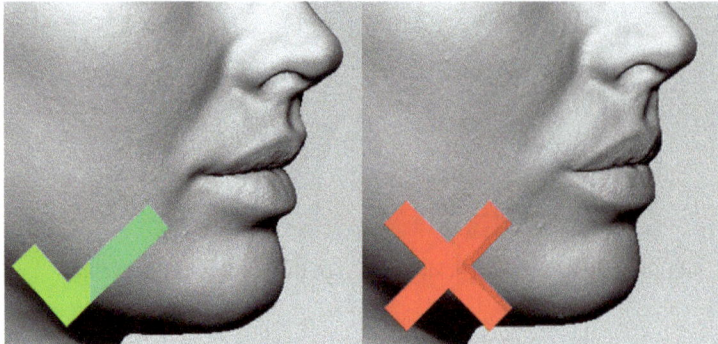

Figure 11.82 – Mistake: Making the lips too flat

This sums up the small list of mistakes that are common in beginner head sculptures. Hopefully you can avoid these and improve the overall quality of your model. If you are new to sculpting a head, keep in mind that it is one of the most challenging subjects in art, and even with the right information and reference material, it will take lots of hours of practice until you can translate what you see and know into a proper sculpture; so, don't be discouraged if the first few attempts do not look spectacular.

Summary

In this chapter, you learned how to sculpt a realistic head in ZBrush.

First, you explored different starting points for sculpting a head and their pros and cons. Then, you dove into the characteristics of different areas of the face and how to sculpt them properly. There, you gained insights into the influence of facial anatomy, such as the skull or facial fat deposits, on the appearance of the face.

After that, you learned how to approach a likeness sculpt, and what kind of reference will help you in achieving a more accurate sculpture. Finally, you explored several common mistakes that beginners make when sculpting a head so you can avoid them and achieve a more realistic result.

The next chapter will cover skin detailing techniques, as well as different ways to create hair so that you can complete your portrait or bust.

12
Adding Skin Detail, Sculpting Hair, and Using FiberMesh

Skin detailing is a crucial skill for many character art and sculpting-related jobs, especially in the VFX and collectibles industries. While it may not be as high on the priority list as some of the previous topics of this book, it should not be overlooked and deserves the attention it gets in this chapter. Equally important, and in even higher demand, is hair creation, which we will also cover.

In the first section of this chapter, you will explore various tools and techniques for creating photorealistic skin detail that will bring the portrait that you started in the previous chapter to life. Here, you will learn how to use scan data for an extremely efficient workflow that produces some of the most realistic kinds of skin detail a 3D artist can create. Then, you will use NoiseMaker, as well as various brushes and Alphas, to create skin detail that comes close to, and even surpasses, 3D scans in some aspects.

In the second section, you will learn how to sculpt hair using basic sculpting brushes, as well as the unique and powerful IMM Curve brush, which can be used to create stylized hair. Plus, you will add eyebrows and eyelashes to complete the portrait.

The last section of this chapter focuses on FiberMesh for hair creation, which is a powerful tool for creating hair that can be exported as curves to render realistic hair in 3D software such as Maya or Blender.

These skin detailing and hair creation techniques will allow you to finish your portrait with a high level of detail and the type of refinement that will make it a useful addition to any portfolio.

So, this chapter will cover the following topics:

- Applying skin detail using NoiseMaker, brushes, and Alphas
- Sculpting hair, eyebrows, and eyelashes with brushes
- Sculpting hair with FiberMesh

Technical requirements

For the best experience, it is recommended that you have a powerful PC that meets the minimum requirements described in the first chapter's *Technical requirements* section. However, you can work on this chapter with just a mouse, a functional PC setup, and a ZBrush license.

If you followed along with *Chapter 11* and sculpted a head, you can apply the skin detail and add hair to that head. Otherwise, you can pick any other head model you like.

Applying skin detail using NoiseMaker, brushes, and Alphas

In this section, you will learn about various tools and methods for creating skin detail for the head that you sculpted in *Chapter 11*. These details will be the finishing touch that gives your sculpture an extra layer of complexity and adds the kind of realism that is needed in many CG jobs.

Projecting 3D scan details

With the increasing availability, quantity, and quality of 3D head scans, many artists implement them in their workflow, as they are a powerful tool for creating photo-realistic humans. Since these scans capture the skin detail extremely accurately, the realism in the detail is near perfect. As the detail does not have to be sculpted, or manually applied, but can simply be projected on a surface, it is much faster to apply and does not require too much skill.

Scanned skin details can be bought as texture files, which can then be applied in various ways, but an easier way, which is increasingly popular, is to use **Head Scans** as 3D models. Instead of the detail being stored in a grayscale image, the detail is already part of a head model – this has a polycount of 50-70 million polys, which means that there is no lack of detail, even if you plan to render extreme close-up shots.

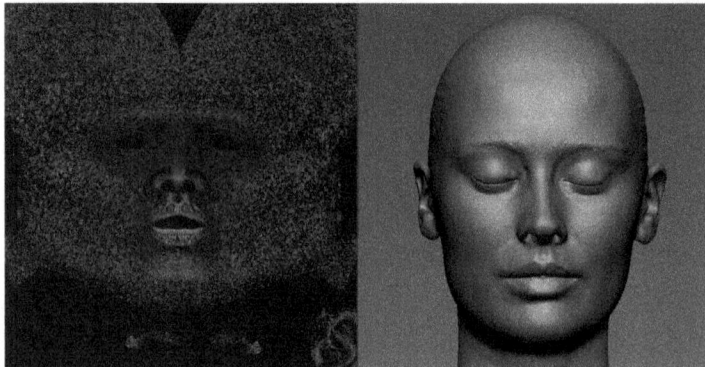

Figure 12.1 – Skin detail on a grayscale image (left) and skin detail as part of a full-head model (right)

Of course, there are cases where textures are going to be useful, but when surfacing a head model, it is more convenient, and easier, to visualize the skin detail as it appears in the 3D space on a head model. For that reason, this section focuses on 3D scan models.

Where to get a 3D head scan

There are several marketplaces that sell 3D scans of heads of high quality. Here are some of the most popular choices:

- **3D Scan Store** (`3dscanstore.com`): This store sells HD head scans that have a very high level of detail, and they come with skin color and other textures. The site also offers one of their female HD head scans for free, so you can try it out.

- **Texturing.xyz** (`https://texturing.xyz/`): This site sells VFace head scan models. With high-fidelity skin detail and matching texture maps, this is another great product.

- **Eisko's Digital Louise** (`https://eisko.com/louise/`): The Digital Louise head scan is available for free, which is great for students and people who just want to experiment with head scans.

How to use a 3D scan to project skin detail

With your model and a 3D scan loaded in your scene, you can use the **ProjectAll** function to transfer detail from the scan onto your head model. But before starting this project, it will be good to retopologize your model if you have a DynaMesh model from the previous chapter.

> **Important note**
>
> A ZRemeshed mesh with subdivision levels lets you have better topology, which makes the sculpting brushes react better with your mesh, and it allows you to work with different resolutions (subdivision levels) based on whether you want to work on detail or make bigger changes to your model. You can find the process for retopologizing a DynaMesh model with ZRemesher in *Chapter 5*.

Once you have a model with good topology and your scan is ready, the process is based on three simple steps: first, matching the scan to your head model; second, preparing both meshes for projection; and third, projecting the skin detail from the scan onto your head model. So, let's take a look at these steps.

Matching the scan to your head model

Here's how to match the scan to your head model:

1. Load the ZTool or other 3D file format that contains the head scan. Here, we will be using Eisko's Digital Louise model. If you downloaded the Digital Louise model, there should be a folder called `Mesh_HD`. In ZBrush, import the model called `Louise_eyesOpened.fbx` from that folder.

Figure 12.2 – Louise_eyesOpened.fbx model

2. This model contains the detail you need, but it does not have subdivision levels. However, ZBrush has a function that lets you restore subdivision levels. Go to **Tool | Geometry** and select **Reconstruct Subiv**, until the lowest subdivision level is restored, and the following message is displayed: **Reconstruction result: Mesh contains triangles, operation canceled**.

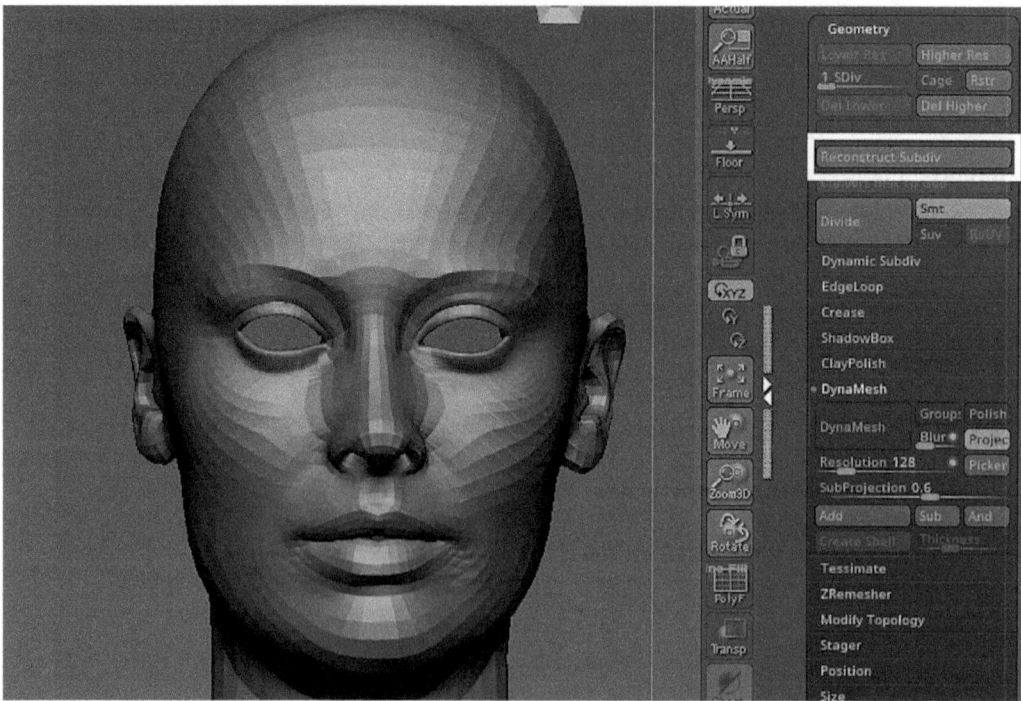

Figure 12.3 – Reconstructing subdivision levels

3. Now load the model that you want to project the skin detail from the scan onto – you can use the head model from *Chapter 11* or you can use another head model if you wish.

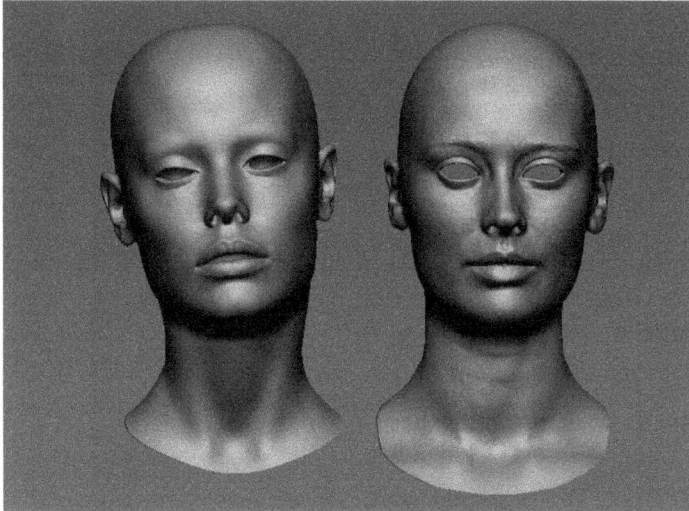

Figure 12.4 – Sculpted head and scan

4. Next, with the **scan** subtool selected, hit *Shift + D*, until you have **Subd level 1** active, and position it on top of your head model. Now, you can use the **Move** brush to match the scan to your head model. This can be seen in the following screenshot, where the **scan** subtool (turquoise) is matched to the head model (gray).

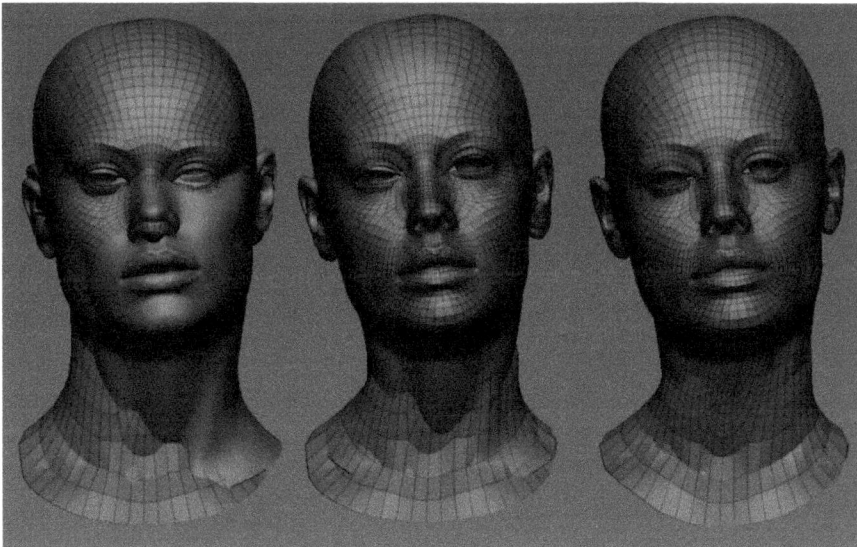

Figure 12.5 – The process of matching the scan to the head sculpture

You don't have to match it perfectly, but make sure to pay extra attention to the eyelids, nose, ears, and lips as these areas have thin shapes that are prone to projection errors when they don't match closely enough.

Now, you should have two models on top of each other, and you can proceed to hide/deselect parts of the meshes that you don't need for the detail transfer.

Preparing the meshes for projection

Here's how to prepare the meshes for projection:

1. Create a PolyGroup for the inside of the mouth and eyes for both the scan and your head model. The PolyGroups will let you deselect those areas since they are not necessary for the projection process.

2. Isolate the face without the eye and mouth interior.

3. Make the same selection for the scan too.

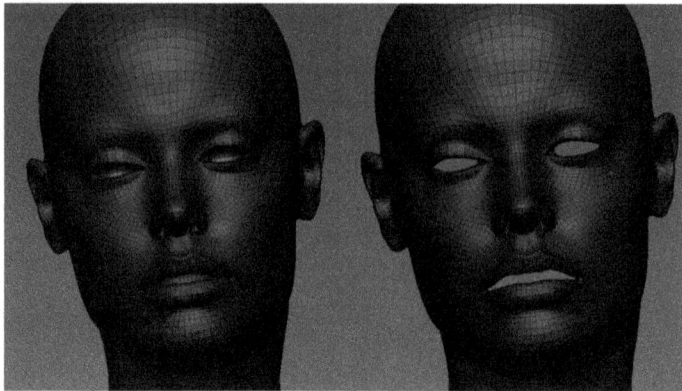

Figure 12.6 – Before and after Isolating the face

4. Now, we can proceed to match the scan even more closely to the head model. Select the scan, go to **Tool | Subtool | Project**, and click **Project All**. Here is the result:

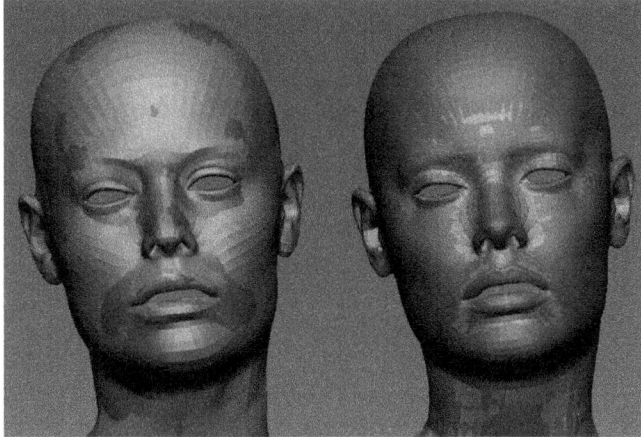

Figure 12.7 – Before (left) and after clicking Project All

5. With the **scan** subtool active, switch to the highest subdivision level, so it shows all the skin detail, which will be projected onto your head model.

At this point, you have both meshes matching closely, unnecessary parts are hidden, and you can start projecting the skin detail.

Projecting skin detail from the scan onto your head model

Here's how to project skin detail from the scan onto your head model:

1. Select your head model and switch to **Subd level 2**, since the lowest levels are already matching.

2. Go to **Tool | Morph Target** and select **StoreMT**. This step is important because it lets you fix potential errors in the projection with the **Morph** brush, which restores the state of the model from before the projection. You may have to delete your current Morph Target if one is stored already.

Figure 12.8 – Morph Target

3. Now you can start projecting the detail from the scan onto your head model. In order to do that, go to **Tool** | **Subtool** | **Project** and click **Project All**. Here is the result:

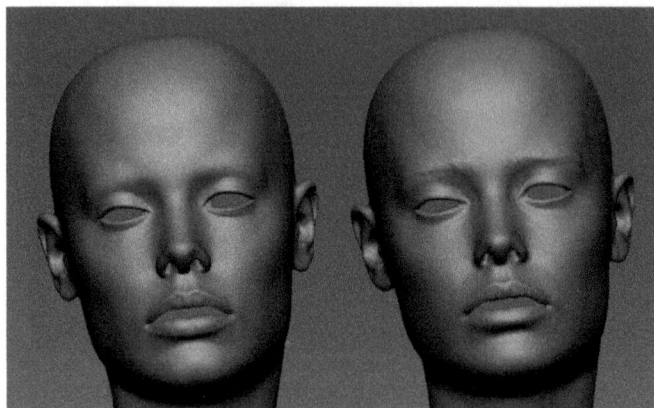

Figure 12.9 – Before (left) and after (right) projecting skin detail onto the head sculpt

> **Important note**
> When you are projecting detail, make sure to start with the lowest subdivision level and work your way up to the highest level. This will prevent projection errors by minimizing the distance between both meshes before projecting the fine detail.

4. Go up to the next subdivision level with your head model.

Repeat these steps until the skin detail on your head model matches that of the scan. This is most likely going to be around 6 to 7 subdivision levels and a polycount of 20 to 50 million polys.

After you have completed this process, you have a fully detailed head, with realistic scan-based detail, without doing a single brush stroke. In many cases, there are still areas that need extra work and enhancement of detail through brushes and Alphas, but this process significantly reduces the time it takes, which makes this a great starting point.

If you have no access to scans or prefer to practice manual skin detailing, the following sections show how to create skin detail from scratch, while showing the detailing techniques based on the unique skin detail of various parts of the face.

To start this process, you will use the NoiseMaker plugin to quickly cover a lot of areas with generic detail, which will serve as a base on which you can apply additional detail to refine it.

Applying tileable skin textures with NoiseMaker

When you are starting to add skin detail to your model, a great way to get started is to use ZBrush's **NoiseMaker** plugin to apply tileable skin textures. You can cover the whole character with detail in a matter of seconds, and if you use this tool in combination with **Morph Target**, you can even control which areas you want to edit.

To decide where to apply this generic skin detail, it is important to take a closer look at the variety of skin detail on a face:

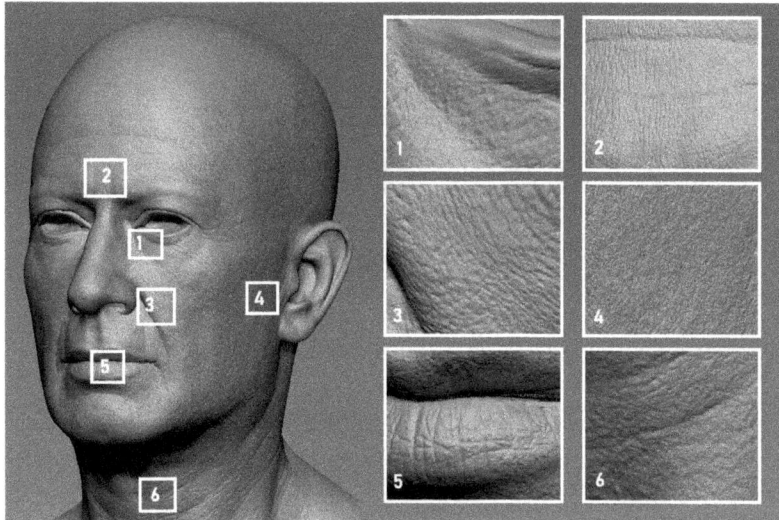

Figure 12.10 – Different kinds of skin detail

As you can see, there are several different kinds of skin detail, based on the area of the face. Of course, the lips are different from the bumpy skin on the neck, but there are even subtle differences between the forehead and the side of the cheeks. Pay attention to these subtle differences when you choose the tools and textures that you use.

> **Important note**
>
> When you have a completely smooth face without any detail, applying a generic, tileable skin texture on the whole face can be a good starting point, as long as you keep it subtle. Then you can apply more intense detail on top of it, depending on the area of the face. This layering of skin detail is a nice way to build complexity and create realistic-looking skin.

Before you start using NoiseMaker though, you need to make sure that your model has UVs.

Checking the UVs

Since we want to apply tileable textures, we need to make sure the model has UVs so NoiseMaker knows how to lay out the detail across the model. If you don't know how to create UVs for your head model, take a look at *Chapter 6*, where this process is explained.

To check whether your model has UVs, go to **Tool | UV Map** and select **Morph UV** (if your model does not have UVs, this button is grayed out):

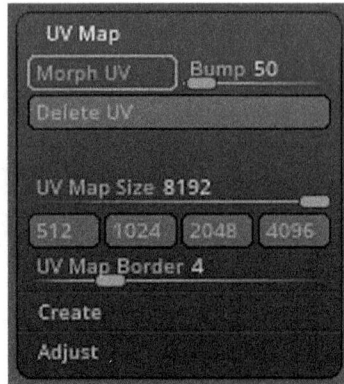

Figure 12.11 – Checking the UVs with Morph Target

If it has UVs, you can now check what the UV layout looks like:

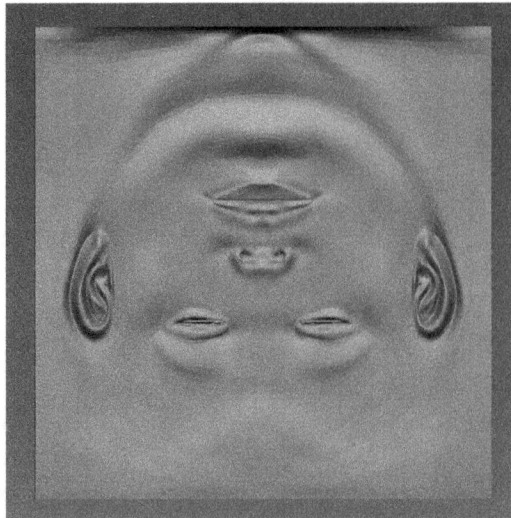

Figure 12.12 – Flattened UV view through the Morph UV command

Don't be surprised that your UVs are displayed vertically flipped. This is something unique in ZBrush and, due to technical reasons we don't need to go into, is to be expected.

When you have a model with UVs, there is still another thing to consider before starting the detailing process, which is adding a layer.

Using layers

If you prefer to work non-destructively, you can add a layer by going to **Tool | Layers** and clicking on the **Add New** symbol, as marked in the following screenshot:

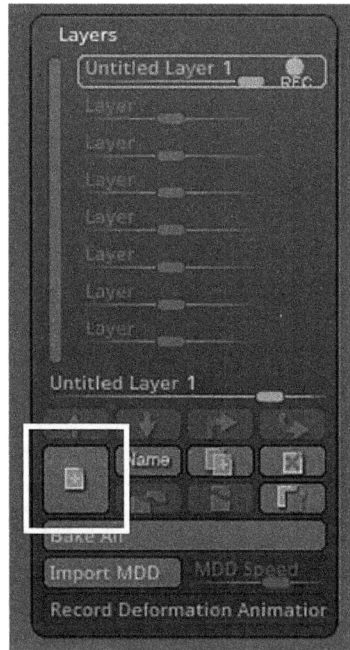

Figure 12.13 – Adding a layer to control the detail

Now, you can turn the layer on and off, and also adjust the strength with the slider, controlling the intensity of the changes you make with the layer active.

However, you need to keep in mind that layers will significantly affect performance, making operations such as undo much slower, which makes the sculpting experience less smooth. This is especially true if your model has a high polycount (more than 10 million).

> **Important note**
>
> You can find more info on layers here: https://docs.pixologic.com/reference-guide/tool/polymesh/layers/.

Let's proceed to add detail with the NoiseMaker plugin.

Using NoiseMaker

Once you have checked for proper UVs (and potentially added a layer), you can apply skin detail with NoiseMaker, as described here:

1. Go to the highest subdivision level and make sure your head model has a polycount of at least 10 million. If it does not, subdivide it to reach or surpass this number. You will be able to see if you have enough resolution once you apply skin detail with NoiseMaker. If the skin detail looks sharp, and there are no seams between polys visible, you have enough resolution. If there is a lack of sharpness, you can simply undo the NoiseMaker operation and add more subdivision levels, as needed.

2. Navigate to **Tool | Morph Target** and click **StoreMT**. Then go to **Tool | Surface** and select **Noise**. This will open the **NoiseMaker** window:

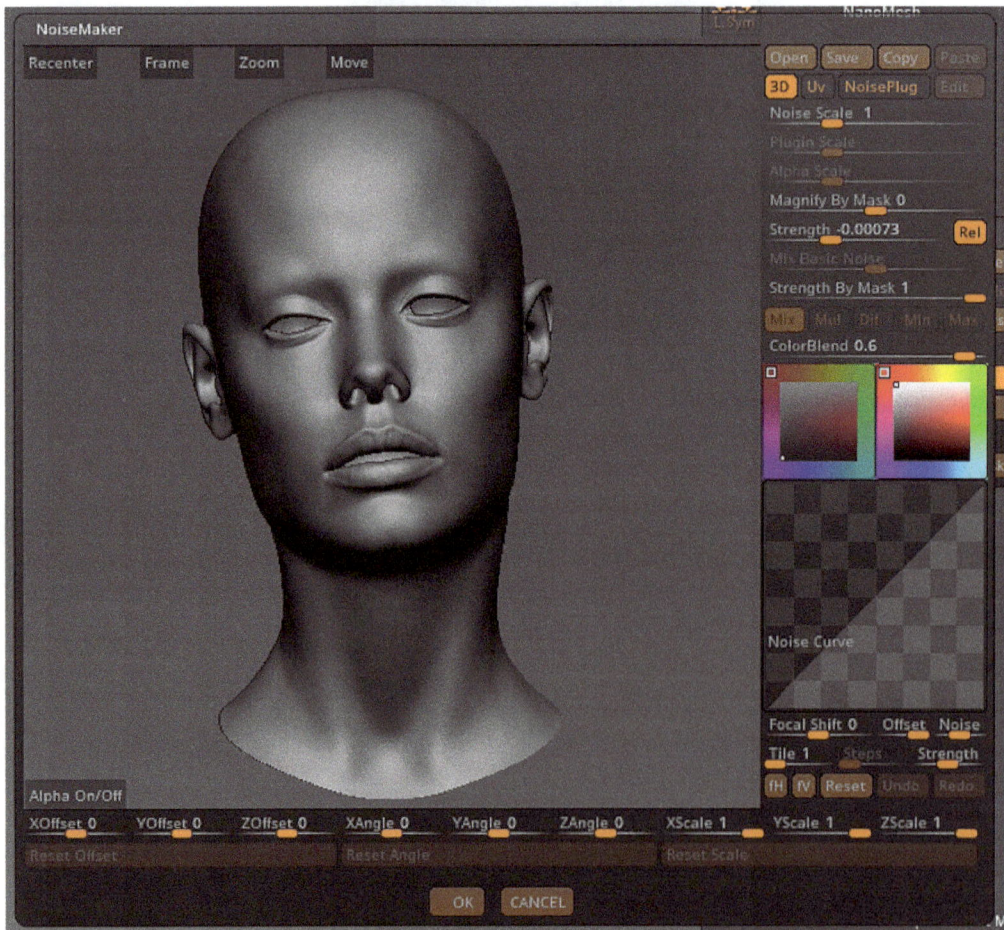

Figure 12.14 – NoiseMaker pop-up window

3. Click on **Alpha On/Off** in the lower-left corner of the window, which will allow you to import your tileable skin texture. You can find these textures on CG sites such as the ArtStation marketplace. For this purpose, make sure to pick a texture with little contrast, and a uniform and subtle effect. That way, it will not look repetitive and will serve as a fitting overall noise that does not interfere too much with additional skin detail that will be added later. Once you load in your texture, check these settings:

I. Enable **Uv** instead of **3D** in the settings, since you want to apply the detail based on the UVs.

II. Set **Mix Basic Noise** to **0**, so you don't mix your imported texture with a generic noise.

III. Finally, adjust **Alpha Scale** and **Strength** so that you have the desired result in the preview window.

Then click **OK**.

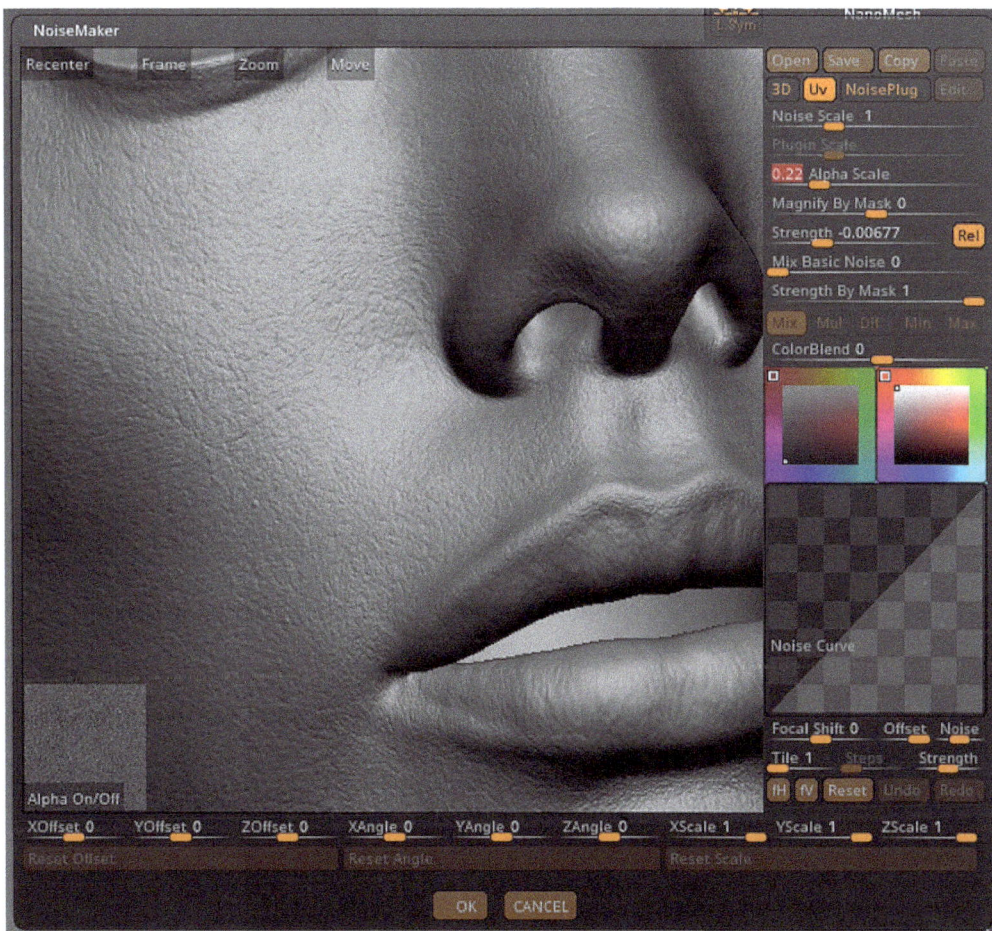

Figure 12.15 – Applying skin detail with NoiseMaker

4.   Now, you can see the effect of NoiseMaker on your model, but it is just a visual preview and not yet applied to your mesh. At this point, you can decide whether it looks right or open NoiseMaker again to make changes to the settings. Then, when you are happy with how the detail looks, click **Apply To Mesh**.

Figure 12.16 – Applying the surface detail

5.   Once you apply the skin detail, the effect on the mesh is a bit weaker than how it looks in **Preview** mode. Luckily, you can easily adjust the strength with the Morph Target that you stored in *step 2*. Go to **Tool | Morph Target** and use the **Morph** slider to increase or decrease the intensity of the skin detail.

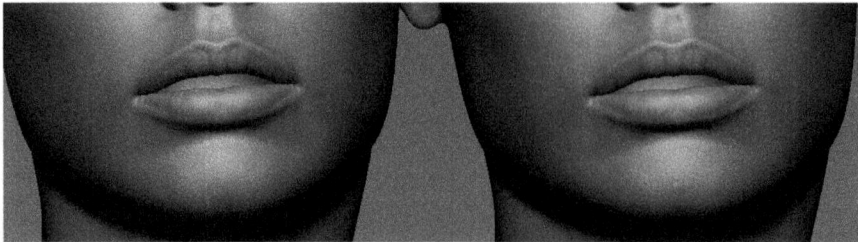

Figure 12.17 – Before (left) and after (right) adjusting the detail intensity with the Morph slider

6.   Maybe you don't want to have this new skin detail on areas like the eyelids or lips, where a different type of detail should be applied. To remove the detail, select the **Morph** brush and apply it to the areas that you do not want to be covered with skin detail – you can remove the detail completely or use a low brush intensity and remove it just slightly. Remember to use reference pictures to see what kind of detail, and what intensity of detail, you need, also based on age, gender, and ethnicity.

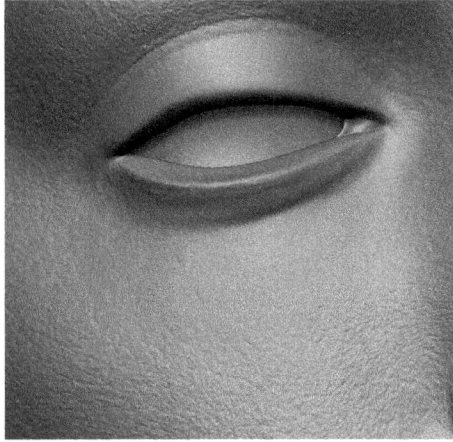

Figure 12.18 – Removing detail with the Morph brush

> **Important note**
> If you want to apply NoiseMaker detail on a small area of the face, make sure to go to **Tool |
> Morph Target**, and click **Switch**. This loads the state of the model, in which the detail is not
> applied yet. Now, select the **Morph** brush to add detail wherever you apply it, instead of using
> it to remove detail. This is a faster way to only add detail in a more limited way.

At this point, the NoiseMaker process is complete. You can continue to apply different textures based
on the part of the face and use the **Switch** function and **Morph** brush to control what area you would
like to affect.

Next, let's take a look at some more precise, manual detailing of different areas of the face.

Adding detail around the eyes

The eyes are the focal point and the most important part of the face, so you should prioritize them
and make sure the detail is accurate and crisp. The different areas around the eye all have their own
characteristic details and need to be sculpted with high attention to detail.

Upper eyelid

Due to its function of shutting and opening the eye, the upper eyelid has many fine wrinkles that
crisscross and create a leathery pattern.

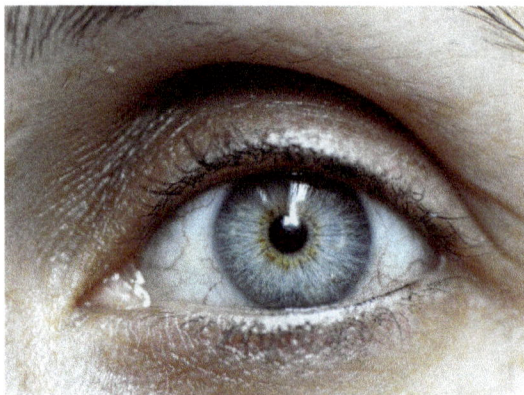

Figure 12.19 – Leathery detail on the upper eyelid

The perfect way to apply detail to the upper eyelid is to use a custom brush with a leather pattern Alpha and **Roll** mode enabled. This mode has a specific effect that is perfect for applying continuous wrinkles.

In order to create this brush, you need to sculpt a wrinkle pattern on a flat plane, like this:

1. Go to **Tool | Subtool | Append** and pick the **Plane3D** model.

2. Then go to the **Geometry** menu, disable **Smt**, and subdivide the plane five times.

3. Sculpt a leathery pattern using the **Dam Standard** and **Standard** brush, with crisscrossed lines like so:

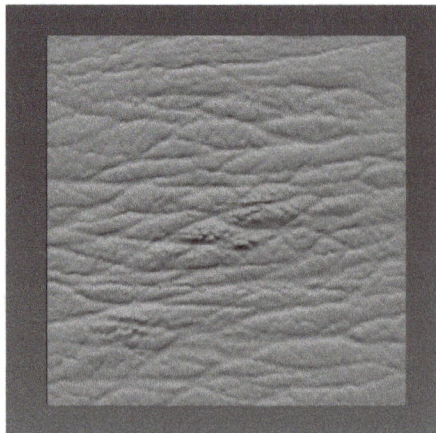

Figure 12.20 – Creating an Alpha for our wrinkles brush

4. Next, go to the **Document** palette and resize the canvas to have square dimensions. This will ensure that you can create a square Alpha that is the preferred dimension of Alphas for brushes.

Figure 12.21 – Resizing the canvas

5. Orient your camera to look straight at the plane that you just sculpted on, and press *F* to center the camera and fill the canvas fully with the plane:

Figure 12.22 – Having the plane fill out the canvas

6. Now select the **Standard** brush, then open the **Brush** menu by pressing *B*, and click **Clone** to create a new brush.

Figure 12.23 – Cloning the Standard brush

7. Click on the **Alpha** slot of your brush, and click **GrabDoc**. Now, the Alpha is attached to the brush like so:

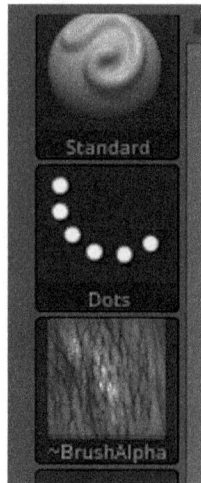

Figure 12.24 – Wrinkles brush

8. Open the **Alpha** palette, navigate to **Modify**, and enable **Surface**, so that the brush applies the texture, based on the grayscale color of the Alpha. This will make the brush change less of the overall volume of the mesh it is applied to while creating sharp detail.

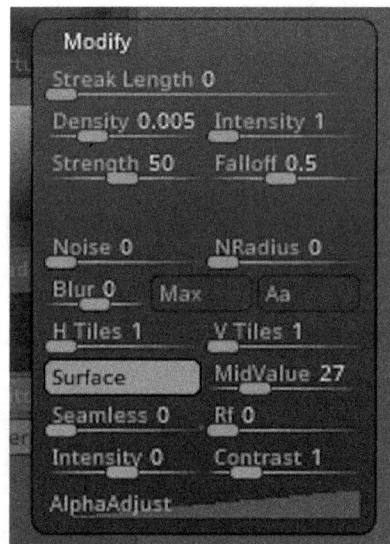

Figure 12.25 – Surface mode in the Alpha palette

9. Finally, go to **Stroke | Modifiers** and enable **Roll**. This will give a continuous effect to the brush, which you will need to sculpt repeated wrinkle patterns in a single brush stroke.

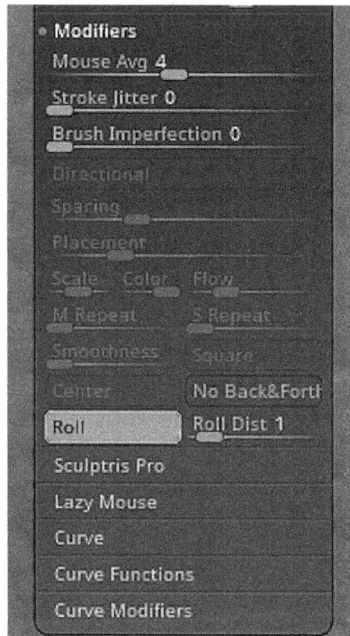

Figure 12.26 – Roll mode in the Stroke palette

Now the wrinkles brush is finished, you can use it to apply a leathery pattern on the upper eyelid. Here is the result:

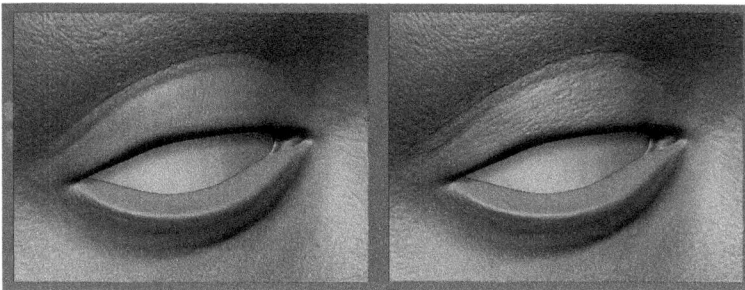

Figure 12.27 – Before (left) and after (right) applying the wrinkles brush

After using our custom wrinkle brush, you can sculpt some more pronounced wrinkles using the **Dam Standard** brush, and then use the **Standard** brush to inflate some of the new shapes.

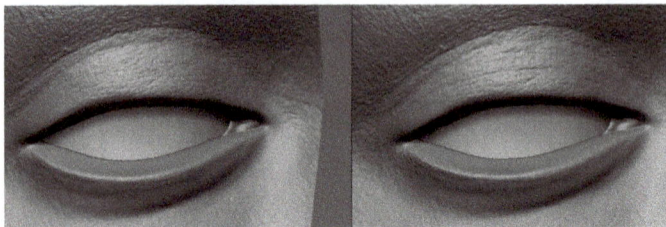

Figure 12.28 – Before (left) and after (right) using the Dam Standard and Standard brush to add detail

This completes the upper eyelid. Next, let's add detail to the lower eyelid.

Lower eyelid

When it comes to sculpting the lower eyelids, make sure to pay attention to the direction and unique pattern of the wrinkles. Sculpting wrinkles in a random position, size, and orientation will lead to unrealistic results. Attention to detail is key here.

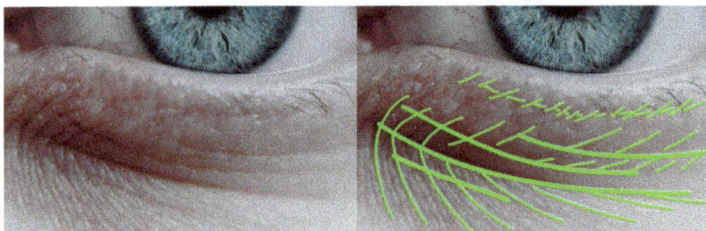

Figure 12.29 – Wrinkle pattern on the lower eyelids

You can use the new custom wrinkles brush for fine wrinkles and the **Dam Standard** brush to sculpt larger wrinkles. The result could look like this:

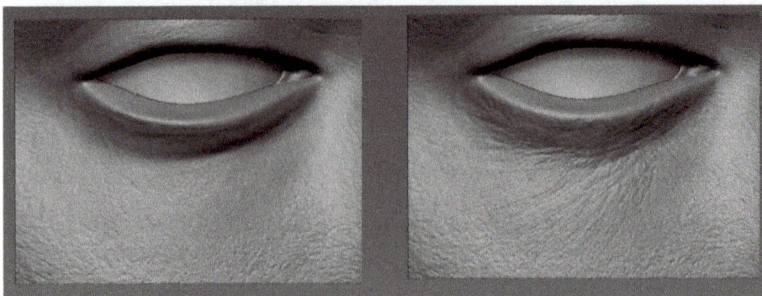

Figure 12.30 – Before (left) and after (right) using the wrinkles
brush and the Dam Standard brush to add detail

You can also use the **Standard** brush with the **DragRect** mode and **Alpha 36** to apply goosebump skin detail, which is typical for the area under the eye.

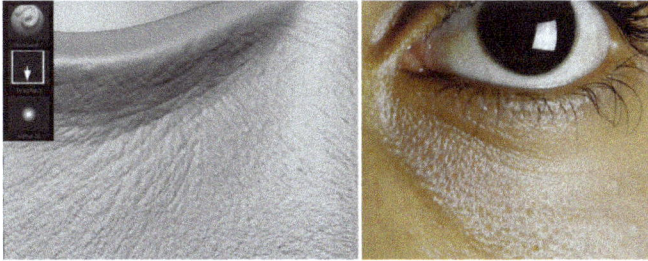

Figure 12.31 – Adding goosebump detail with the Standard brush and Alpha 36

Next, let's move on to the forehead.

Detailing the forehead

Not dissimilar to other types of skin detail, the detail on the forehead changes a lot with increasing age. Of course, there are individual differences but, generally, stronger forehead wrinkles become more pronounced at a certain age. For a young woman, this detail will be much more subtle, but it is good to implement it nonetheless because it creates the natural variation and complexity that gives the head more realism.

Creating the base detail

Before creating the base detail, make sure that you find high-resolution images of the forehead, in which you can see the pores and their size and frequency. It is important to get these characteristics right, so you don't end up with pores that are too small or big. Searching for Face Macro or Face Closeup on websites such as pexels.com or unsplash.com will deliver high-quality, close-up images of faces.

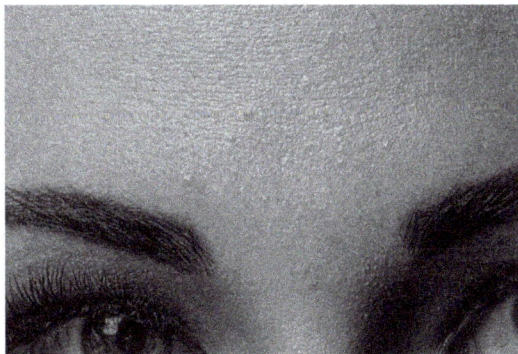

Figure 12.32 – Subtle wrinkles, visible in macro photos

Once you have these references, the first thing to do is to apply a tileable pore texture. Although skin pores on the forehead tend to be very small and subtle, they exist, so the first pass on the forehead should be done with NoiseMaker.

To do this, go through the steps of the previous subsection, but instead of applying a generic skin texture, apply a tileable pore texture. Again, you can find this type of texture in CG stores, such as the Artstation marketplace.

After you have applied pores on the forehead, it could look like this:

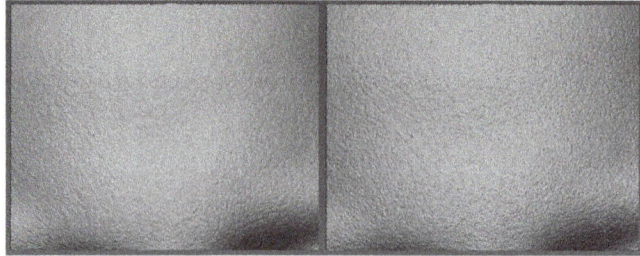

Figure 12.33 – Before (left) and after (right) applying a tileable pore texture to the forehead

With this base established, you can move on to add subtle wrinkles to refine the forehead further.

Creating wrinkles

The forehead is one of the areas that has a lot of wrinkles since a lot of facial expressions use the stretching and compression of the forehead through the facial muscles. Before you start sculpting the wrinkles, there are two important points to consider for the placement and appearance of those wrinkles:

- Pay attention to the placement of the main wrinkles, as they typically have a relatively even distance between each other. Reference photos are key for getting the most accurate placement possible.

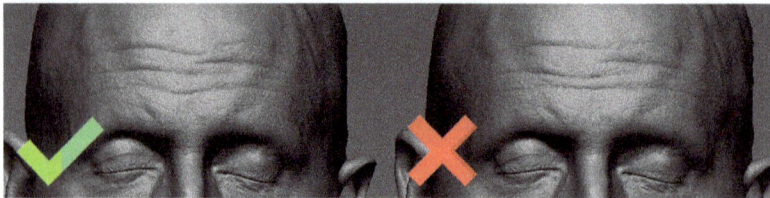

Figure 12.34 – Placement of the forehead wrinkles should not be random

- Do not sculpt long, continuous wrinkle lines. Wrinkles tend to consist of multiple lines, that roughly intersect. Beginners tend to forget about this breakup into smaller shapes.

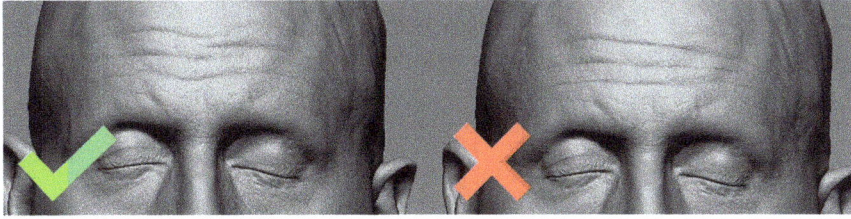

Figure 12.35 – Avoid long, continuous wrinkles without breakup

Now, you can go ahead with the **Dam Standard** brush and sculpt fine wrinkles. Since we are sculpting a young woman, we only need a very subtle indication of these wrinkles; reducing **Z Intensity** to around **20** can be a good strength for the type of subtle wrinkles that are needed here.

As you're sculpting, try to keep the previous two rules in mind and sculpt multiple fine lines with some crisscrossing lines, aiming to create a natural look with variation in strength and orientation.

You can "connect" pores by making the wrinkles pass through the pores instead of making them pass between them. This requires a bit of time, but it is the realistic way to do it.

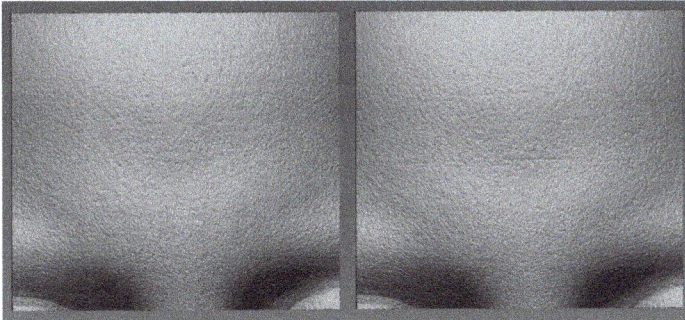

Figure 12.36 - Before (left) and after (right) sculpting subtle wrinkles with the Dam Standard brush

After you are done sculpting the subtle wrinkles, you can use the **Standard** brush with **Intensity** lower (around **10** to **14**) to inflate the area around the wrinkles to increase their volume and visibility (though make sure to keep it subtle, based on the gender and age of your subject).

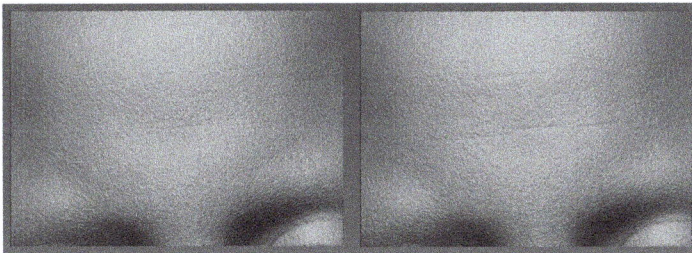

Figure 12.37 – Before (left) and after (right) using the Standard brush to add volume around the wrinkles

Finally, you can use the **Standard** brush and sculpt bumps, pimples, and small cuts. These imperfections will add realism and bring the details to life.

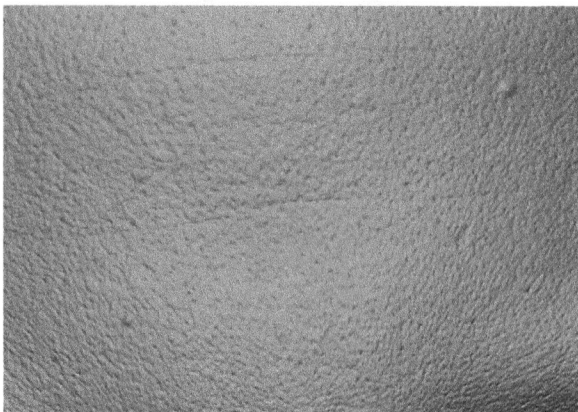

Figure 12.38 – Adding small pimples and imperfections with the Standard brush

The next area we will look at is the nasolabial folds, which are the lines extending from the sides of the nose to the corners of the mouth.

Adding pores to the cheeks and nasolabial folds

The cheeks usually have the biggest, most noticeable pores in the face. The more you want to push the realism and quality of the skin, the more important it is to pay attention to detail; here it means that you need to make sure to use a good quality pore alpha texture.

NoiseMaker versus a custom pore brush

When adding bigger pores to the face, you could use NoiseMaker again without much effort. However, a more precise, though time-intensive, way is to add them manually with a pore brush. This gives you more control, as you can adjust the scale of the pores based on the stretching and compression of the skin.

Another benefit of placing the pores manually is that it creates a less repetitive look with more variation in size and density, which can make the model feel more natural.

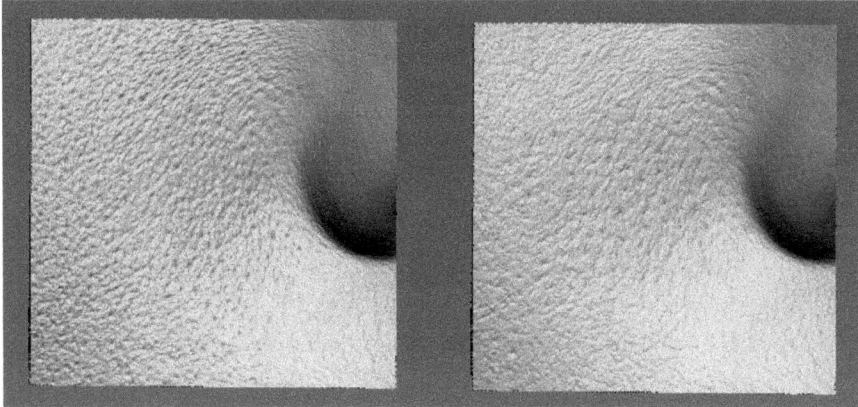

Figure 12.39 – Tileable pore texture versus manually added pores

That being said, some areas of the face have very subtle pores, in which case they can be created with NoiseMaker without noticeably inferior results.

Creating a custom pore brush

If you decide to create a pore brush, you need to sculpt an individual pore on a plane first, similar to the way you created a **Roll** brush earlier when adding wrinkles to the eyelids.

To do this, sculpt several pores on the plane so that you can add detail more quickly, since placing individual pores would take an absurd amount of time. You may create both an Alpha with a single pore and one with a bunch of pores. Make sure to sculpt some radial wrinkles around the pore, using the **DamStandard** brush. That way, they integrate better into the skin you apply them to.

Figure 12.40 – Sculpting a bunch of pores

Also, when you sculpt the pores, instead of sculpting a small hole that has harsh borders, make sure to give them a gradual fall-off. You can achieve this by applying the **Dam Standard** brush with a large brush radius on a centered point, automatically creating the needed effect.

Figure 12.41 – Using the Dam Standard brush to give the pores a proper fall-off

Then extract an Alpha from your sculpted pore plane(s) and create a **DragRect** brush from it. The full process is described in *Chapter 4*.

Now, you can place the pores on the model. You may choose to add some of them with your custom brush, and then fill out the rest, using NoiseMaker. **Morph Target** and **Morph Brush** will help to control the effect.

Next, you can move on to another important area with unique detail: the lips.

Creating the detail of the lips

When detailing the lips, you can use the **Dam Standard** brush and the Alphas that you bought or created yourself. Here, we will just focus on the characteristics of lips, instead of focusing on one of these ways, so that you can use your preferred tools to achieve the desired result.

If you look at close-up photos, you can observe the typical leathery pattern (which you find in other skin details as well). The border of the lips has small vertical wrinkles at a higher frequency, while the wrinkles toward the inner lips are fewer and of increasing size.

Figure 12.42 – The skin detail and pattern of lips

When you sculpt the wrinkles, make sure to add variation to avoid a repetitive look:

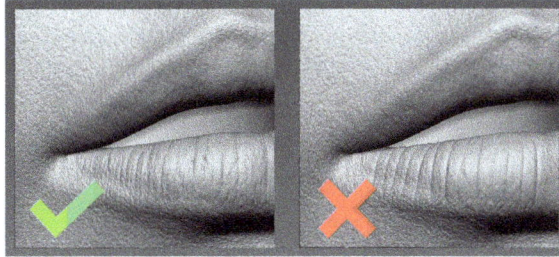

Figure 12.43 – Avoiding a repetitive look

Important note

It is important to not neglect small details. While it seems like they are unimportant and barely visible from a distance, leaving them out will reduce the realism, which is especially noticeable when rendering your model with a proper skin shader that has a subsurface Scattering effect. A lack of sculpted detail will create a plastic or rubber look instead of a skin look.

Now we will focus on the neck, which needs slightly adjusted detailing as well.

Creating the skin detail of the neck

The skin on a person's neck has a slightly different look than the rest of the skin – the base texture has a typical leather-like structure, and on top of that, there is a bumpy effect.

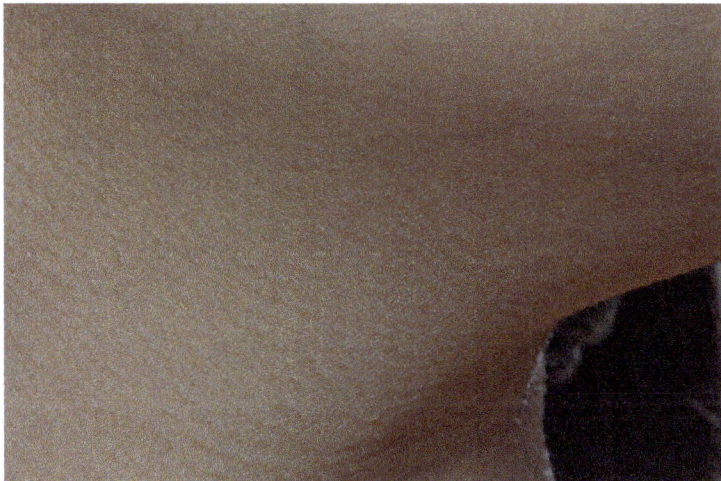

Figure 12.44 – Neck skin detail

To add this detail to the neck, a great solution is using a custom **Roll** brush, again, like you did when you detailed the upper eyelid. First, you need to sculpt the details on a plane with the **Dam Standard** brush. Here, adding some directional wrinkles as the base, similar to the eyelid detail brush, will be a good starting point. Then, you can use the **Standard** brush on a low intensity, to sculpt a bumpy effect, as shown in the previous screenshot. The following screenshot shows the result:

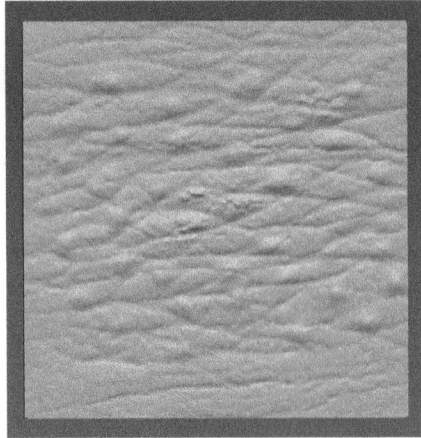

Figure 12.45 – Sculpting the neck detail

From here, proceed with the instructions under the *Adding detail around the eyes* subsection to create a custom brush. After using this brush on the neck, it could look similar to this:

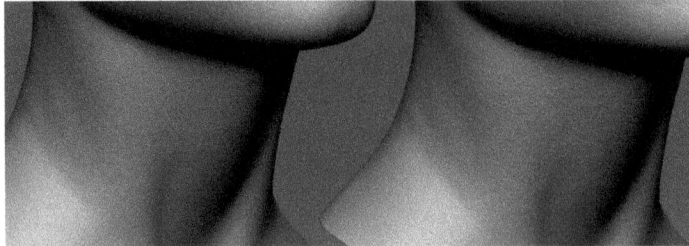

Figure 12.46 – Skin detail on the neck, before (left) and after (right) using a custom brush

Now, you have a fully surfaced head, with the appropriate variation in detail based on the area of the head. As a final step, you can increase its realism by adding some more imperfections.

Finalizing the skin detail with imperfections and surface variation

The head model is now in a place where the details look nice and crisp, however, it looks too smooth and uniform overall. A final pass of subtle sculpting is needed to add the sort of imperfections that will bring the realism to the next level.

Before you get start with this sculpting pass, make sure to go to **Tool | Morph Target** and click **StoreMT**, so you can adjust the strength of this sculpting pass after applying it.

This pass can be done with just the **Standard** brush on a low intensity, although you are of course free to choose any brush or tool that you like. Here, you can sculpt small pimples, bumps, small scars, and wrinkles. Additionally, you can inflate bigger areas slightly to create more surface variation. For that, it is important to stick to reference images and do it while paying attention to anatomy, so it does not look too random, which would make it counter-productive.

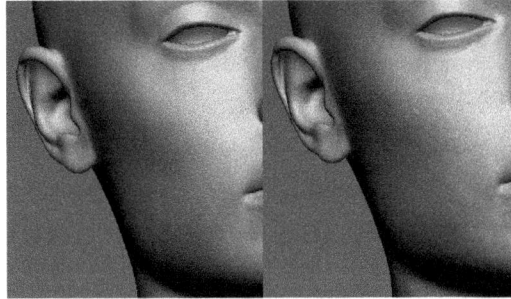

Figure 12.47 – Using the Standard brush to add surface variation and imperfections

This sculpting pass can make the model appear slightly older, so keep this in mind when you add this detail. At the same time, the 3D model will not display all of this detail once it is rendered with a skin shader with subsurface scattering, so the detail needs a certain intensity to even be visible. Keep this in mind if you are planning to render your model in software such as Maya or Blender.

After completing this sculpting pass, you can use the **Morph** slider in the **Morph Target** menu to reduce the effect again, if it is too intense. After using the **Morph Target** slider with a value of **50** to reduce the sculpting pass to 50%, the final result of the skin detailing work looks like this:

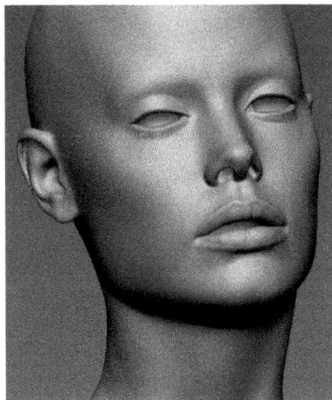

Figure 12.48 – The result of the manual skin detailing process without scan data

> **Important note**
>
> Just like there is a lot of variation in the proportion and appearance of faces, there is not only one type of skin detail and there are big differences between individuals. Older people especially will require a much more intense kind of detail that includes more, and more defined, wrinkles. Gathering high-resolution reference images that show the skin detail close up will give you the information needed to create a more realistic and believable sculpture.

This completes the skin detailing part of this chapter. As you will see, realistic and great-looking skin detail can be achieved without scan data, using some Alphas and basic brushes. In the end, it is up to you which tools you would like to use, how much scan data is helpful, and where you need to achieve specific effects with your own sculpting.

The last thing this head needs is some hair, which you will add next.

Sculpting hair, eyebrows, and eyelashes with brushes

Sculpting hair can be challenging since you always need to apply some kind of stylization. This makes it less intuitive to sculpt than the head, but at the end of the day, the design rules that apply to sculpting the head, or any other thing, apply to the hair as well.

In this section, you will learn about two ways of creating hair – with basic brushes and, specifically, with the **IMM Curve** brush – as well as adding eyebrows and eyelashes. Then, in the final section, you will learn a third way of creating hair: with FiberMesh.

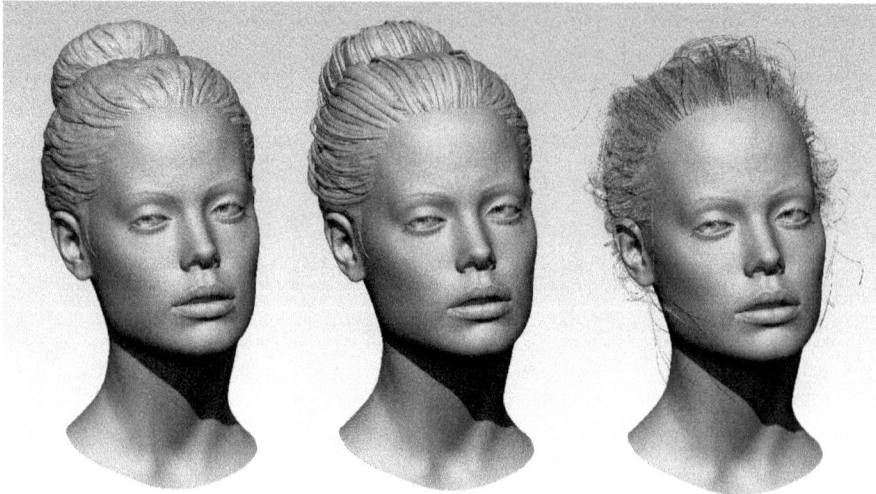

Figure 12.49 – Sculpted hair, IMM Curve brush hair, and FiberMesh

Sculpting the hair with basic brushes

If you're creating a project focused on sculpting and want to create a classic sculpture look, sculpting the hair with basic brushes such as **ClayBuildup**, **Dam Standard**, and the **Standard** brush will be your best choice. Here is a simple workflow:

1. Go to **Tool | Subtool | Append** and pick the **Sphere3D** shape, to use it as a starting point for blocking out the hair.

Figure 12.50 – Using a Sphere3D shape to block out the hair

2. Scale the sphere with the Gizmo and position it inside of the head model, so it covers the area where you want to sculpt the hair.

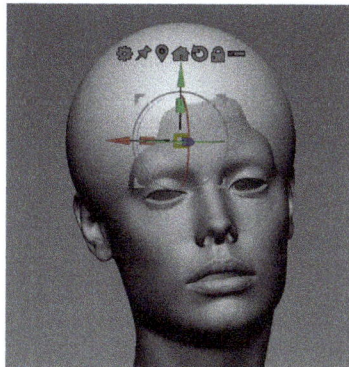

Figure 12.51 – Scaling and positioning the sphere

3. Now, you can use the **Move** brush to push the sphere into position, creating the proper hairline. You can also use the **ClayBuildup** brush to establish the overall shape of the haircut you want to sculpt.

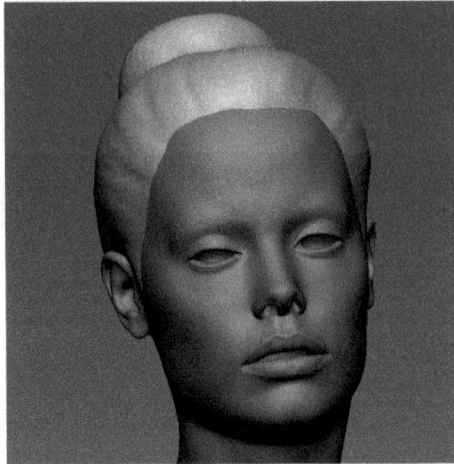

Figure 12.52 – Establishing the rough hair shape

4. Go to **Tool | Geometry | ZRemesher** and click **ZRemesh**, making sure **Target Polygons Count** is **5**. This will give you more even topology, so you can sculpt properly without having areas that lack resolution.

Figure 12.53 – Using ZRemesher to create even topology

5. At this point, you can subdivide the model 2-3 times for now. You may add more subdivisions later, as you run out of resolution to sculpt the detail.

6. Next, you can start to refine the hair a bit, creating points of higher contrast with the **Dam Standard** brush to create sharp lines, and the **Standard** brush to add volume to the hair. Make sure to have a variety of areas with little detail and areas that have more detail.

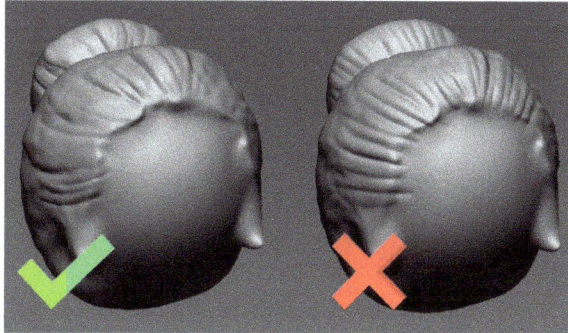

Figure 12.54 – Avoid a repetitive look by sculpting hair strands of varying size and intensity

7. As you create a more detailed mesh, you can define the individual hair strands more, and even define very small strands of hair with the **Dam Standard** brush. Just make sure to not get carried away detailing too much, ending up with a high frequency of hair clumps everywhere – some clumps can be combined and merged into one bigger shape, while other hair clumps can overlap each other.

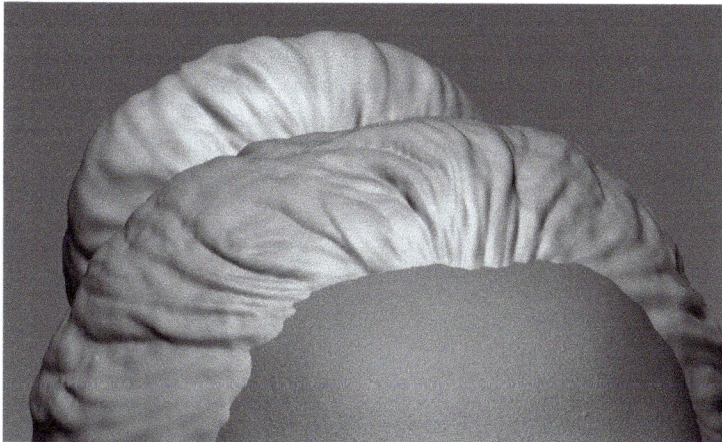

Figure 12.55 – Refining hair strands and creating an overlap effect

8. Add more subdivision levels, until you can sculpt clean details. Use the **Dam Standard** brush to sculpt in some finer hair strands and sharp lines, keeping the bigger picture in mind and not applying the details too uniformly.

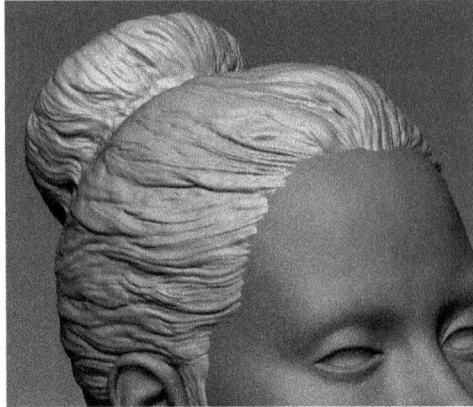

Figure 12.56 – Adding fine details to the hair sculpt

Now, you have a way of creating hair that fits nicely for a classic sculpted look. However, there is another interesting, unique way to create hair in ZBrush that is especially suited for creating stylized hair, which is using an **IMM Curve** brush. You will learn about this workflow next.

Creating hair with an IMM Curve brush

IMM Curve hair brushes let you create hair strands along a brush stroke, giving you simple and clean geometry, making it a great fit for stylized characters that are simple. If you decide to create this kind of hair, you could buy a custom brush on sites such as Gumroad or the ArtStation marketplace, but if you want to create your own hair strand brush, here is how to do it:

1. Go to **Tool | Subtool | Append** and pick any shape, for example, **Cube3D**.

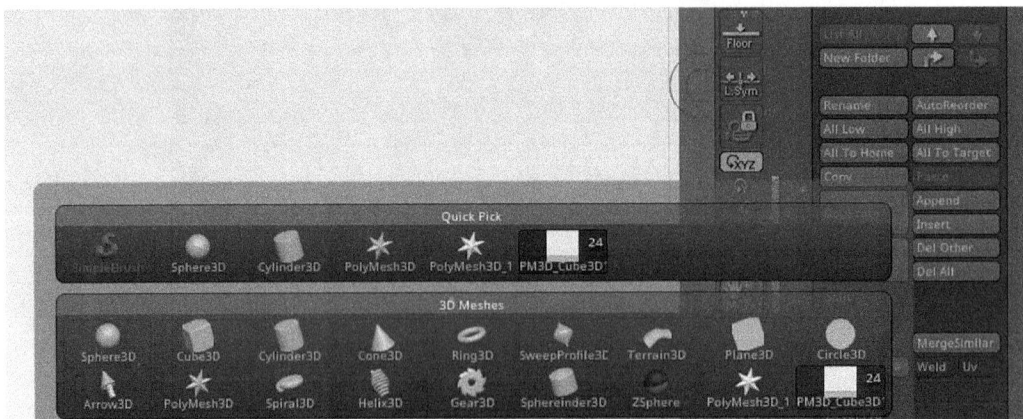

Figure 12.57 – Appending Cube3D

2. In the **Tool** palette, navigate to the **Initialize** menu, and click **QCyl Y** to transform the cube into a cylinder.

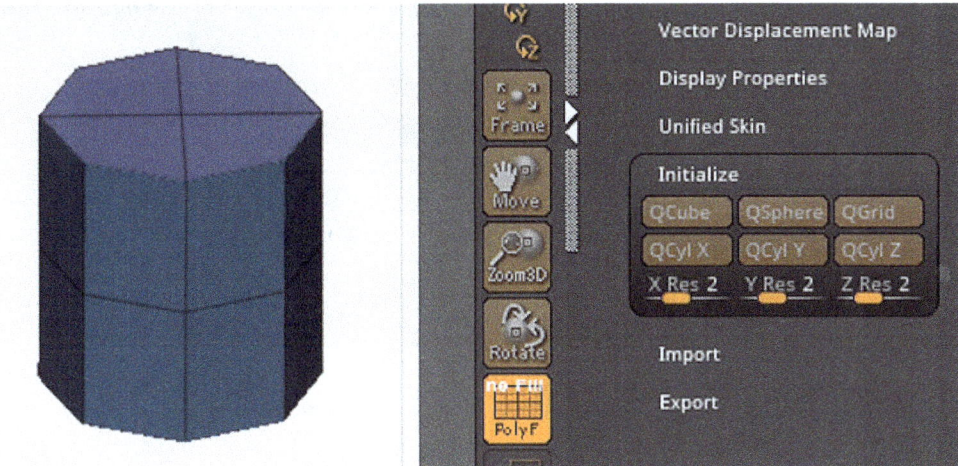

Figure 12.58 – Initializing your mesh

3. Go to **Tool | Subtool** and select **Duplicate** at least 3-4 times. These duplicates will be the different hair strands. Then use the **ZModeler** brush to create variations with different widths, thicknesses, and profiles, while keeping the length and position the same. After you are done, make sure to merge the meshes into one subtool. The result could look something like this:

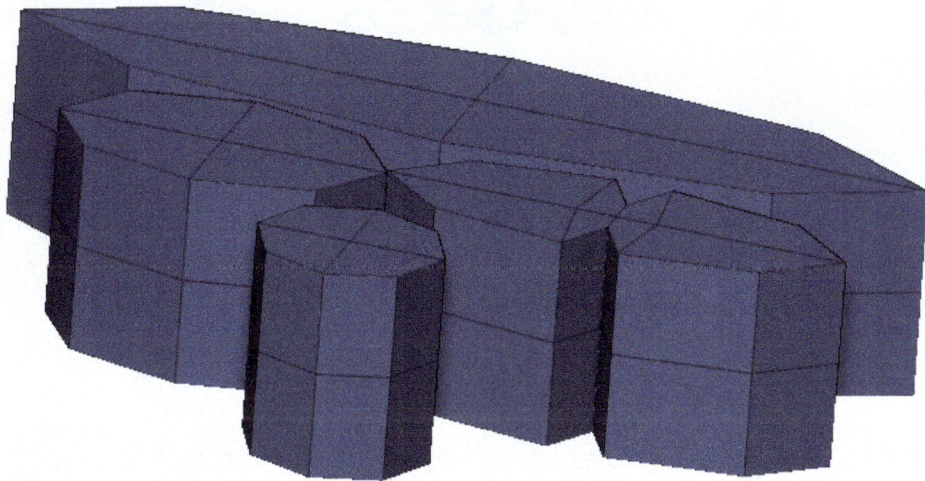

Figure 12.59 – Hair strands

4. Now select the **IMM Basic** brush, open the **Brush** menu, and select **Create InsertMesh**. A pop-up window opens, asking you if you would **Like to APPEND the active mesh to this brush or create a NEW brush** – here, you need to select **New** to create a new **InsertMesh** brush based on your hair strands model.

5. Next, open the **Stroke** palette, go to the **Curve** menu, and enable **Curve Mode**.

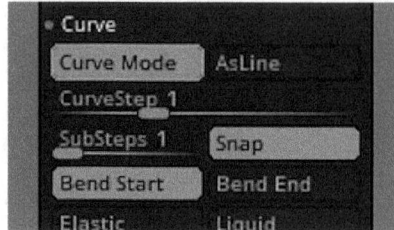

Figure 12.60 – Enabling Curve Mode

6. Next, navigate to the **Brush** palette, click **Modifiers**, then enable **Weld Points** and **Stretch** to make sure the brush will create hair strands as one continuous mesh, instead of many separate pieces.

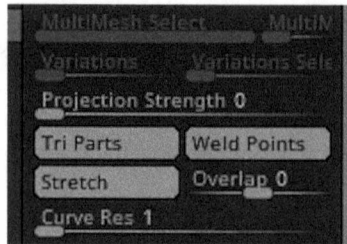

Figure 12.61 – Adjusting the Brush modifiers

7. Go to the **Stroke** palette, open the **Curve Modifier** menu, and enable **Size**. This will give the hair strands a tapering effect. You can adjust the curve profile, by dragging a dot on the graph, to test the brush and see the different effects it has. In the following screenshot, the hair strands on the left were created without **Size** enabled, and on the right, they were created with **Size** enabled.

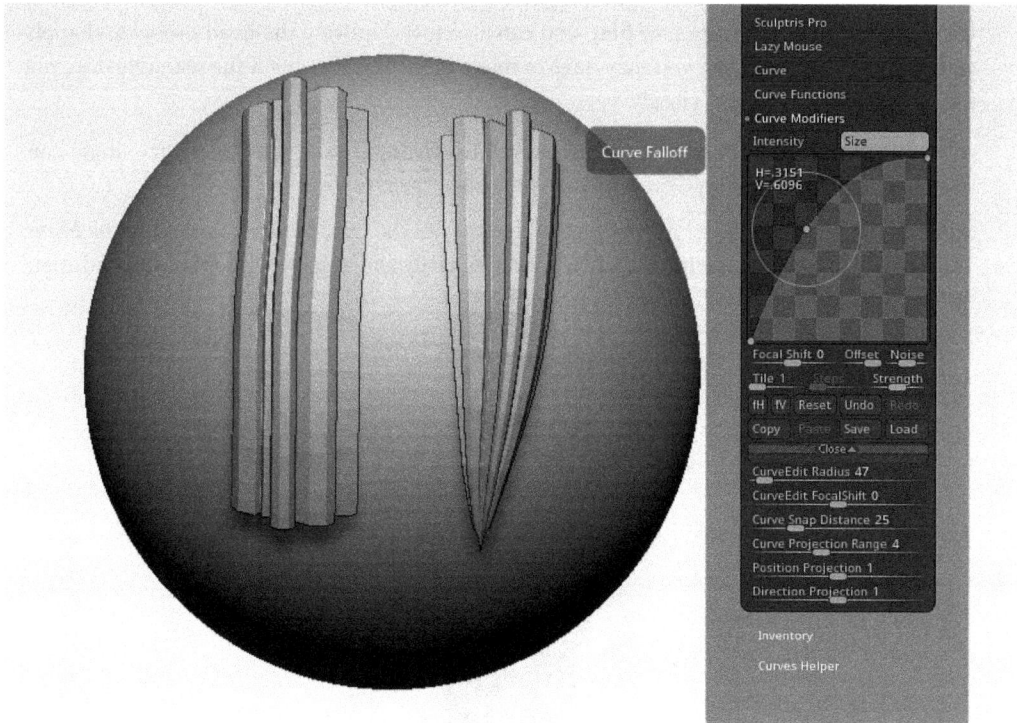

Figure 12.62 – Using Size to give a curve brush a tapering effect

8. Now, with the **IMM Curve** brush created, you can go ahead and apply the brush on the hair, though make sure to delete the subdivision levels of the mesh, or you will not be able to apply the **IMM Curve** brush. For this workflow, a very rough sculpt is enough and will be a useful guide to apply your **IMM Curve** brush and achieve the hairstyle you aim for.

Figure 12.63 – Adding hair strands with a custom IMM Curve brush

If you do not want to sculpt hair first, you can also just duplicate the head model and apply your brush there. However, you may have to make more adjustments if the hairstyle does not follow the head shape very closely.

9. After that, go to **Tool | PolyGroups** and click **Auto Groups**. Now you have every single hair strand as an individual PolyGroup so you can isolate them easily.

10. Separate the hair strands and move them so that they fit the head better. You can use the **Move Topological** brush to move individual hair strands easily and quickly without having to isolate them. The result could be similar to this:

Figure 12.64 – Hair created by IMM Curve brush

Now you know a great way to create stylized hair and have the ability to create custom curve brushes, which can come in handy in many scenarios.

At this point, we still don't have eyebrows and eyelashes, so let's tackle those next.

Sculpting the eyebrows

Sculpting eyebrows is simple, and you can get great results with just a couple of brushes and the right references. Particularly if this is your first time, it is worth gathering some high-resolution images of eyebrows and observing the orientation of the hair:

Figure 12.65 – Hair direction of the eyebrows

With the hair orientation in mind, you can go ahead and sculpt the eyebrows on your head model:

1. Start by storing a Morph Target, so you can bring back skin detail on areas that you don't end up covering with hair.

2. Then use the **Smooth** brush to get rid of the skin detail on areas where you plan to sculpt the eyebrows.

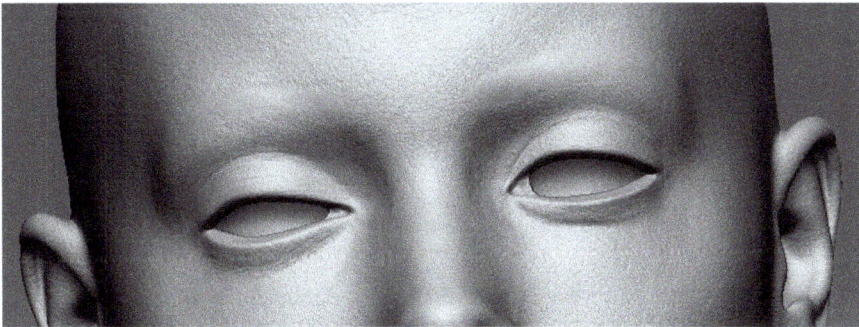

Figure 12.66 – Using the Smooth brush to remove skin detail

3. Use the **ClayBuildup** brush to sculpt in some rough forms, defining the overall shape and position of the eyebrows.

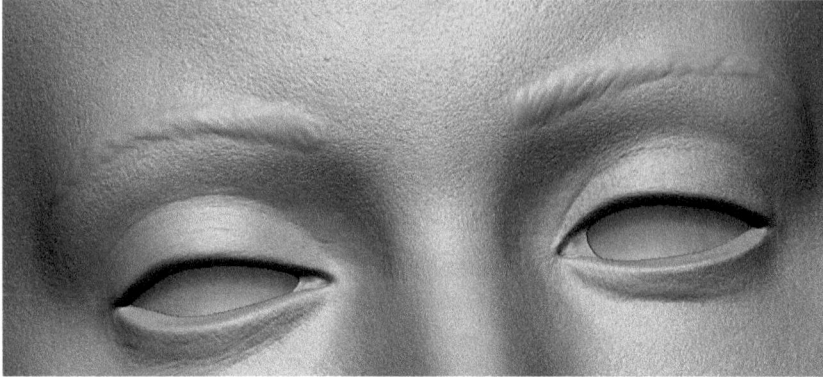

Figure 12.67 – Blocking out the eyebrows with the ClayBuildup brush

4. Next, use the **Dam Standard** brush on both eyebrows, as well as using the **ZAdd** and **ZSub** modes to sculpt individual hairs. Try to sculpt in a variety of forms, creating a balance of defined individual hairs and smaller clumps.

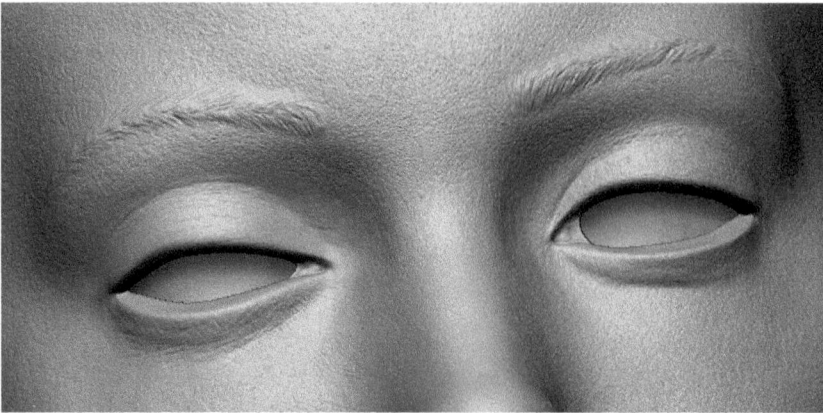

Figure 12.68 – Refining and detailing the eyebrows

Now that you have added the eyebrows, you can add the last missing piece: the eyelashes.

Adding eyelashes

Eyelashes are a small but important detail that you can't miss if you aim to create a realistic head. Once again, make sure to take a quick look at some photos before getting started, so you can get familiar with the characteristics of eyelashes. Here, you can observe the size and quantity of the eyelashes, what curvature they have, and how they tend to clump together:

Figure 12.69 – Curvature and clumping of eyelashes

From the **Curve** and **InsertMesh** brushes to **FiberMesh**, there are several ways in which you can create eyelashes. An easy and convenient way is to create an individual eyelash first and then copy it until the eyelashes are complete. Here is how to do it:

1. Go to **Tool | Subtool | Append** and pick the **Cylinder3D** shape.

Figure 12.70 – Using a Cylinder3D shape to create the eyelash shape

2. Use the Gizmo to scale up the cylinder, so that it has the appropriate length and width for eyelashes.

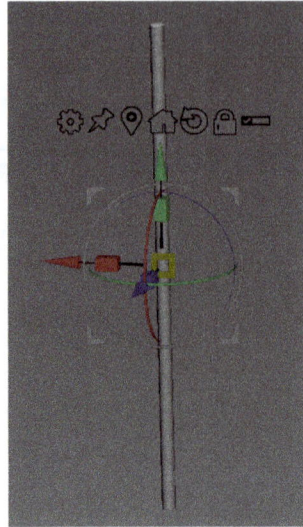

Figure 12.71 – Scaling with the Gizmo

3. In the **Gizmo** menu, open the **Deformation** menu, and select the **Taper** deformation. Then at the end of the cylinder, taper the top.

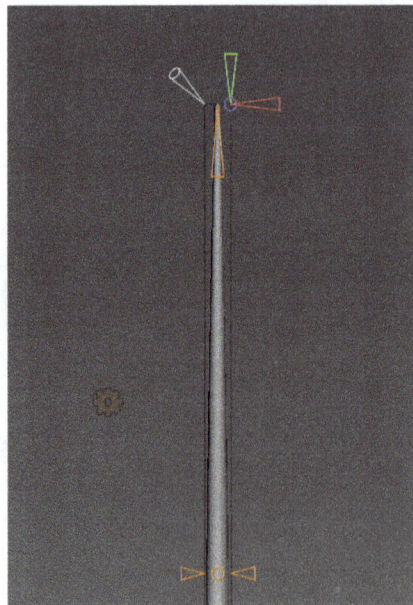

Figure 12.72 – Using the Taper deformation

4. Next, use the **Bend Arc** modifier and give the mesh some curvature.

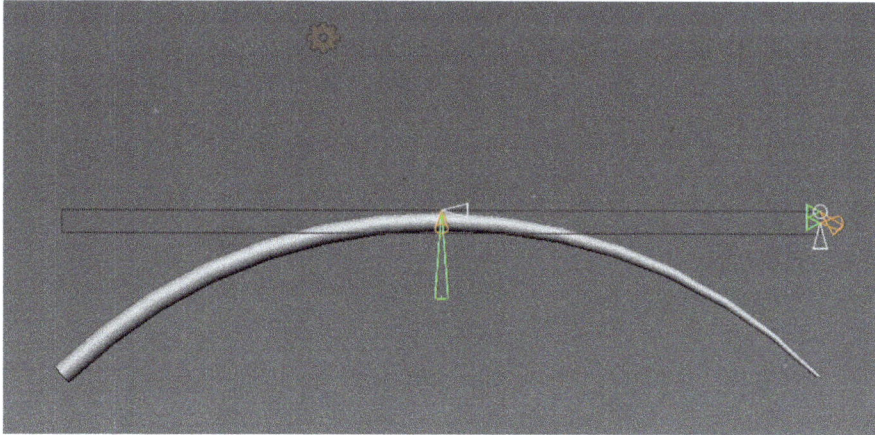

Figure 12.73 – Using the Bend Arc modifier

5. Copy the mesh to create a clump of eyelashes, and then copy that clump along the entire lower eyelid. You can do this by pressing *W* to access the Gizmo, then pressing *Ctrl + Left-click*, and dragging your cursor.

6. Use the **Move Topological** brush to move individual eyelashes, creating more variation between the clumps.

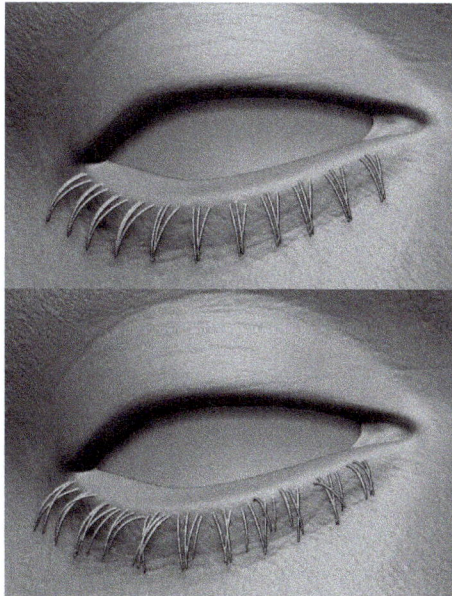

Figure 12.74 – Before (upper) and after (lower) adding variation in the eyelashes

7. Now go to **Tool** | **Subtool** | **Duplicate**, rotate the copy of the eyelashes, and place them on the upper eyelid.

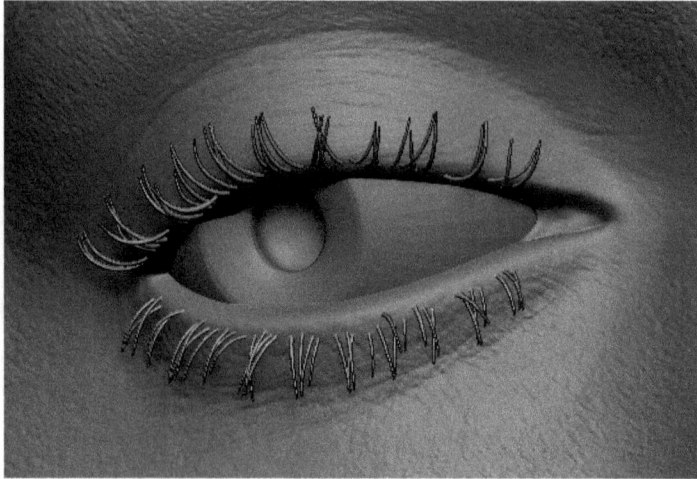

Figure 12.75 – Adding the upper eyelashes

8. Next, go to **Tool** | **Geometry** | **Modify Topology** and click **Mirror and Weld** to mirror the eyelashes onto the other eyes. If your model has an asymmetrical pose, you will need to use the Gizmo to rotate and move the eyelashes to match the eyelid shape.

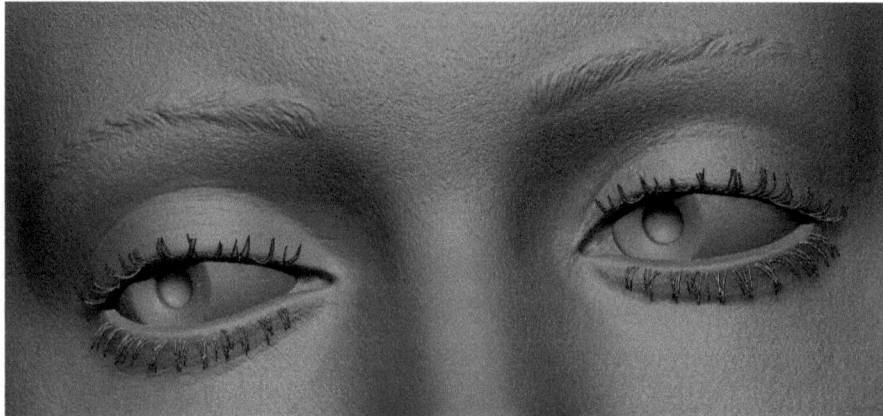

Figure 12.76 – Mirroring the eyelashes to the other side of the face

Now you know how to sculpt hair with different brushes, you will learn about FiberMesh next, ZBrush's hair system that can be used to create photorealistic CGI hair.

Sculpting hair with FiberMesh

If your goal is to create a classic sculpture, sculpting the hair is the way to go. But perhaps you want to create a realistic render with a 3D render engine outside of ZBrush. In that case, you can use ZBrush's tool for hair creation, FiberMesh.

> **Important note**
>
> While FiberMesh can be used to create realistic hair in software such as Maya, it is not an industry-standard tool. Grooming artists mainly use XGen, Yeti, or Houdini – they offer a variety of sophisticated tools that let you create a variety of different hairstyles with many more settings, but more importantly, the workflow is less manual and less destructive. In ZBrush, the grooming process will require using grooming brushes, and there is no option to get automatic effects. Tools such as XGen, on the other hand, let you add effects such as clumping and noise through simple sliders and values. In case you do not have access to any other hair tool, FiberMesh can still be useful and get the job done.

FiberMesh modifiers

FiberMesh can not only be used for hair. You can create a variety of different plants and objects with it, based on modifiers that let you determine shape, size, color, and other characteristics. Here you will learn how to adjust the modifiers so the fibers are easy to groom:

1. Create a mask on the head model, based on where you want the hair to be.

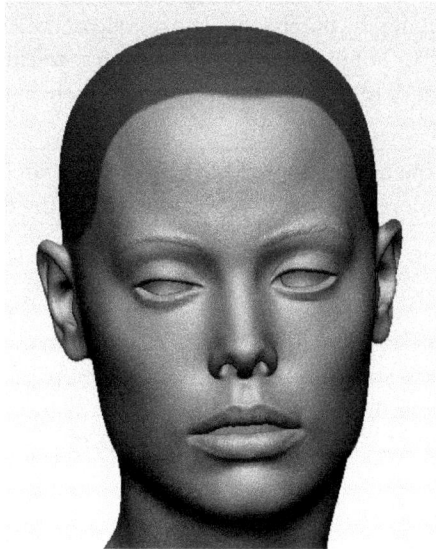

Figure 12.77 – Applying a mask to the head

2. Go to **Tool** | **FiberMesh** and click on **Preview**.

3. Open the **Modifiers** tab and adjust the options so you have the right starting point. Here are some of the important values:

Modifier	Description
MaxFibers	This will determine how many fibers/hairs will cover the masked area. Choose this according to your specific preference.
ByMask	This will make sure that coverage of fibers will occur based on the mask, so this should be set to **1**.
Length	This determines the hair length. Again, choose this according to your specific preference.
Coverage	This sets the thickness of the fibers. Make sure that the value is not too low so that the fibers are not too thin and are clearly visible.
Scale Root/Scale Tip	This determines the size of the root and tip. You do not have to change the default values.
Gravity	This sets the gravity effect. A value of **0** removes gravity, which makes the hair stand up straight, while a value of **1** makes the hair point downward. For grooming the hair, a value of **0** gives a better starting point because hair strands can be isolated more easily.
Profile	This determines the number of sides. A higher number can help you see the hair more clearly. A profile of **1** is just a thin plane, while a value of **2** or **3** will turn it into a shape with thickness, making it easier to visualize it from certain angles.
Segments	This determines the number of segments along the length of the hair. Long hair needs a higher number, so there are enough segments to bend the hair without running out of resolution. For long and curly hair, the maximum value of **50** is ideal.
Base/Tip	This gives the base and tip of the hair color. It can be good to use bright and saturated colors so the hair is clearly visible during the grooming process.

Figure 12.78 – FiberMesh modifiers

All the other options can also be useful to create variation, but they are more suited for other things, such as grass, plants, and anything that does not require grooming. Since you will need to manually groom and style the hair, these options do not matter as much. Before proceeding, make sure **ByMask** is set to **1** and everything else is set to **0**.

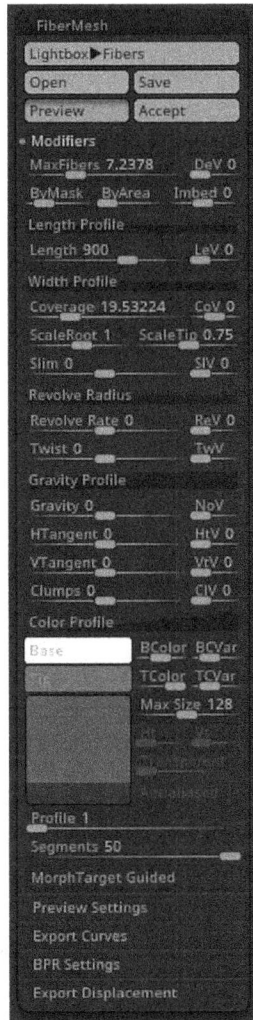

Figure 12.79 – FiberMesh basic hair settings

1. Once you have everything entered, click **Accept**. ZBrush may ask if you want to use **FastPreview rendering**, which improves performance when working with a large number of fibers, which you can accept too.

Important note

FastPreview rendering helps to improve performance when you have a dense hairstyle with lots of fibers and segments. This mode will reduce the hair to a smaller percentage, which you can groom and still create the overall hairstyle. This does not only improve performance, but it can also help simplify the process, by letting you focus on smaller portions of hair at a time.

2. If you enabled **FastPreview rendering**, navigate to the **Preview Settings** menu, and adjust the **PRE Vis** value according to the percentage of hair you want to have displayed (this value depends on the total amount of fibers you created and personal preference).

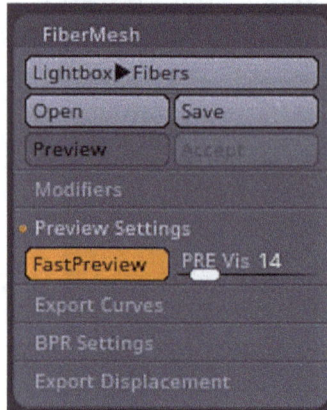

Figure 12.80 – FastPreview mode and PRE Vis value

At this point, your model should look like this:

Figure 12.81 – The starting point of the grooming process

Now, you can start to style the hair. Let's look at the brushes we'll use to do this.

Grooming FiberMesh brushes

Grooming fibers and creating hairstyles is relatively simple and straightforward. In ZBrush's default **Brush** library, there is an extra category of brushes just for grooming, which all start with **Groom**, however, the most effective brushes for longer hair are, in fact, four regular brushes. These are the ones that we will look at here.

SelectRect

SelectRect is the first brush you will be using, allowing you to move hair strands into position, without affecting all the hair at once. A good practice is to group the hair strands that you isolate into PolyGroups, making them easier for you to access.

Figure 12.82 – Using the SelectRect brush to isolate parts of the FiberMesh

Move brush

The **Move** brush is perfect for moving hair strands. You can increase **Z Intensity** to **100** to make moving the hair easier since the default strength of **51** does not move hair as quickly as it does for normal meshes.

Before you start using the brush, go to the **Brush** palette, click **FiberMesh**, and set **Preserve Length** to **100** – this will ensure that the hair keeps its length when you move it (a lower value will cause the **Move** brush to stretch the hair out, which means that you will have inconsistent hair length).

Then, a bit further down in the same menu, set **Front Collision Tolerance** to a value of **5** or lower. This setting determines the collision between FiberMesh and objects in the ZTool scene. High values will cause a glitching effect when the hair comes close to the head, which you want to avoid when creating the hairstyle, which is why we chose a lower value.

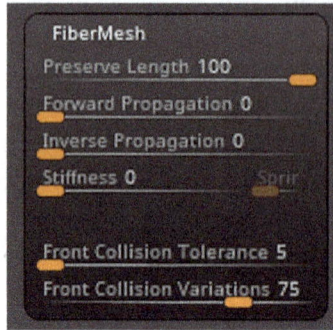

Figure 12.83 – FiberMesh brush settings

Keep in mind that these settings are only applied to the selected brush, so you will have to apply them to any other brush that you are using.

> **Important note**
> Working with FiberMesh and trying to create a good-looking haircut can be a tedious and tricky task if you are trying it out for the first time. This is completely normal, and with some practice you will get a better understanding of how the fibers react to your brushes.

Here is the result of moving fibers with the **Move** brush:

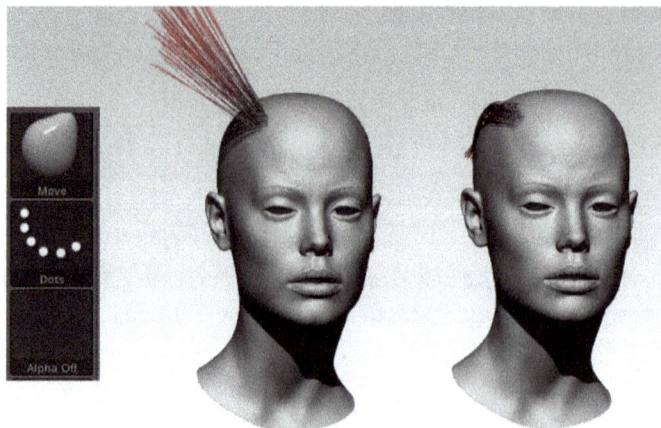

Figure 12.84 – Using the Move brush to shape hair strands

Pinch brush

The **Pinch** brush is great for creating clumps because it will pull the hair within the brush radius together. Before using the brush, it is best to isolate the hair you want to clump together, so you don't affect hair that is not supposed to be part of the clump. Then make sure to set **Preserve Length** to **100**, as you did with the **Move** brush – although this brush will not cause hair stretching as much, it is better to change this setting anyway if you want to keep the hair length consistent.

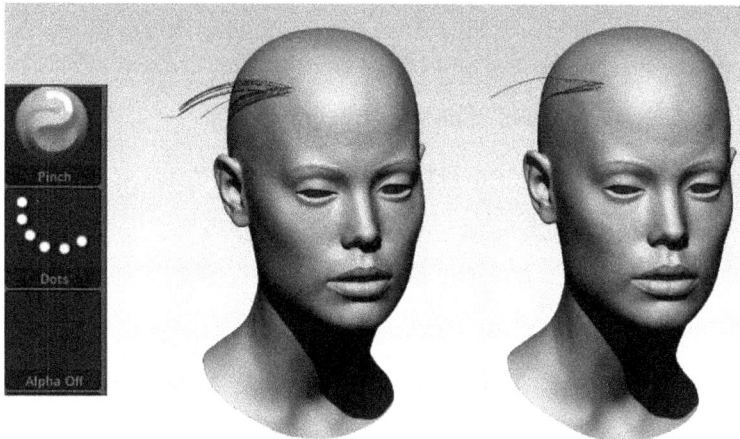

Figure 12.85 – Using the Pinch brush to create hair clumps

Smooth brush

The **Smooth** brush is essential when working with FiberMesh. Hair becomes distorted in the process of creating the hairstyle, and the **Smooth** brush can soften up jagged hair quickly.

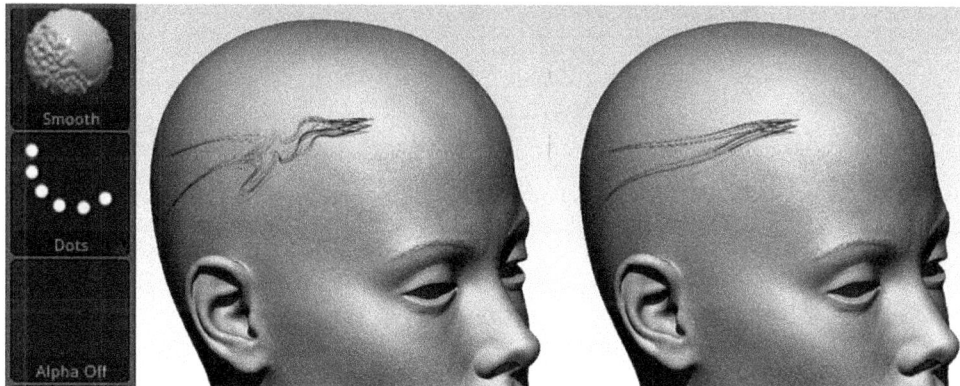

Figure 12.86 – Using the Smooth Brush to soften up jagged hair

The **Smooth** brush can also be used to shorten the hair, if it is applied to the tips of hairs, given that **Preserve Length** is set below **100**.

With these brushes in your arsenal, you can go ahead and start creating hair clumps and styling the hair, based on your concept.

Creating a hairstyle

At this point, you can go from hair strand to hair strand, using **SelectRect** to isolate them, and then shape the hair with the **Move**, **Pinch**, and **Smooth** brushes until you create your hairstyle.

The evolution of the hairstyle could look something like this:

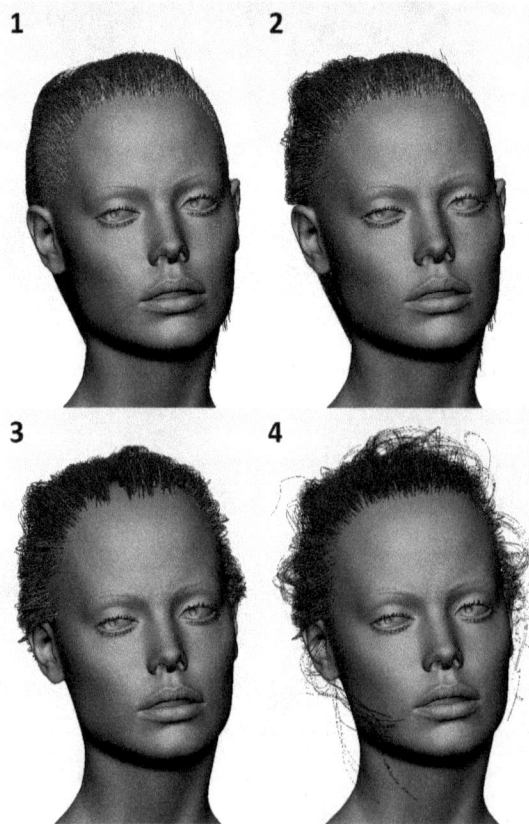

Figure 12.87 – Evolution of the hairstyle

Let's break down the screenshot:

1. First, you can slick back all the hair, so that it is not so messy and you get a better overview of which clumps you already worked on.

2. Then, you can start to isolate small parts of the hair to clump them together, as described before.

3. Eventually, you should have worked on all the hair so that you have a nice variation of clump size and orientation.

4. Finally, you can make an even smaller selection of hairs, and move them a bit further apart to create so-called flyaway hair.

When it comes to creating a great CG hairstyle, the two main things are patience and attention to detail, so make sure to gather some reference pictures and play some music or a good podcast to manage the boredom!

Most likely, the first attempt at creating a hairstyle will not give the desired result when implementing it in your 3D software of choice, and it can be a good idea to test this early, so you can make adjustments, and proceed according to the results.

Here is the result of the FiberMesh grooming:

Figure 12.88 – The result of creating a hairstyle with FiberMesh

Now, you know how to create a simple female hairstyle with a small selection of effective brushes. Finally, in the last section, you will learn how to export the hair as curves, so you can import it into other 3D software that lets you use curves to create realistic hair.

Exporting FiberMesh as curves

If you plan to export your fibers to use them to create hair with software such as XGen, you will have to export them as curves. These curves can then be converted to hair directly or serve as guides to set the hair direction of the clumps of the hairstyle. Here's how to do this:

1. Go to **Tool** | **FiberMesh** | **Export Curves**.

Figure 12.89 – Export Curves

2. If the number of curves exceeds 1,000, you will get a message asking you if you want to adjust the number of curves through the **Preview** mode option.

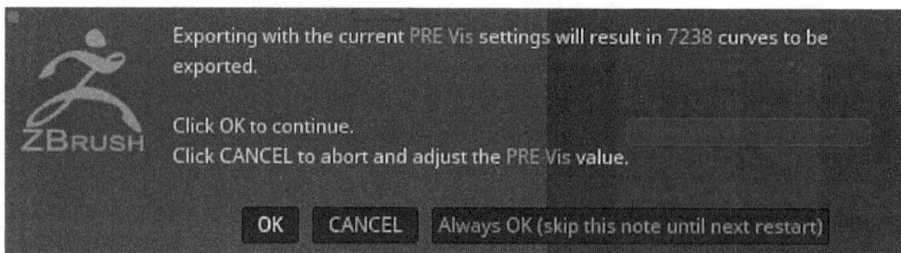

Figure 12.90 – Export warning message

Depending on how you plan to use the curves, you may only need a limited number of curves. For example, if you plan to use them as guides for XGen, you may only need something in the 10s to 100s, since they will be converted to guides, rather than to hair. One guide can determine the shape and profile of multiple hair strands, which is why you need far fewer than if you convert them to hair directly.

Also, If the curve amount exceeds a couple of thousand, it may cause performance issues in your 3D software. Figuring out the number of curves needed requires experience with the appropriate hair creation tool, and probably some trial and error as well. If you have the number of curves you like, click **OK**.

3. If you export the curves for use in Maya/XGen, make sure to save the curves as .ma (because .obj will not be recognized by Maya).

After that, you can proceed in Maya, importing the curves, and using them with XGgen, nHair, or other hair creation tools, to create realistic CG hair. If this interests you, here is the documentation for one of the more common uses, which is the conversion of curves to XGen guides: `https://help.autodesk.com/view/MAYAUL/2024/ENU/?guid=GUID-1511C398-13D1-4D6D-917E-8E5A0A9C3C3A`.

Summary

In this chapter, you added skin details to the head that you sculpted in the previous chapter. By doing so, you explored several types of skin details, and the best tools to create them, in order to apply the right kind of details based on the area of the face. This included sculpting techniques with basic brushes, as well as using projections of 3D scan data, NoiseMaker, and custom **Roll** mode brushes. Then, you added hair, eyebrows, and eyelashes to the head, learning how to sculpt "classic" hair with basic brushes, a custom IMM Curve brush for creating stylized hair, and FiberMesh for more advanced looks.

Unfortunately, it is not enough to be proficient with ZBrush to make a living as a digital sculptor. Regardless of whether you are looking to get work as a character artist, modeler, or any other digital sculpting-related role, there will be a lot of skilled competition out there, so it is essential that you maximize the effectiveness of your portfolio and leverage social media to get the kind of visibility that gets you job offers and opportunities. This will be the topic of the next, and final, chapter.

13
Building a Portfolio and Leveraging Social Media

The idea of the "starving artist" was popularized in the 19th century, describing an artist who had to sacrifice their financial and materialistic pursuits in order to dedicate themselves to their craft, hoping for a breakthrough in the art field.

Nowadays, you can be an artist without the "starving" part, if you put in the effort and apply some strategic thinking. Of course, there is no lack of competition, so creating a portfolio that stands out and lets you land jobs will take some time and luck. At the end of the day, it always comes down to the quality of your artwork, but you also need to get your artwork in front of recruiters, too.

This chapter aims to showcase ways to generate attention for your artwork. First, we will do this by exploring the most effective subjects to choose for your portfolio. Not every topic is popular, and picking the right one can make the difference between going viral or drowning in a seemingly endless sea of artwork.

Then you will learn tips and practices for presenting your work so that it highlights the strengths of your artwork and shows the key skill sets that you want recruiters to see.

Finally, you will learn how to leverage social media, so that you can get as much visibility for your artwork as possible, increasing the chances of attracting clients.

This chapter will cover the following topics:

- Picking your subjects
- Presenting your artwork
- Sharing your art on social media

Picking your subjects

Whenever you start a personal project, whether just for practice or with the goal of creating a new portfolio piece, there are endless options for the subject of the artwork. Naturally, there are certain topics that will interest you more than others, but if you treat your art like a professional would treat their art, your personal interests should not be the only consideration for deciding what to create.

Ultimately, you are a service provider and you have to fulfill a demand – this is part of the mindset you need to have while building your portfolio. Here are some of the things to consider when you pick the subject for a project.

Can it generate attention?

Ultimately, some subjects are more popular than others. If your subject is not popular, even if it is executed flawlessly, you will miss out on attention. Here are some of the more popular themes:

- **Fanart**: If you create artwork based on a TV show, game, or other medium, the fans of the title will relate to your artwork more than with any other type of project.

- **Male versus female**: If you sort ArtStation projects by likes, you will find that high-ranking projects with female subjects are much more numerous than those with a male subject.

Figure 13.1 – Male versus female

- **Character art versus environments**: Characters and creatures tend to be a bit more popular than environments or props on ArtStation because they are usually more catchy and make for a more engaging visual. If you want to build your portfolio around props or environment, it just means that you have to be extra mindful in creating a striking composition that can compete with vampires, dragons, and all living things.

Figure 13.2 – Character/creature versus object

- **Anatomy**: This is a popular theme in art in general, and it enjoys great popularity in the 3D community as well. Characters showing more detailed skin will draw more attention, and if executed well, they will enjoy greater popularity than characters covered with more clothing. Other anatomical elements such as bones and skulls can enhance designs easily too.

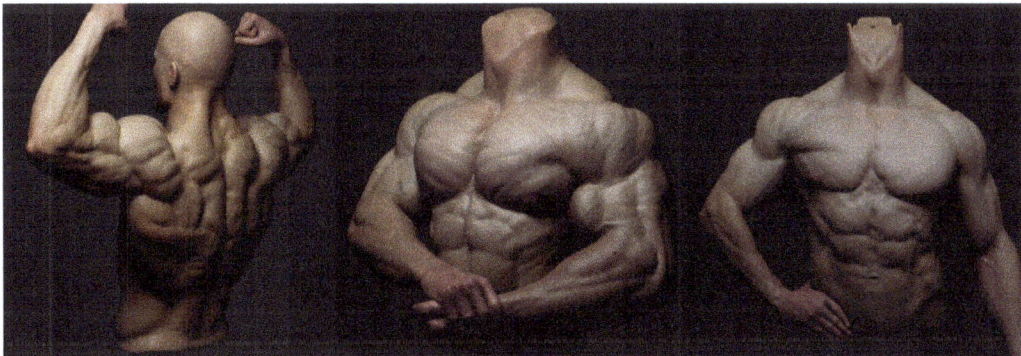

Figure 13.3 – Anatomy

- **Full character versus upper body**: The thumbnails of the most popular artwork on ArtStation have either a crop of the upper body or a close-up of the face. Including the full body would mean that the head will take up very little space in the image, but since we connect most to faces, this is not ideal. It may also miss out on all of the face modeling details, too. You might still decide to create a full character, but in terms of generating a catchy thumbnail, focusing on the upper body would be enough.

Figure 13.4 – A close-up will be more striking as a thumbnail

Does it showcase key skills?

Depending on the job opportunities you want to attract, some subjects make more sense than others. Even if your artwork goes viral, it does not matter if it does not stir the interest of recruiters in the target industry.

Here are some of the job roles that require digital sculpting, and the key skills that you should aim to display in your portfolio to qualify as a candidate.

Character modeler

For this position, anatomy plays a big role because it is an essential skill that does not leave much room for a lack of skill, and it cannot be learned in a short period of time.

Since the head is one of the most important elements of any character design, you need strong head sculpting skills, facial anatomy knowledge, and a strong skin detailing ability. If your project displays muscles or other anatomical elements such as bones and the skeleton, that will be a bonus.

Other than that, the overall quality of the design and composition should be good so it shows that the artist can create characters that look appealing and have a strong silhouette and good shape language. That is why it is important that you pick a concept with a strong design.

Figure 13.5 – Character modeling

Finally, good surfacing skills will also be appreciated. Detailing different surfaces, such as fabrics, metal, and skin, will be an important part of the high poly creation, so make sure to treat the surfacing of your character with attention to detail.

> **Important note**
>
> Since character modelers are often not expected to create the designs themselves, but instead work with designs from a concept artist, you will not have to come up with the design of the project yourself. Instead, look for successful concept art that has a lot of likes, and create your 3D model based on that. This will ensure that you have a concept that is proven to be liked by many people, which makes it more likely that you also get a strong response and visibility for your artwork and portfolio.

Creature modeler

If you are looking to enter the VFX industry as a creature modeler, you need a portfolio that displays strong animal anatomy skills. Ideally, you would show skill in creating a range of different animal anatomy, from four-legged carnivores such as lions and bears to reptilians, equidae (horses, zebras, etc.), and primates.

When you create sculptures of those creatures, you do not have to limit yourself to the actual animals, but rather you can sculpt fantasy designs based on the anatomy of animals. Here, it is less important to create the most anatomically accurate model but rather to make the creature look believable and well designed overall.

Figure 13.6 – Creature modeling

Generally, you should aim to showcase your overall sculpting skills, in which you implement strong primary, secondary, and tertiary shapes. Furthermore, Creature Modelers have to create fine detail and make sure the creature looks hyperrealistic so they can look believable in the end. Make sure to showcase the realistic detailing of various surfaces, from skin and scales to horns and fleshy wounds, to score extra points.

To help out, make sure to take inspiration from skilled creature artists on ArtStation to see the quality bar of professionals.

Digital sculptor for collectibles

Similar to the previous two roles, anatomy is key once again. Since many collectibles and figures are muscular superheroes from comics and movie franchises, sculpting muscles and human anatomy is a crucial aspect.

Picking an already popular collectibles concept is a good starting point, but beyond that, you can pick a design that shows a lot of skin and body definition so that you show your ability to create all kinds of characters. Based on their size, many 3D-printed figures can have a very high level of detail, so it is essential that you put a lot of time and effort into the surfacing of your character.

Now that you have an idea about picking a good subject in terms of popularity and key skills involved, let's look at another consideration when it comes to picking subjects.

Is it easy, fast, and fun to create?

Let's say you pick a popular concept that shows all the right skills for the job you are trying to get. If the creation of this artwork takes you three months, and you don't enjoy the process, it is probably not the best project to do. In a worst-case scenario, the project might even fall short of your expectations, and you wasted an opportunity to create something with a bigger impact.

Here are some things to watch out for when picking a concept:

- Some concepts are harder to translate into 3D than others. If the 2D concept is not very refined, and there is a lack of definition and clarity, it will be hard to get great results, so make sure you pick an artwork that is crisp and easy to read.

- As you progress as an artist, naturally some skills will come easier to you and you will develop certain strengths. Building on these strengths can help you stand out, so you should try to incorporate these skills in your project if possible.

Important note

If you are a complete beginner, it is a great time to experiment with different themes and styles, but eventually, you will have to choose a couple of these themes and styles to hone in on so that you can build them to a degree that makes you competitive.

- If you are working on a very unusual design, it will be hard to find reference pictures and similar artwork for inspiration, which will make the process unnecessarily hard for you. Something simple and mainstream might not be so exciting, but it allows you to do the simple things right and even small successes can help build momentum.

Of course, there is also a great learning experience in a challenging project, but especially for beginners, picking an easier concept can help to keep motivation and output high.

With these tips for picking a subject in mind, you can create artwork that has the potential to reach a wider audience. Once you created your model, and are about to finish a project, it is time to think about proper presentation, as this can have a significant impact, especially if you want to impress recruiters.

Presenting your artwork

After you put in hours and hours of hard work into creating a stunning and detailed 3D model, it is time to think about proper presentation so you can get the most out of your work and present your model in the best light possible. Let's explore practices for optimum presentation.

Picking the thumbnail

When you share your artwork on ArtStation, you have to pick a thumbnail for your project that is as catchy as possible, so it can stand out against all the other thumbnails. If you follow the tips about picking the subject in the first section of this chapter, you already have a great starting point for creating an effective thumbnail, but to get even more out of them, here are the main things to keep in mind:

- The focus should be on close-up images. Given that the quality is consistent, an upper-body will outperform a full-body shot, and a head will outperform an upper-body shot. Of course, there are exceptions, but as a general rule, using a closer shot is superior.

- It needs to have enough brightness and contrast. A common mistake is having too much shadow area and a lack of bright highlights in your artwork, so make sure that your model is sufficiently lit. Use a relatively high contrast, and if you have strong shadows, make sure that they are not so dark that the design cannot be read anymore.

Figure 13.7 – Ensuring sufficient contrast and light

- Study successful artwork. You can always browse ArtStation and filter the site by likes. This helps you find the most popular artwork, and you can get an idea of some of the common characteristics that you can apply to your own artwork.

Adding images

After the thumbnail, the main content of your project usually consists of multiple renders of your model. Here are some considerations you should make here.

The main/first image

When it comes to ArtStation, or any other portfolio or social media site, the strongest image should come first. This image is often what you used in the thumbnail, but it doesn't have to be. Since viewers have already engaged with your artwork, there is less need to have something catchy, but you can impress with something more subtle and complex instead.

Additional angles

Viewers will be interested in seeing more angles of your 3D model. It is normal for you to just focus on one – usually the best – angle of your model, but you should spend some time on the other angles to make sure that they look good as well. Then you can include these extra angles in your project; this is especially important if the pose plays a bigger role in your project, as it will be essential to be able to see how it works from more than one angle.

Figure 13.8 – Including side view or angles of your model

Additional lighting

Showing your model with different light setups can add some interesting variation to your project. Obviously, it allows you to show your lighting skills, but it also lets you show your shading and look development skills.

> **Important note**
>
> Look development is the process of establishing the style of the project. It includes the use of lights, textures, and shaders to establish mood, color schemes, and the overall design and style. For 3D artists, this process also includes the proper use of render settings.

Creating a model that looks good with various light setups speaks to the quality of your model, and it is especially relevant for the games and VFX industries where this kind of quality is needed.

Close-ups

If you have a detailed model with crisp textures, you will want to show off some of these details. Attention to detail is appreciated in general, but recruiters will take extra notice.

Figure 13.9 – Adding close-up renders to your project

Educational content

Sharing the "how-to" information, and any kind of educational content, usually enjoys great popularity, especially amongst beginner artists. It can be anything from software tips and explanations of a specific workflow to a simple breakdown of your artwork and the steps that were involved in the creation of it. This can help your artwork get some extra engagement, which will improve the visibility and reach of your post.

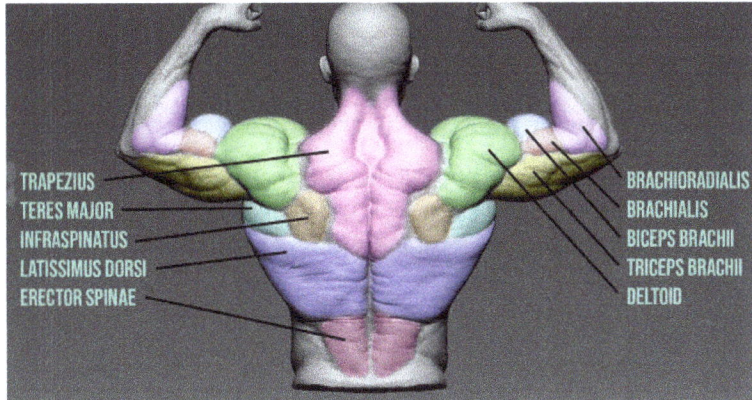

Figure 13.10 – Adding an educational aspect to your project

Work-in-progress images

Not every project needs to be super refined and sophisticated. The presentation part of the project can take up a significant amount of time, and it is not always needed to get a good response. **Work-in-progress** (**WIP**) posts, which show the model in an unfinished state, can sometimes get an even bigger response than the finished piece.

Especially if you want to showcase your sculpting and design skills, a screengrab from ZBrush can be enough to create a viral post. Since it is easy and fast to do, there is not really any downside to sharing a WIP post, as long as it shows a certain level of quality. Once you are done, you can still share the finished project, and now you have two or more opportunities to share your work with your audience.

Figure 13.11 – Creating two posts, WIP and finished, out of one character model

Adding videos

Videos are a great medium for showcasing your model, and with the tools in ZBrush's **Movie** palette at your disposal, you can get some great effects easily and quickly.

If you have rigged and animated your model, it will show a completely different skillset and make for a very impressive portfolio piece, but a simple turntable (rotation of, or around, your model) is already great, as it lets the viewer assess your model in a way that is not possible with images.

Take a look at *Chapter 7* to learn how to create turntable videos in ZBrush!

Showcasing key skills

If your project was created with the intention of getting a certain job or showcasing a certain skill set, your project should highlight these key skills.

For instance, if you are interested in a job in the games industry, you should have images showing the topology of your model (using wireframe rendering), as well as information about UVs, textures, and in the best case, even the implementation in a game engine such as Unreal Engine or Unity.

Figure 13.12 – Including a wireframe render showing the character's topology

Similar technical and artistic skills need to be highlighted for jobs in the VFX and collectibles industry. Depending on the job role, not only does this include anatomy and organic sculpting, but also grooming, surfacing, and technical skills. Do some research on some of the top artists of the desired job roles to see how they present their work and what they focus on.

Linking your social media

When you share your project, it is a great opportunity to link to other social media profiles. This can help get traffic to these sites and build your following there. You can also use the attention you get from your artwork to promote digital products and services.

Simply add a note in the project description, including the link to the social media platform, and optionally a note on what people can expect/gain from clicking on the link.

Social media is one of the most powerful tools for marketing your art, so let's take a closer look at some of the best apps for 3D artists and some quick tips on how to use them effectively.

Sharing your art on social media

When it comes to making a living as an artist, creating great art is only half the work. If your work does not reach a wider audience, nobody will even learn about your talent and skills, and you will miss out on commissions, job offers, and other opportunities.

This section will focus on the best platforms for artists that let you share your artwork and build a following so that you can gain momentum and visibility over the long term.

ArtStation

ArtStation is not only the most popular portfolio hosting site for 3D artists, but it is also a social media platform in which artists share their work, write blogs, sell products, and more.

As a portfolio website, it is the most common way of sharing an overview of your best art when you are applying for jobs, but it is also an opportunity to be found by recruiters and other people interested in your artwork. After publishing a new piece of artwork, it will be seen by people for a short duration, making it an opportunity to connect with clients and build your following, since some of the people who like your artwork will click the follow button to see future artwork.

A relatively new feature on the site is the ArtStation Marketplace, which allows 3D artists to sell their models, video tutorials, or any kind of digital product, making it a great opportunity to build an extra stream of income.

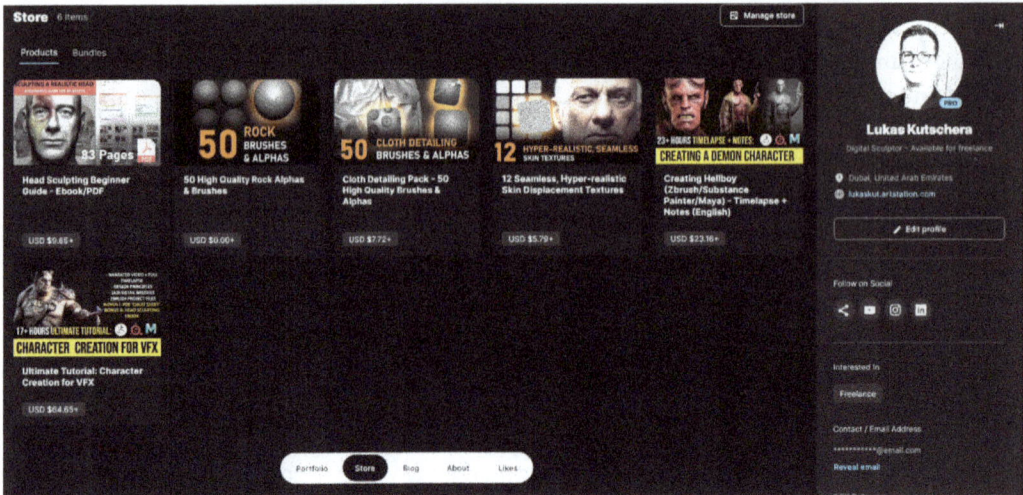

Figure 13.13 – Selling digital products on the ArtStation Marketplace

There is also the option to write a blog, which can be a useful way to generate extra attention for your artwork and digital products.

It is important to note that the more people that click the like button or leave a comment, the higher the artwork will rank on the ArtStation trending page, increasing the visibility of your artwork further. However, with over 3 million monthly active users, there is a lot of traffic on the site, which also means a lot of artists competing for attention. This makes it essential that your artwork has a thumbnail that stands out and can successfully compete for attention.

Here are some tips for creating a strong thumbnail:

- Make sure to have enough contrast in your thumbnail.

- Make sure it has enough brightness and the model is not covered too much in shadows. Of course, there are always exceptions, but often you see artwork that would benefit from increased brightness.

- If you publish a character, create a crop that focuses on the upper body or face. Unless you have elements of your character design that are important to include for the design to work, having a smaller crop generates more attention than a bigger one.

- Make sure that the thumbnail is aligned with the tips on popular themes, described at the beginning of the chapter.

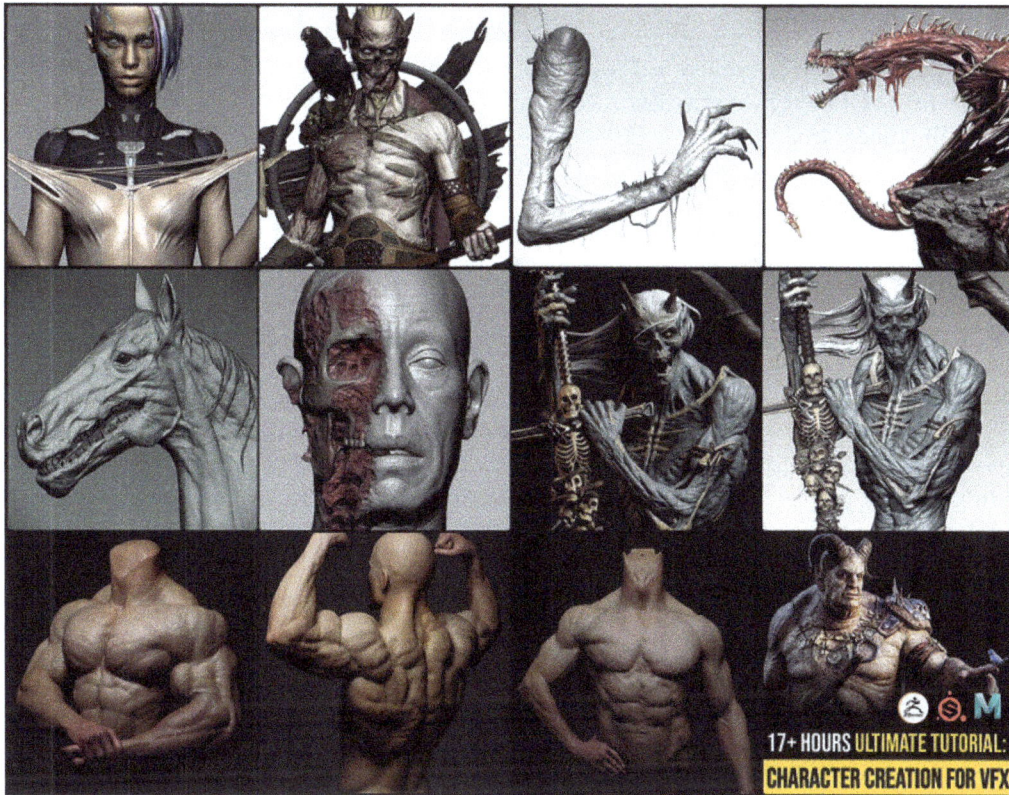

Figure 13.14 – Project thumbnails on ArtStation

Now that you know how to create catchy thumbnails, here are some more tips to get the most out of your ArtStation projects:

- Pay attention to the best time to post. This will require some experimentation and research, but generally, a good time is during the week when it is the evening for your main audience and they are off work. If you post at a bad time, you might not get the initial traction that is needed to get high on the trending page, which means it might never get the attention it deserves based on its quality.

- Shortly after you publish your ArtStation project, make sure to share it on other social media platforms and include a link to your ArtStation project. This will generate extra traffic to your ArtStation project, which will boost the ArtStation algorithm and push it higher on the trending page, exposing it to a wider audience.

- Following on from the previous tip, you can use the project description of your artwork to include a link to one or more of your other social media platforms in order to build up an audience for these platforms. For example, you could link to your Instagram profile or embed a YouTube video.

At the end of the day, ArtStation can be very competitive for beginners, but you should not let it discourage you; rather, you should use it as motivation to improve your art and compete with some of the best sculptors and modelers in the world.

Instagram

Since Instagram is focused purely on pictures and videos, it is a no-brainer for artists. However, on this platform, you do not only compete with other artwork but with all kinds of content, which makes it very hard to reach larger audiences, especially when starting. This makes it vital to be aware of how the Instagram algorithm works, and you need to have a strategy for posting your content accordingly.

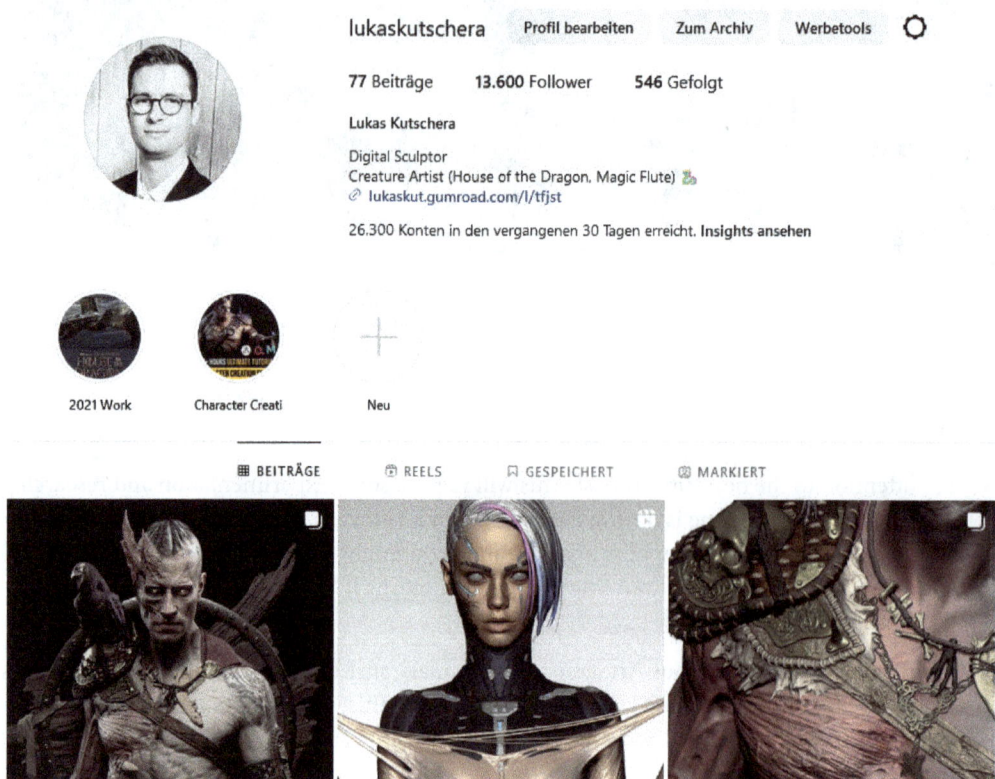

Figure 13.15 – Instagram

Similar to ArtStation, the virality and reach of your posts depend mostly on the quality of your artwork. On Instagram, you can post much more frequently because there are many types of content that you can post besides finished artwork.

Here are some content ideas:

- **WIP posts**: You can share everything from the early beginning stages of a project to something more refined. A WIP post does not need to look very refined, but of course, it needs to look promising and impressive regardless if you want to reach more people.

- **Videos**: You could share any kind of video recording, such as a turntable or animation of your model, but a recording of you sculpting in ZBrush or any behind-the-scenes videos can make for engaging content as well.

- **Old artwork**: You may share old projects that you feel are worth sharing. Instagram even has the popular #Throwbackthursday, which encourages people to use old content.

- **Non-art content**: Of course, you can also share other types of content, such as memes, your breakfast, or something inspirational. While that can also be a good chance to share something more personal and connect to your followers, the performance of these posts will usually be a bit inferior.

You will find that it requires some experimentation with different posts to find what resonates with your audience and what leads to the biggest gain in followers. So, when you start on Instagram, there will probably be mixed results at first, but eventually, you should find your niche and be able to build your following.

Just as important as your content is your consistency in posting, so make sure to post regularly and avoid not posting anything for too long. If you are serious about the growth of your following, you should try to post daily.

Here are some more tips for better results on Instagram:

- When you post artwork on other platforms, you can provide a link to your Instagram account so you can drive some traffic there. This will help boost the performance of your posts so you can start gaining followers organically. Instagram is one of the harder platforms to gain traction on and gain followers, so the importance of this can not be overstated.

- When you start from zero, a more tedious but popular method is to engage with other artist's content and comment on their posts to encourage them to return the favor. It is not a very efficient and sustainable technique, but it can help get the ball rolling.

- Pay attention to the best time to post. Just like on ArtStation, picking the time at which your audience is active allows you to get more interaction with your post early on, which helps to boost the post's performance.

- Many people on Instagram follow hashtags or get posts recommended based on hashtags. This makes it important to use the right hashtags. There are many free tools for finding good hashtags, such as `https://keywordtool.io/instagram`.

  - Make sure to stay up to date with the latest Instagram algorithm trends. The algorithm is constantly changing, and some forms of content can be superior for some time, then suddenly not be. There are several marketing websites that track these sorts of changes, including `https://neilpatel.com/blog/`.

  - Don't buy followers or likes. While it seems like a cheap and easy solution, this method brings useless, fake followers that add little value. Additionally, Instagram will notice suspicious activity and it will result in a decreased performance of your posts.

To sum up, a good strategy for Instagram is to post good art often, at the right times, and with the right hashtags. While there are many marketing services and shady profiles promising quick results, the reality is that there are no shortcuts and the biggest factor of success will be the quality of your artwork. Of course, there are many content marketing strategies that can work, such as collaborating with other creators or joining competitions or events, so if you want to maximize your results, there is a ton of information on content marketing and social media strategies available online.

Facebook groups

Facebook groups are a great place to share your art for self-promotion, but at the same time, they are supportive communities, ideal for beginner artists who want to get feedback, learn, and improve their art.

The content strategy here needs to be a little bit different from ArtStation and Instagram. Since Facebook groups are of a much smaller size, self-promotion should not be done excessively, and the frequency of posts has to be appropriate for the size of the group. This depends also on the kind of groups, as some groups are specifically meant for self-promotion while others are communities of enthusiasts.

Here is how to get the most out of Facebook groups for self-promotion:

- Make sure to post your best artwork for the highest engagement. Similar to the thumbnails for ArtStation, the main artwork should be catchy enough to get attention, since you will not be the only one sharing your artwork.

- As with any social media platform, it is best to use a time when the majority of the audience is active so that you can get a good initial engagement with your artwork.

- One of the best uses of Facebook groups is to promote artwork on other social media platforms. For example, you could link to your ArtStation project, right after publishing it, so you can boost the ArtStation algorithm. You can also link to your Instagram, YouTube, or any other account you want to drive traffic to, and vice versa.

As you can see, Facebook groups are a great addition to your social media promotion program, and they can be especially useful to kick-start growth and boost performance on other platforms.

Additional platforms

While the prior three platforms alone will deliver solid results, there are many other platforms and communities that can deliver equally good results. If you have time and would like to focus more on your marketing efforts, any one of the following platforms can work very well if you use them consistently (just make sure to do some research, so that you know the best practices for promoting your artwork and get most of them):

- **LinkedIn**: The Facebook for business, this platform can generate a lot of attention for 3D art, and you can also gain followers, which makes it similar to the previous platforms. As always, the better your artwork, the better the response and the more effective your marketing efforts will be. Since the traffic can be significant, LinkedIn can also be used to link to other platforms and drive traffic there.

- **Reddit**: The so-called subreddits on Reddit are communities for certain topics. There is a ZBrush subreddit, a 3D modeling subreddit, and more digital art-related ones in which you can share your artwork. These have a relatively small community size, but it is worth sharing your art there anyway, and if you are a beginner, you will also find that you can get great constructive feedback on your projects if that is something you are interested in.

- **Discord**: Discord is a communication platform, mainly used for chatting, but there are also groups and communities, called servers, in which you can connect to fellow artists, and share your artwork as well. Like any community of enthusiasts, this is probably not the place for using a sophisticated marketing strategy, but rather a place to chat and share a common interest.

- **X/Twitter**: This is another platform that artists and studios use to promote their artwork or projects. Similar to Instagram, it can be slow to begin with, so you might want to consider driving traffic from other platforms where you have a following already. Proper use of hashtags will be essential to be seen by the right audience, or, in your case, the 3D artist community.

- **YouTube**: This platform is very different from the other sites in many regards. Creating videos takes a lot of effort, and reaching a wider audience also requires a good strategy for picking the subject, as well as additional marketing efforts to drive traffic to YouTube. This, and the fact that results come much slower, is also an advantage because it means that there is far less competition than on other social media platforms. This is probably not for everyone, but if you enjoy creating videos, then it could be worth looking into.

Now, you have an overview of some of the most effective social media platforms, and you can start trying out some of them. While there are definitely certain tips and tricks that can help get a bit more out of them, the most important factor remains the quality of your artwork, which means that you can spend most of your time creating the art to reap the benefits. The mentioned sites do not require much time or effort, but it is important that you are consistent in posting if you want to see steady growth in your following and a higher frequency in commissions and job offers.

Summary

In this chapter, you learned about a strategic approach to building a portfolio that gets attention and, with some luck, attracts job offers and other opportunities. In the first section, you explored different criteria for picking a good subject for your artwork, which increases the odds for your social media posts to get viral and to get a lot out of it in general. Then you learned about the proper presentation of your work, and what to include in the project, so you can get the most out of the artwork you already invested the time and energy in. Finally, you got an overview of various social media platforms, their benefits, and how to use them.

This wraps up the book, which aims to give an insight into the most effective tools and techniques in ZBrush. It should make you a better sculptor and prepare you for a variety of tasks in the VFX, games, and 3D printing industries. If you were a complete beginner, some lessons might have been challenging and some steps require some time and repetition until they feel natural, but you can be proud of finishing the book. You will find that, even though ZBrush has an unusual UI and a steep learning curve at first, if you keep practicing, things will become intuitive and smooth very quickly.

If you want to dive deeper into any topic and learn more about ZBrush functionality, the single best resource available is the YouTube channel of artist and ZBrush instructor Michael Pavlovich (`https://www.youtube.com/channel/UCWiZI2dglzpaCYNnjcejS-Q`).

At this point, I would like to thank you for reading! Feel free to message me on LinkedIn (`https://www.linkedin.com/in/lukas-kutschera-289686181/`) or ArtStation (`https://www.artstation.com/lukaskut`) if you have any questions or comments about the book!

Index

Symbols

3D model
managing 74
subtools 74, 76
Symmetry mode 76-78

3D scan details
3D head scan 403
matching, to head model 403-406
meshes for projection, preparing 406
need for 403
projecting 402
skin detail, projecting
onto head model 407, 408

3D scan store
reference link 403

3D Sculpting Brushes menu 94, 95

A

Alphas 106
custom Alpha, creating
for demon bust 111, 112
modifying 107-111
used, for applying skin detail 402

Ambient modifier 145, 146

ambient occlusion (AO) 195

anatomy
base mesh, correcting 225-228

blocking 221
fat deposits, adding 250-252
hands and feet sculpture 249
lower body sculpture 243, 244
refining 230
skeleton 221, 222
skeleton, adjusting 222-225
upper body sculpture 233

anatomy references
preparing 214
resources, loading 217, 218

anatomy references, types 214
bodybuilder and athlete pictures 215
muscles and skeleton 216, 217
proportions and measurement charts 214

anatomy, sculpting brush 230
ClayBuildup 230
DamStandard 231
Move and Move Topological 232
StandardBrush 232

anatomy sculpture 253
asymmetry, adding 255, 256
model pose 254, 255
shapes transition, working 256, 258

anti-aliasing (AA) 195

ArtStation 42
reference link 42

‹packt›

Other Books You May Enjoy

If you enjoyed this book, you may be interested in these other books by Packt:

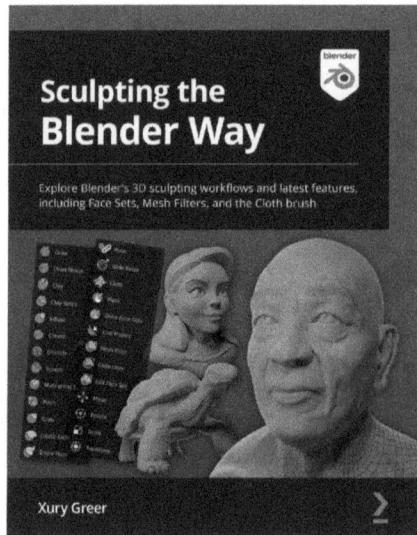

Sculpting the Blender Way

Xury Greer

ISBN: 978-1-80107-387-5

- Configure your graphics tablet for use in 3D sculpting.
- Set up Blender's user interface for sculpting.
- Understand the core Blender sculpting workflows.
- Familiarize yourself with Blender's basic sculpting brushes.
- Customize brushes for more advanced workflows.
- Explore high-resolution details with brush alphas and Multiresolution.
- Try out the all-new Cloth brush.
- Render your finished artwork for and make it portfolio-ready.

Packt is searching for authors like you

If you're interested in becoming an author for Packt, please visit `authors.packtpub.com` and apply today. We have worked with thousands of developers and tech professionals, just like you, to help them share their insight with the global tech community. You can make a general application, apply for a specific hot topic that we are recruiting an author for, or submit your own idea.

Share Your Thoughts

Now you've finished *Sculpting in ZBrush Made Simple*, we'd love to hear your thoughts! Scan the QR code below to go straight to the Amazon review page for this book and share your feedback or leave a review on the site that you purchased it from.

`https://packt.link/r/1-803-23576-4`

Your review is important to us and the tech community and will help us make sure we're delivering excellent quality content.

Download a free PDF copy of this book

Thanks for purchasing this book!

Do you like to read on the go but are unable to carry your print books everywhere?

Is your eBook purchase not compatible with the device of your choice?

Don't worry, now with every Packt book you get a DRM-free PDF version of that book at no cost.

Read anywhere, any place, on any device. Search, copy, and paste code from your favorite technical books directly into your application.

The perks don't stop there, you can get exclusive access to discounts, newsletters, and great free content in your inbox daily

Follow these simple steps to get the benefits:

1. Scan the QR code or visit the link below

https://packt.link/free-ebook/978-1-80323-576-9

2. Submit your proof of purchase
3. That's it! We'll send your free PDF and other benefits to your email directly